# Contributions from Biology Education Research

Contributions from Biology Education Research (CBER) is the international book series of the European Researchers in Didactics of Biology (ERIDOB). The series includes edited collections of state-of-the-art papers presented at the ERIDOB international conferences, and monographs or edited collections of chapters by leading international scholars of the domain. The aim of the series is to shed light on global issues and trends in the teaching and learning of biology by gathering cutting edge research findings, theoretical views, and implications or concrete suggestions for everyday school practice regarding biology. The books may serve as resources for (a) getting informed about the most recent findings of biology education research to possibly integrate them in new personal research, and (b) studying about the teaching and learning of biology as a pre-service or in-service biology teacher. So, the main audiences for the series range from senior to early career biology education researchers and pre- or in-service biology teachers working at all educational levels. Book proposals for this series may be submitted to the Publishing Editor: Claudia Acuna E-mail: Claudia.Acuna@springer.com

More information about this series at https://link.springer.com/bookseries/16350

Konstantinos Korfiatis • Marcus Grace
Editors

# Current Research in Biology Education

## Selected Papers from the ERIDOB Community

Springer

eridob
european researchers
in didactics of biology

*Editors*
Konstantinos Korfiatis
Department of Education
University of Cyprus
Nicosia, Cyprus

Marcus Grace
School of Education
University of Southampton
Southampton, UK

ISSN 2662-2319 ISSN 2662-2327 (electronic)
Contributions from Biology Education Research
ISBN 978-3-030-89479-5 ISBN 978-3-030-89480-1 (eBook)
https://doi.org/10.1007/978-3-030-89480-1



This Springer imprint is published by the registered company Springer Nature Switzerland AG
The registered company address is: Gewerbestrasse 11, 6330 Cham, Switzerland

# European Researchers in Didactics of Biology (ERIDOB)

# Preface and Introduction

This is a book produced by members of the community of European Researchers in Didactics of Biology (ERIDOB), at a time when communication between researchers has been severely affected by the COVID-19 pandemic.

Since 1996, researchers in biology education have come together at the biennial conference to exchange ideas, research questions, designs, instruments, results, interpretations, and conclusions. However, the usual way of face to face communication at conferences was disrupted by the pandemic, and although colleagues had submitted research papers for the conference, the ERIDOB2020 conference never happened. So instead, the ERIDOB Academic Committee wanted to provide the opportunity for biology education researchers to communicate their research through this publication. An invitation was distributed to all the colleagues whose papers had been accepted for presentation at the ERIDOB2020 conference, asking them to submit full papers for publication in this book.

The submitted papers were double-blind peer reviewed by two anonymous experienced reviewers, and all ensuing corrections were further reviewed by at least one member of the ERIDOB Academic Committee. The result is the present book with 23 selected papers from 9 countries.

We believe that this edited volume has managed to capture the diversity and quality of current biology education research being carried out across Europe and beyond.

Today's challenges faced by humanity highlight the importance of biological education for equipping young people with the tools and resources needed to become scientifically literate, critical thinkers, and socially responsible citizens, and these concerns are highlighted in this book.

Throughout the parts, new chapters on biology education and up-to-date research regarding biology topics are presented, with the goal of shedding light on the important role of biology education – from addressing critically socio-scientific issues to innovative teaching approaches and other emerging topics where further research is needed.

The first group of chapters (Part I: Socio-scientific Issues, Nature of Science and Scientific Thinking) focus on socio-scientific issues and nature of science. The first chapter is by Ayelet Baram-Tsabari, who had very kindly agreed to be a keynote speaker at the conference. Her chapter exactly corresponds to the need to establish suitable and efficient ways to communicate science with bigger audiences as a necessity to tackle conspiracy theories, fake news, and socially dangerous attitudes and behaviours. This chapter explores the complex ways that science literacy supports people in spotting fake news and misinformation and could help them make informed decisions about socio-scientific issues that affect their lives.

The chapter by Blanca Puig and Noa Ageitos highlights the importance of health literacy and critical thinking skills in building arguments against the anti-vaccination movement.

Julia Holzer and Doris Elster used the psychological theory of planned behaviour to develop a teaching unit that could positively influence the intention for stem cell donation. The results presented in the chapter show that the teaching unit has been effective in raising the awareness of this important health issue.

Andreani Baytelman, Kalypso Iordanou, and Costas Constantinou found that grounded scientific knowledge and informed epistemic beliefs do play a role in the construction of reasonable arguments on health socio-scientific issues.

The final two chapters of this part deal with aspects of the nature of scientific inquiry and the nature of scientific knowledge. The chapter by Corinne Charlotte Wacker, Marius Barth, Christoph Stahl, and Kirsten Schlüter presents the results from the construction and evaluation of an instrument for assessing biology students' knowledge of the nature of scientific inquiry. The chapter by Marida Ergazaki and Aggeliki Laourdeki shows that even elementary school children as young as 7 years old can improve their knowledge on the nature of science by following a learning intervention based on the study of scientists' biographies.

The second part of the book (Part II: Teaching and Learning in Biology) includes eight chapters that evaluate learning interventions and learning environments. Two of them examine digital environments.

The chapter by María Napal Fraile, María Isabel Zudaire Ripa, Irantzu Uriz Doray, and Lander Calvelhe presents a programme for training teachers in the implementation of digital storytelling.

Lars Emmerichs, Meike Mohneke, Sandra Hofhues, and Kirsten Schlüter compare the use of digital mobile technologies in education with the typical 'paper and pencil' method. Interestingly enough, their results do not show any explicit advantage of either of the two methods in assisting learning gains.

Another comparison between the more contemporary inquiry-based teaching with the traditional lecturing is made in Yvette Samaha and Assaad Yammine's chapter, and their findings clearly indicate that the jury is still out over this! Emanuel Nestler, Carolin Retzlaff-Fürst, and Jorge Groß highlight, after an experimental intervention study, the importance of biology-related quality mentoring for the professional training of pre-service biology teachers.

The next two chapters of this parts assessed specifically designed educational proposals. The chapter by Patricia Esteve-Guirao, Isabel Banos-González, and Magdalena Valverde evaluates two alternative proposals for teaching genetics to tenth graders, while Athanasios Mogias, Eleftheria Peskelidou, and Theodora Boubonari used the Ocean Literacy Framework to support vocational students' understanding of marine life.

Johannes Zang and Marcus Hammann report on a trait-specific approach to promote an understanding of relational causality in the context of genetic education, and especially in understanding gene-environment interactions.

Leroy Großmann and Dirk Krüger end this part highlighting the importance of taking into account students' conceptions and misconceptions about biology when trainee teachers design their biology lessons.

The third part of the book (Part III: Perceptions of Biology and Biology Education) consist of five chapters presenting and discussing students' views on biology, or on various aspects of teaching biology. Annkathrin Wenzel and Norbert Grotjohann explore students' ideas concerning interdisciplinary teaching, while Vida Lang and Andrej Šorgo examine how the popularity of secondary school biology is connected with biology knowledge and its perceived importance for personal well-being.

The chapter by Vanda Janštová, Natálie Tichá, and Petr Novotný examines the perception of biological disciplines by upper secondary school students within the Czech educational context.

The correlation between students' conceptions on modelling, scientific reasoning, and cognitive abilities is the focus of Maximilian Göhner and Moritz Krell.

Andrej Šorgo and Vida Lang discuss students' opinions on the advantages and disadvantages of various representational tools in teaching biology.

In the next part of the book (Part IV), we have included two chapters dealing with the interesting topic of textbook analysis. Both of them refer to Greek textbooks. The first one, by Georgios Ampatzidis and Anastasia Armeni, explores the content of Greek secondary education textbooks on human reproduction from late nineteenth century to present. The second chapter by Akrivi Christidou, Penelope Papadopoulou, and Anastasia G. Stamou presents a critical discourse analysis of the gene concept, as it is defined and explained in two biology lower high school textbooks.

The book ends with two chapters in Part V confirming the close interrelationship of biology education with outdoor and environmental education. Richard Sheldrake and Michael Reiss present the results of a survey of young children in England on their connectedness with nature and their interest in learning about nature.

The chapter by Marianna Kalaitzidaki, Katerina Mantadaki, and Aris Tsantiropoulos uses a 'photovoice' participatory research method approach in order to allow university students to express what they appreciate from their urban environment.

We would like to thank Stella Petrou, Vasiliki Vasileiou, and Rafaella Malouppa for helping with the logistics of the submission and review process. We are also extremely grateful to all the colleagues in the field of biology education who kindly took part in the peer reviewing of the chapters in this book:

| | |
|---|---|
| Jeremy Airey | Mats Lundström |
| Indira Banner | Birgitta McEwen |
| Andri Christodoulou | Christina Ottander |
| Justin Dillon | Patricia Schneeberger |
| Marida Ergazaki | Penelope Papadopoulou |
| Daniela Fiedler | Michael Reiss |
| Charles Frédéric | Bernard Sandie |
| Marcus Hammann | Martin Scheuch |
| Michal Haskel-Ittah | Gregor Torkar |
| Lissy Jäkel | Dina Tsybulsky |
| Moritz Krell | Micha Ummels |
| Dirk Krüger | Michal Zion |


Nicosia, Cyprus — Konstantinos Korfiatis

Southampton, UK — Marcus Grace
April 2021

# Contents

**Part I Socioscientific Issues, Nature of Science and Scientific Thinking**

1 **The Relevancy of Science Education to Public Engagement with Science** ... 3
Ayelet Baram-Tsabari

2 **Engaging in Argumentation as Critical Evaluation of the Anti-vaccination Movement** ... 19
Blanca Puig and Noa Ageitos

3 **Theory of Planned Behavior in the Context of Stem Cell Donation** ... 31
Julia Holzer and Doris Elster

4 **Prior Knowledge, Epistemic Beliefs and Socio-scientific Topic Context as Predictors of the Diversity of Arguments on Socio-scientific Issues** ... 45
Andreani Baytelman, Kalypso Iordanou, and Costas P. Constantinou

5 **Toward a Questionnaire to Assess Biology Student Teachers' Knowledge of the Nature of Scientific Inquiry (NOSI)** ... 59
Corinne Charlotte Wacker, Marius Barth, Christoph Stahl, and Kirsten Schlüter

6 **Introducing Primary School Students to Aspects of the Nature of Scientific Knowledge** ... 71
Marida Ergazaki and Aggeliki Laourdeki

**Part II Teaching and Learning in Biology**

7 **Digital Narratives for Biology Learning** ... 87
María Napal Fraile, Isabel Zudaire Ripa, Irantzu Uriz Doray, and Lander Calvelhe

8 **Digital First? Effects of Digital and Analogue Learning Tools on the Plant Knowledge Acquisition of Future Biology Teachers** . . . . . 99
Lars Emmerichs, Meike Mohneke, Sandra Hofhues, and Kirsten Schlüter

9 **Comparing Inquiry-Based Learning and Interactive Lectures While Teaching Physiology to Undergraduate Students** . . . 111
Yvette Samaha and Assaad Yammine

10 **Train the Trainer in the Jigsaw Puzzle of Biology Education: Effects of Mentor Training on Teaching Quality** . . . . . 127
Emanuel Nestler, Carolin Retzlaff-Fürst, and Jorge Groß

11 **The Assessment of an Educational Proposal to Address the Relationship Between Genetic Information and Protein Synthesis** . . . . . 141
Patricia Esteve-Guirao, Isabel Banos-González, and Magdalena Valverde

12 **Evaluating the Effectiveness of a Teaching Intervention in a Marine Biology Course: The Case of Greek Vocational Students** . . . . . 155
Athanasios Mogias, Eleftheria Peskelidou, and Theodora Boubonari

13 **Promoting Students' Understanding of Gene-Environment Interaction in Genetics Education** . . . . . 167
Johannes Zang and Marcus Hammann

14 **Students' Conceptions as a Neglected Perspective in Trainee Teachers' Biology Lesson Plans** . . . . . 181
Leroy Großmann and Dirk Krüger

**Part III Perceptions of Biology and Biology Education**

15 **Students' Opinions About Interdisciplinary Lessons** . . . . . 197
Annkathrin Wenzel and Norbert Grotjohann

16 **Correlation Between the Popularity and Difficulty of Secondary School Biology and Perceived Importance of Knowledge Acquired for Personal Wellbeing** . . . . . 209
Vida Lang and Andrej Šorgo

17 **Perception of Biological Disciplines by Upper Secondary School Students** . . . . . 219
Vanda Janštová, Natálie Tichá, and Petr Novotný

**18 Analyzing the Relationships Between Pre-service Biology Teachers' Modelling Processes, Scientific Reasoning Competencies and General Cognitive Abilities** .................. 231
Maximilian Göhner and Moritz Krell

**19 The Pitfalls of Using Presentation Technology in the Biology Classroom** .................................. 245
Andrej Šorgo and Vida Lang

**Part IV Textbook Analysis**

**20 Human Reproduction in Greek Secondary Education Textbooks (1870s to Present)** ................................. 257
Georgios Ampatzidis and Anastasia Armeni

**21 Exploring Logico-Semantic Relations of Historical Gene Models in Similar Genetic Content in Greek Biology School Textbooks** ..................................... 269
Akrivi Christidou, Penelope Papadopoulou, and Anastasia G. Stamou

**Part V Outdoor and Environmental Education**

**22 Children's Interests and Orientations Towards Nature: Views from Young Children in England** ....................... 285
Richard Sheldrake and Michael Reiss

**23 Picturing the Urban Environment: Using Photovoice to Bring University Students' Views and Voices into Urban Environmental Education** .......................... 301
Marianna Kalaitsidaki, Katerina Mantadaki, and Aris Tsantiropoulos

# Contributors

**Noa Ageitos** Department of Experimental Sciences, University of Santiago de Compostela, Santiago de Compostela, Spain

**Georgios Ampatzidis** Department of Chemistry, University of Patras, Patras, Greece

**Anastasia Armeni** Department of Obstetrics and Gynecology, Medical School, National and Kapodistrian University of Athens, Athens, Greece

**Isabel Banos-González** Departamento de Didáctica de las Ciencias Experimentales – Facultad de Educación, University of Murcia, Murcia, Spain

**Ayelet Baram-Tsabari** Faculty of Education in Science and Technology, Technion-Israel Institute of Technology, Haifa, Israel

**Marius Barth** Research Methods and Experimental Psychology, Department of Psychology, University of Cologne, Cologne, Germany

**Andreani Baytelman** Learning in Science Group, Department of Education, University of Cyprus, Nicosia, Cyprus

**Theodora Boubonari** Department of Primary Education, Democritus University of Thrace, Alexandroupolis, Greece

**Lander Calvelhe** Department of Humanities and Educational Sciences, Universidad Pública de Navarra (UPNA), Campus Arrosadía, Pamplona, Spain

**Akrivi Christidou** Department of Early Childhood Education, University of Western Macedonia, Florina, Greece

**Costas P. Constantinou** Learning in Science Group, Department of Education, University of Cyprus, Nicosia, Cyprus

**Irantzu Uriz Doray** Department of Sciences, Universidad Pública de Navarra (UPNA), Campus de Arrosadia, Pamplona, Spain

**Doris Elster** Institute for Science Education – Biology Education, University of Bremen, Bremen, Germany

**Lars Emmerichs** Institute of Biology Education, University of Cologne, Cologne, Germany

**Marida Ergazaki** Department of Educational Sciences and Early Childhood Education, University of Patras, Patras, Greece

**Patricia Esteve-Guirao** Departamento de Didáctica de las Ciencias Experimentales – Facultad de Educación, University of Murcia, Murcia, Spain

**María Napal Fraile** Department of Sciences, Universidad Pública de Navarra (UPNA), Campus de Arrosadia, Pamplona, Spain

**Maximilian Göhner** Department of Biology, Chemistry, Pharmacy, Freie Universität Berlin, Berlin, Germany

**Jorge Groß** Department Science Education, Otto-Friedrich-University Bamberg, Bamberg, Germany

**Leroy Großmann** Department of Biology, Chemistry, Pharmacy, Freie Universität Berlin, Berlin, Germany

**Norbert Grotjohann** Biology Didactics, Faculty of Biology, Bielefeld University, Bielefeld, Germany

**Marcus Hammann** Centre for Biology Education, University of Münster, Münster, Germany

**Sandra Hofhues** Fakultät für Kultur- und Sozialwissenschaften, FernUniversität in Hagen, Hagen, Germany

**Julia Holzer** Institute for Science Education – Biology Education, University of Bremen, Bremen, Germany

**Kalypso Iordanou** School of Sciences, University of Central Lancashire UCLan, Pyla, Cyprus

**Vanda Janštová** Department of Biology Education, Faculty of Science, Charles University, Prague, Czech Republic

**Marianna Kalaitsidaki** Department of Primary Education, University of Crete, Crete, Greece

**Moritz Krell** Department of Biology, Chemistry, Pharmacy, Freie Universität Berlin, Berlin, Germany

**Dirk Krüger** Department of Biology, Chemistry, Pharmacy, Freie Universität Berlin, Berlin, Germany

**Vida Lang** Faculty of Natural Sciences and Mathematics, University of Maribor, Maribor, Slovenia

Department of Biology, University of Maribor, Maribor, Slovenia

**Aggeliki Laourdeki** Department of Educational Sciences and Early Childhood Education, University of Patras, Patras, Greece

**Katerina Mantadaki** Department of Primary Education, University of Crete, Crete, Greece

**Athanasios Mogias** Department of Primary Education, Democritus University of Thrace, Alexandroupolis, Greece

**Meike Mohneke** Institute of Biology Education, University of Cologne, Cologne, Germany

**Emanuel Nestler** Department Biologiedidaktik, University of Rostock, Rostock, Germany

**Petr Novotný** Department of Biology Education, Faculty of Science, Charles University, Prague, Czech Republic

**Penelope Papadopoulou** Department of Early Childhood Education, University of Western Macedonia, Florina, Greece

**Eleftheria Peskelidou** Department of Primary Education, Democritus University of Thrace, Alexandroupolis, Greece

**Blanca Puig** Department of Experimental Sciences, University of Santiago de Compostela, Santiago de Compostela, Spain

**Michael Reiss** UCL Institute of Education, London, United Kingdom

**Carolin Retzlaff-Fürst** Department Biologiedidaktik, University of Rostock, Rostock, Germany

**Yvette Samaha** Faculty of Pedagogy-Dean's Office, Lebanese University, Beirut, Lebanon

**Kirsten Schlüter** Institute of Biology Education, University of Cologne, Cologne, Germany

**Richard Sheldrake** UCL Institute of Education, London, United Kingdom

**Andrej Šorgo** Faculty of Natural Sciences and Mathematics, University of Maribor, Maribor, Slovenia

**Christoph Stahl** Research Methods and Experimental Psychology, Department of Psychology, University of Cologne, Cologne, Germany

**Anastasia G. Stamou** Department of Linguistics and Language Teaching, Aristotle University of Thessaloniki, Thessaloniki, Greece

**Natálie Tichá** Department of Biology Education, Faculty of Science, Charles University, Prague, Czech Republic

**Aris Tsantiropoulos** Department of Philosophy and Social Studies, University of Crete, Crete, Greece

**Magdalena Valverde** Departamento de Didáctica de las Ciencias Experimentales – Facultad de Educación, University of Murcia, Murcia, Spain

**Corinne Charlotte Wacker** Institute of Biology Education, University of Cologne, Cologne, Germany

**Annkathrin Wenzel** Biology Didactics, Faculty of Biology, Bielefeld University, Bielefeld, Germany

**Assaad Yammine** Faculty of Pedagogy-Dean's Office, Lebanese University, Beirut, Lebanon

**Johannes Zang** Centre for Biology Education, University of Münster, Münster, Germany

**Isabel Zudaire Ripa** Department of Sciences, Universidad Pública de Navarra (UPNA), Campus de Arrosadia, Pamplona, Spain

# Part I
# Socioscientific Issues, Nature of Science and Scientific Thinking

# Chapter 1
# The Relevancy of Science Education to Public Engagement with Science

**Ayelet Baram-Tsabari**

> *"The wise man's eyes are in his head, But the fool walks in darkness. Yet I myself perceived, That the same event happens to them all"*
>
> *Ecclesiastes 2:14*

For the past few decades science education aimed to prepare non-scientists to make sense of science[1] in their daily lives, to be critical consumers of science information, and to make informed decisions about scientific issues (Roberts & Bybee, 2014). This goal of inclusive science education for non-scientists is termed "science literacy" (Hurd, 1958). "Science education for all" can be justified if it offers value for all rather than just for the minority who will become scientists (Osborne & Dillon, 2008). What is this value?

The US National Academies' report *Science Literacy: Concepts, Contexts, and Consequences* (Snow et al., 2016) provides four broad rationales for science literacy. The first is economic, and states that advanced economies require a scientifically and technologically skilled population to remain competitive in a global market. The second is cultural and considers that science is one of the greatest achievements of humankind which as part of a liberal education, offers ways to understand the world. The third is democratic and recognizes that citizens in a

[1] In accordance with the scope of science education, 'Science' refers here to STEM: Science, technology, engineering, and mathematics. Science includes a body of knowledge (content and theories), a way of knowing the world (processes and dispositions), as well as aspects of human endeavour (relevant institutions and people).

A. Baram-Tsabari (✉)
Faculty of Education in Science and Technology, Technion-Israel Institute of Technology, Haifa, Israel
e-mail: ayelet@technion.ac.il



K. Korfiatis, M. Grace (eds.), *Current Research in Biology Education*, Contributions from Biology Education Research,
https://doi.org/10.1007/978-3-030-89480-1_1

democracy participate directly and indirectly in decision-making about science related policies (e.g. regulation of technology, allocation of funding), and that democracies function better when their citizens are informed. Finally, the personal rationale for the importance of science literacy is that it "helps people respond to issues and challenges that emerge in their personal and community context" (Snow et al., 2016, p. 24), such as deciding whether to vaccinate against COVID-19, buy an electric car or use a cell phone in an elevator.

According to this rationale, science and technology are intertwined with the lives of people in the twenty-first century, and therefore understanding some science and an ability to interact with it helps people make better decisions and engage in informed actions that lead to richer and healthier lives. This vision of science education corresponds to Robert's (2007) vision II[2] of science literacy which refers to a curriculum looking 'outside of science' (Roberts & Bybee, 2014) by including skills and knowledge that lay people need to have to deal with science in their daily lives, such as Ryder's (2001) 'functional science literacy', and Fensham's idea of 'connoisseurs of science' that emphasize citizens' capacity to make judgments (Fensham, 2015).

The OECD's framework for internationally testing youth for science literacy (Organisation for Economic Co-operation Development [OECD], 2016, p. 18) argues that:

> Given that knowledge of science and science-based technology contributes significantly to individuals' personal, social, and professional lives, an understanding of science and technology is thus central to a young person's "preparedness for life". (p. 18)

My research in recent years explores this "given".

Education is ultimately about the transfer of what was learned to relevant contexts (Bransford & Schwartz, 1999; Perkins & Salomon, 1988; Sadler & Donnelly, 2006). However, in their *Science* paper, Feinstein et al. (2013) bluntly state:

> With the exception of modest and isolated gains in conceptual knowledge, it is not clear that these campaigns [educational efforts to enhance science literacy] have enhanced people's ability to function in a world where conflicting health advice clutters the Internet, research is filtered through political screens, and the media strip context from scientific claims. (p. 314)

Aikenhead (2006) tells us that most students tend not to integrate science into their everyday thinking, and Rudolph and Horibe (2016) reprimand:

> It appears to be a matter of faith among policymakers and researchers that high-quality science education…will… result in producing students – and thus citizens – who can successfully navigate science-related civic issues as they arise later in their lives. (p. 806)

Although some people are interested in science for its own sake, many engage with science in response to situation-specific needs and tend to be interested in science only when it supports their goals (Feinstein et al., 2013), and when they see it

[2] Vision I of science literacy offers a solid comprehensive foundation of literacy within science (Roberts, 2007).

as useful in light of their pre-existing commitments and motivations (Feinstein, 2011). Feinstein proposed that scientifically literate people are *competent outsiders* with respect to science; i.e., people who can recognize the moments when science has some bearing on their needs and interests and interact with sources of scientific expertise in ways that help them achieve their own goals (Feinstein, 2011; Fensham, 2015).

In this context, personal usefulness can be understood as instances where:

> …science education can help people solve personally meaningful problems in their lives, directly affect their material and social circumstances, shape their behavior, and inform their most significant practical and political decisions. (Feinstein, 2011, p. 169)

Personal usefulness can be related to both individual concerns (e.g. losing weight) as well as to widespread public concerns involving civic engagement (e.g. becoming an environmental activist). The distinctions are used as analytical and explanatory tools. Naturally, a personal decision can also be political – think about one's personal choice not to wear a face mask during a pandemic.

I think about usefulness as a form of transfer; i.e. people's abilities to learn new information in knowledge-rich environments and relate their learning to previous experiences (Bransford & Schwartz, 1999). Science education becomes valuable if people believe it is relevant to their needs, and choose to employ the content (facts, principles) and practices (including the importance of evidence and questioning) learned in science education to answer novel science-related questions in an effective manner.

"Science education" usually refers to K-12 formal education, higher education, and varied informal learning experiences in designed and everyday settings. In fact, lifelong, self-directed science learning process are dominated by out-of-school, free-choice learning experiences (Falk & Dierking, 2012). People learn science in non-school settings (Bell et al., 2009) and develop interests, understandings and identity related to science through an accumulation of non-school experiences (Falk et al., 2016). But when I am trying to empirically relate science education to public engagement with science in adult life, I focus mainly on mandatory formal science instruction. The reason is that access to informal science education is confined by socioeconomic, cultural, ethnic, historical and systemic factors (Bell et al., 2009; Dawson, 2017; Feinstein & Meshoulam, 2014).

Most lifelong learners in personal (non-professional) contexts pursue knowledge in an area of personal interest (Horrigan, 2016), and people who are interested and curious about science will be more likely to seek informal science learning opportunities than those who are not (Falk et al., 2016). The uninterested are not a negligible group. Almost half of all Europeans are not interested in developments in science and technology (46%) according to a 2013 Eurobarometer (European Commission, 2013), and there is a strong correlation between interest and feeling informed (European Commission, 2013). Science education is considered personally useful for all, not just the individuals who are interested.[3] Therefore, I mainly examine the personal usefulness of mandatory science education to non-scientists adults.

## 1.1 Problematizing Science Literacy

The EU Framework of Science Education for Responsible Citizenship (Hazelkorn et al., 2015) identified the main issues involved in

> …helping all citizens acquire the necessary knowledge of and about science to participate actively and responsibly in, with and for society, successfully throughout their lives. (p. 7)

Its first recommendation is that science education should be an essential component of a learning continuum for all, from pre-school to active engaged citizenship. Science education is seen in this report as vital to promoting a culture of scientific thinking and dispositions, and inspiring citizens to use evidence-based reasoning for decision making. In our editorial "Bridging science education and science communication research" (Baram-Tsabari & Osborne, 2015) Osborne and I put forward an observation: Science education makes the optimistic assumption that if science is taught properly, people who encounter science in their everyday lives can acquire the necessary knowledge to reach an informed, scientifically-based decision. However, empirical evidence from the field of public engagement with science indicates that people make meaning of the science they encounter in their lives using different narratives based on culturally relevant prior knowledge that may or may not include science (e.g. Carrion, 2017; Hine, 2012; Laslo et al., 2011; Layton et al., 1993; Feinstein, 2014).

Science literacy plays out in a social context in that social, cultural, and demographic differences influence how people engage with science, and what concerns and resources they have. People's own values, interests, and non-scientific knowledge shape their perspective. The weight people attach to underlying scientific vs. non-scientific knowledge when making decision is a function of this individual perspective (Scharrer et al., 2016).

Science literacy can also be seen as a characteristic of the community with its division of cognitive labor (Scharrer et al., 2016), rather than as an individual characteristic. In Feinstein's (2015) account of the Lippmann-Dewey debate, both scholars agreed that the ignorance of individual citizens was an inevitable by-product of specialization, and the distracting demands of everyday life. Dewey's (1927) solution to healthy and informed civic participation in democracy was to address the level of communities rather than that of individuals in a

[3] I adhere to a quote attributed to the Danish physicist Niels Bohr, who was asked why a brilliant scientist like himself would nail a horseshow above his door for good luck. He replied "I understand it brings you luck, whether you believe in it or not".

> …well-functioning community an individual does not need to know about a particular topic as long as she is meaningfully connected to someone who does. (Feinstein, 2015, p. 156)

This idea is echoed in writings about adult education that try to create an 'autonomous agent in a collaborative context' (Mezirow, 1997, p. 8).

To be scientifically literate one needs to have the ability to make thoughtful decisions, which includes a critical assessment of scientific claims (e.g., Kolstø et al., 2006; Norris & Phillips, 1994; Ryder, 2001), especially in the media (McClune & Jarman, 2010). However, Norris (1995), based on Hardwig (1985) argues that non-scientists cannot avoid some epistemic dependence[4] (Pritchard, 2015) on experts: "non-scientists' belief or disbelief in scientific propositions is not based on direct evidence for or against those propositions, but, rather, on reasons for believing or disbelieving the scientists who assert them" (p. 206). This epistemic dependence is further explained by Bromme (Bromme & Goldman, 2014; Bromme et al., 2010) who noted that laypeople frequently have to judge the validity of scientific knowledge claims that are of great relevance to their lives, but they lack the necessary epistemic capabilities to make such judgments adequately. Therefore, the ability of laypeople to reach informed decisions is usually based not on a first hand evaluation of what is true but on a second hand evaluation of "who knows what and whom to trust" (Bromme & Goldman, 2014). These and other studies led Bromme and Goldman (2014) to argue based on the theory of bounded rationality (Kahneman, 2003) that the public's limited understanding of science leads people to make science related decisions mostly by using quick heuristics, a phenomenon they called a "bounded understanding of science".

To complicate things even further, when science is popularized and seems easy to understand, the ease of processing leads readers to underestimate their dependence on experts (Scharrer et al., 2016). Furthermore, depending on the scientific topic, people appear to distinguish between what they judge to be credible and what they personally believe to be true (Bromme et al., 2015). It is not clear how much school science instruction improves children's pre-existing reasoning skills to make judgments about science expertise (Sandoval et al., 2014). Other studies, however, show that competent outsiders are able to make sophisticated judgments about the credibility of scientific claims based on cues such as professional reputation, publication venue, institutional affiliation, and potential conflicts of interest, even when they do not understand the technical nuances of experimental designs or laboratory techniques (Feinstein et al., 2013).

A potential bridge between public engagement with science scholarship and science education goals and methods is research concerning socio-scientific issues (SSI), a pedagogical movement consistent with Vision II of science literacy. SSI refers to complex, ill structured and controversial social issues that are often value-laden with conceptual, procedural, or technological ties to science on which competing views are held (Simonneaux, 2008). Features that distinguish them from

[4] The dependence of one's knowledge on factors outside one's cognitive agency, e.g. one does not possess direct evidence that the Earth circles the sun, but believes so nonetheless.

traditional school science are uncertainty and acceptance of different types of knowledge, rather than only scientific knowledge (Christensen & Fensham, 2012). There is broad support within the science education community for the inclusion of decision-making in the context of SSI as an integral component of scientific literacy (Sadler & Donnelly, 2006) as well as its evaluation (Allchin, 2011; Romine et al., 2017).

## 1.2 Public Engagement with Science: How Do People Use Science Knowledge in Real Life Situations?

On the one hand, there are cases where laypeople have developed science-related expertise (Epstein, 1995), use science to become expert consumers and effective caregivers or patients (Feinstein, 2014), and participate in scientific research (such as citizen science) (Bonney et al., 2016; Golumbic et al., 2020; Lewenstein, 2016). An eminent case study (Wynne, 1996) of farmers' responses to scientific advice show laypeople capable of

> ...extensive informal reflection upon their social relationships towards scientific experts, and on the epistemological status of their own local knowledge in relation to science as an 'outside' knowledge. (p. 281)

It also demonstrated how the public uptake of science depends primarily upon the trust and credibility public groups place in scientific institutions and representatives (Wynne, 1996).

On the other hand, this literature points to the marginality of science knowledge to non-scientists' decision making. In the foundational study *Inarticulate Science* (Layton et al., 1993), four cases in which laypeople had to cope with science-related situations were analyzed. The authors found that in all four cases, participants were rarely inclined to frame their challenges in terms of science. The authors concluded that in order to become practical, science knowledge needs to be relevant to the person concerned, align with personal experience, relate to other social knowledge, and derive from a trustworthy source.

In the last 10 years, the internet in general, and social networking sites in particular, have become a primary source for science and technology related information (National Science Board, 2020), discussion and deliberation (Brossard, 2013; Brossard & Scheufele, 2013), and a place to go to for advice and emotional, social and psychological support (Zillien et al., 2011). Many of the insights regarding public engagement with science are replicated online (e.g. Betten et al., 2017). However, the new media are also characterized by some features that shape the how public engagement with science happens (Brossard, 2013; Brossard & Scheufele, 2013; Peters et al., 2014). The new media landscape is characterized by an abundance of content, interactivity, mobility, and multimediality (Schejter & Tirosh, 2016). Each of these is a double-edged sword enabling new affordances, while making it harder for a non-expert audience to reach an informed science related decision

(Baram-Tsabari & Schejter, 2019), especially in a context where false news spreads further and faster than truth (Vosoughi et al., 2018) and science and health communication is being weaponized for political reasons (Broniatowski et al., 2018).

It is tempting to assume that individuals who do not accept the scientific consensus are anti-science or simply uninformed, but empirical data do not support this claim. Studies in the US context have shown that when ideology-related science controversies are studied, gaps in opinions are typically larger among individuals with more years of formal education (e.g. Kahan et al., 2012). A secondary analysis of a large nationally representative US based survey revealed that where religious or political polarization existed (stem cell research, the Big Bang, human evolution, and climate change), it was greater among individuals with more general education and among individuals with greater scientific knowledge, as measured by both science course attainment and scores on a science literacy test (Drummond & Fischhoff, 2017a).

This phenomenon is typically attributed to motivated reasoning, in which people seek, evaluate, interpret, and recall information that supports their beliefs and commitments (Kunda, 1990). Education and specific knowledge of the content area give people a more elaborate toolbox to interpret evidence in support of their preferred conclusions. Better educated people may be more likely to know when political or religious communities have chosen sides on an issue, and hence what they should think about it. Finally, educated individuals may also have greater confidence in their own knowledge (Drummond & Fischhoff, 2017b).

To conclude with the sobering words of the adult education researcher Mezirow (1990):

> In reality…we never have complete information, are seldom entirely free from external or psychic coercion of some sort, are not always open to unfamiliar and divergent perspectives, may lack the ability to engage in rational and critically reflective argumentation…and only sometimes let our conclusions rest on the evidence and on the cogency of the arguments alone. (p. 7)

This depiction is far from the optimistic view that if we teach science well, people will make evidenced-based decisions about science related issues.

## 1.3 The Relevancy of Research in Public Engagement with Science to Science Education

The relevance of formal science instruction and its goals to adult life can be illuminated through an examination of public engagement with science from a science literacy theoretical perspective. The remainder of the chapter reviews several such insights based on our research over the past decade.

1. **Is science knowledge useful for advocacy?**

   Our study of parents of hearing-impaired children explored their use of science knowledge in advocating for their children's rights (Shauli & Baram-Tsabari, 2019).

Hearing impaired child's potential of integrating into hearing society largely depends on developing efficient forms of communication (Most et al., 2012). Achieving that depends to some degree on parental dedication to realizing child's potential. The vast majority of parents (90%) of deaf children are not hearing impaired, and have little or no prior science knowledge of audiology (Northern & Downs, 2002). Therefore, parents are required to learn vast amounts of new science knowledge in the context of hearing impairment in a very short time. We called that 'Contextual science knowledge'. Parents also need to speak and act on behalf of their child to help address his/her preferences, strengths, and needs (Wolfensberger, 1977) – this is termed 'Advocacy', and we viewed it as a type of personal usefulness.

Shauli and I found that parents sometimes used general science knowledge as a background for understanding contextual science terms (e.g. frequency), procedures (e.g. tympanometry) and possible outcomes (e.g. fluid accumulation). A lack of general science information hinders this process (e.g. not knowing what soundwaves are). We then investigated the quantitative interactions between general scientific knowledge, contextual scientific knowledge in the field of hearing, and parents' advocacy knowledge and attitudes. Based on 115 parents' questionnaires, general science knowledge was a predictor of contextual science knowledge, and parents who displayed higher contextual scientific knowledge emerged as having only slightly better advocacy attitudes and knowledge. The small explained variance (5.5%) is consistent with previous research on knowledge and the prediction of behavior (Ajzen et al., 2011; Kaiser & Fuhrer, 2003). Possibly, our operationalization of usefulness was not comprehensive enough to include the range of situations in which science knowledge becomes a strong tool for parents. Improved measures might reveal a stronger link between context-specific scientific knowledge and practical real-world outcomes.

2. **Are science education background and knowledge relevant to COVID-19 decision making?**

At the time of writing, we are exploring the transferability of science instruction and knowledge to the understanding of and informed decision making in the context of the COVID-19 pandemic. In a series of survey-based studies we found very weak indications for the support of science education or knowledge in conformity with COVID-19 guidelines among diverse groups of sports fans, Ultraorthodox Jews, and the general public. Science education was, however, correlated with understanding the science underlying COVID-19 and relating health guidelines and with the ability of participants to justify their decisions (Baram-Tsabari et al., 2021; Taragin-Zeller et al., 2020). Similarly, Math education was a strong predictor of understanding quantitative information in the news among the general Israeli public, but did not affect the decisions people took – only their ability to justify it (Heyd-Metzuyanim et al., 2021). However, findings are not clear cut. Sometimes knowledge level increases risk perception and motivation for risk-reducing behavior. Such a connection was recently demonstrated in the context of the understanding of exponential growth and COVID-19 social-distancing regulations (Lammers et al., 2020). Indeed, the pandemic is a prime opportunity to study the relevance of science education and science knowledge to adult informed reasoning and decision making, and to examine the role of science literacy in real life context.

3. **Is "Engaging in argument from evidence" a realistic goal during social media discussions?**

In the end of May 2013 traces of the Polio virus were found in the sewage of a southern town in Israel. In August parents to young children in Israel were called to re-vaccinate their children to Polio (OPV), since the original vaccination protected them, but did not prevent them from spreading the virus to risk groups once infected. As one can imagine, parents on social media had a lot to say about that, e.g.:

> Why should I, as a mother of a daughter who has been vaccinated… have to give her the OPV too so… she won't affect other children who have not been vaccinated… why do I need to take responsibility for others?

Our study of Polio vaccination discussions in a *Facebook* group of parents and experts found that although half the items addressed scientific or medical issues as their primary topic, most items (96%) did not present any evidence to support their arguments. That was also true for the physicians who took part in the discussion. Reasoning with evidence did not seem to be a natural part of the online informal discussion, even for scientifically literate people, although it mainly revolved around science (Orr & Baram-Tsabari, 2018). This raises issues regarding the attainability of the learning goal "engaging in argument from evidence" outside the classroom.

The medical experts who were involved in the discussions indeed tried to establish epistemic trustworthiness, competence and to counter misinformation, but they also tried hard to establish benevolence while maintaining integrity and clarity. This mix of goals, constrains and considerations did not leave much place for using evidence in the online dialogue that was created (Sharon & Baram-Tsabari, 2020b).

4. **Does a mandatory science curriculum prepare people to follow science in the media and discussions in the reader comments section?**

Using animals in research is a contested socio-scientific issue. Our analysis (Laslo & Baram-Tsabari, 2021) of expressions of science literacy in readers' comments about online coverage of animal experimentation found that over half of the scientific concepts used by the commentators were at the high school or academic level, in which science is elective. Thus, in order to participate in the discussion or even just follow it passively, members of the public need to learn many new science concepts independently. Usually, their teacher is the media. Similarly, Shea (2015) found that the knowledge needed to understand news articles about genetics exceeds what is expected by the learning progression in school curricula. The finding that the majority of the science concepts introduced by readers and media were beyond the compulsory science curriculum level highlights the importance of the ability and motivation for independent lifelong learning as an outcome of science education.

5. **Do people who refer more to science knowledge and practices side more with the scientific consensus?**

   Not as much as you might expect. In several of our studies on online authentic discussions we found that expressions of science literacy did not necessarily go hand-in-hand with the scientific consensus. In fact, scientific knowledge was often used to support beliefs that were at odds with the scientific consensus (Asakly et al., 2016; Dalyot et al., 2019; Laslo & Baram-Tsabari, 2021; Orr & Baram-Tsabari, 2018). Knowledge is power, but that includes the power to convince yourself that your opinion is supported by evidence and that that your side in right. These findings challenge the 'deficit model', which sees scientific illiteracy as the root cause of opposition to the scientific consensus.

   But can science literacy support people in spotting fake news and misinformation? That would help people to form an informed opinion on science-related public controversies. That depends on which components of science literacy one emphasizes. Of the seven components of science literacy identified by the National Academies report (Snow et al., 2016), we argue that four are most likely to help individuals identify misinformation in everyday life: (1) Understanding of scientific practices; (2) Identifying and judging appropriate scientific expertise, (3) Epistemic knowledge, and (4) Dispositions and habits of mind, e.g., inquisitiveness and open-mindedness. However, apart from scientific practices, these are not the components that commonly used in definitions or teaching of science literacy in schools (Sharon & Baram-Tsabari, 2020a).

## 1.4 Concluding Remarks

Science education focuses on K-12 students, higher education, and learning in formal and informal environments. Science communication focuses on interactions between the scientific community and other publics and science in society. The two domains share the goal of engaging the public in informed discussions on science-related issues. However, their evolution as independent academic fields have impeded the identification of shared methods and ideas (Baram-Tsabari & Osborne, 2015). In this chapter I asked – what is the relevance of science education to public engagement with science? The main goal of science education in recent decades is educating scientifically literate people, who make informed science-based decisions in their personal and social lives. However, as I demonstrated, today this is still more of a normative statement than supported by evidence. Yet, bridging science education and science communication research informs us about the indirect and complex ways in which science education supports public engagement with science in adult life, and about the relevance of the different goals of formal science instruction.

## References


Aikenhead, G. S. (2006). *Science education for everyday life: Evidence-based practice*. Teachers College Press.

Ajzen, I., Joyce, N., Sheikh, S., & Cote, N. G. (2011). Knowledge and the prediction of behavior: The role of information accuracy in the theory of planned behavior. *Basic and Applied Social Psychology, 33*(2), 101–117. https://doi.org/10.1080/01973533.2011.568834

Allchin, D. (2011). Evaluating knowledge of the nature of (whole) science. *Science Education, 95*(3), 518–542.

Asakly, D., Orr, D., & Baram-Tsabari, A. (2016). *Characteristics of an authentic scientific discourse in social networks: The case of drinking water fluoridation*. Paper presented at the Public Communication of Science and Technology (PCST) conference, Istanbul.

Baram-Tsabari, A., & Osborne, J. (2015). Bridging science education and science communication research (Editorial). *Journal of Research in Science Teaching, 52*(2), 135–144.

Baram-Tsabari, A., & Schejter, A. (2019). The double-edged sword of new media in supporting public engagement with science. In Y. Kali, A. Schejter, & A. Baram-Tsabari (Eds.), *Learning in a networked society* (Computer supported collaborative learning (CSCL) book series). Springer.

Baram-Tsabari, A., Bronshtein, J., Rozenblum, Y., Barel-Ben David, Y., & Swirski, H. (2021). *Sports fans' science knowledge is relevant to their stance on COVID-19 guidelines, but only if they don't care who wins*. Paper presented at the Public Communication of Science and Technology (PCST) 2020+1 Conference, Aberdeen, Scotland, United Kingdom.

Bell, P., Lewenstein, B., Shouse, A. W., & Feder, M. A. (2009). *Learning science in informal environments: People, places, and pursuits*. Retrieved from Washington, DC http://www7.nationablacademies.org/bose/Learning%20Science%20in%20Informal%20Environment.html

Betten, A. W., Broerse, J. E. W., & Kupper, F. (2017). Dynamics of problem setting and framing in citizen discussions on synthetic biology. *Public Understanding of Science*, 0963662517712207. https://doi.org/10.1177/0963662517712207

Bonney, R., Phillips, T. B., Ballard, H. L., & Enck, J. W. (2016). Can citizen science enhance public understanding of science? *Public Understanding of Science, 25*(1), 2–16.

Bransford, J. D., & Schwartz, D. L. (1999). Chapter 3: Rethinking transfer: A simple proposal with multiple implications. *Review of Research in Education, 24*(1), 61–100.

Bromme, R., & Goldman, S. R. (2014). The public's bounded understanding of science. *Educational Psychologist, 49*(2), 59–69.

Bromme, R., Kienhues, D., & Porsch, T. (2010). Who knows what and who can we believe? Epistemological beliefs are beliefs about knowledge (mostly) to be attained from others. In L. D. Bendixen & F. C. Feucht (Eds.), *Personal epistemology in the classroom: Theory, research, and implications for practice* (pp. 163–193). Cambridge University Press.

Bromme, R., Scharrer, L., Stadtler, M., Hömberg, J., & Torspecken, R. (2015). Is it believable when it's scientific? How scientific discourse style influences laypeople's resolution of conflicts. *Journal of Research in Science Teaching, 52*(1), 36–57. https://doi.org/10.1002/tea.21172

Broniatowski, D. A., Jamison, A. M., Qi, S., AlKulaib, L., Chen, T., Benton, A., Quinn, S. C., & Dredze, M. (2018). Weaponized health communication: Twitter bots and Russian trolls amplify the vaccine debate. *American Journal of Public Health, 108*(10), 1378–1384.

Brossard, D. (2013). New media landscapes and the science information consumer. *Proceedings of the National Academy of Sciences, 110*(Suppl 3), 14096–14101.

Brossard, D., & Scheufele, D. A. (2013). Science, new media, and the public. *Science, 339*(6115), 40–41.

Carrion, M. L. (2017). "You need to do your research": Vaccines, contestable science, and maternal epistemology. *Public Understanding of Science*. https://doi.org/10.1177/0963662517728024

Christensen, C., & Fensham, P. J. (2012). Risk, uncertainty and complexity in science education. In B. J. Fraser, K. Tobin, & C. J. McRobbie (Eds.), *Second international handbook of science education* (pp. 751–769). Springer Netherlands.

Dalyot, K., Sharon, A. J., Orr, D., Ben-David, Y. B., & Baram-Tsabari, A. (2019). Public engagement with science in everyday life: Perceptions of Wi-Fi radiation risks in schools. *Research in Science Education*, 1–20. https://doi.org/10.1007/s11165-019-09894-w

Dawson, E. (2017). Social justice and out-of-school science learning: Exploring equity in science television, science clubs and maker spaces. *Science Education, 101*(4), 539–547. https://doi.org/10.1002/sce.21288

Dewey, J. (1927). *The public and its problems*. Holt.

Drummond, C., & Fischhoff, B. (2017a). Individuals with greater science literacy and education have more polarized beliefs on controversial science topics. *Proceedings of the National Academy of Sciences, 114*(36), 9587–9592.

Drummond, C., & Fischhoff, B. (2017b). Individuals with greater science literacy and education have more polarized beliefs on controversial science topics. *Proceedings of the National Academy of Sciences*, 201704882.

Epstein, S. (1995). The construction of lay expertise: AIDS activism and the forging of credibility in the reform of clinical trials. *Science, Technology & Human Values, 20*(4), 408–437. https://doi.org/10.1177/016224399502000402

European Commission. (2013). *Responsible Research and Innovation (RRI), science and technology*. https://op.europa.eu/en/publication-detail/-/publication/ee9bacdf-fdad-46eb-8cd8-32879e310191/language-en

Falk, J. H., & Dierking, L. D. (2012). Lifelong science learning for adults: The role of free-choice experiences. In B. J. Fraser, K. Tobin, & C. J. McRobbie (Eds.), *Second international handbook of science education* (pp. 1063–1079). Springer Netherlands.

Falk, J. H., Dierking, L. D., Swanger, L. P., Staus, N., Back, M., Barriault, C., Catalao, C., Chambers, C., Chew, L.-L., Dahl, S. A., Falla, S., Gorecki, B., Lau, T. C., Lloyd, A., Martin, J., Santer, J., Singer, S., Solli, A., Trepanier, G., … Verheyden, P. (2016). Correlating science center use with adult science literacy: An international, cross-institutional study. *Science Education, 100*(5), 849–876. https://doi.org/10.1002/sce.21225

Feinstein, N. (2011). Salvaging science literacy. *Science Education, 95*, 168–185.

Feinstein, N. (2014). Making sense of autism: Progressive engagement with science among parents of young, recently diagnosed autistic children. *Public Understanding of Science, 23*(5), 592–609.

Feinstein, N. W. (2015). Education, communication, and science in the public sphere. *Journal of Research in Science Teaching, 52*(2), 145–163.

Feinstein, N. W., & Meshoulam, D. (2014). Science for what public? Addressing equity in American science museums and science centers. *Journal of Research in Science Teaching, 51*(3), 368–394. https://doi.org/10.1002/tea.21130

Feinstein, N., Allen, S., & Jenkins, E. (2013). Outside the pipeline: Reimagining science education for nonscientists. *Science, 340*(6130), 314–317.

Fensham, P. J. (2015). Connoisseurs of science: A next goal for science education? In *The future in learning science: What's in it for the learner?* (pp. 35–59). Springer.

Golumbic, Y. N., Fishbain, B., & Baram-Tsabari, A. (2020). Science literacy in action: Understanding scientific data presented in a citizen science platform by non-expert adults. *International Journal of Science Education, Part B, 10*(3), 232–247.

Hardwig, J. (1985). Epistemic dependence. *The Journal of Philosophy, 82*(7), 335–349.

Hazelkorn, E., Charly, R., Yves, B., Constantinos, C., Ligia, D., Michel, G., & Welzel-Breuer, M. (2015). Science education for responsible citizenship. In *Report to the European Commission of the expert group on science education*. European Commission.

Heyd-Metzuyanim, E., Sharon, A., & Baram-Tsabari, A. (2021). Public understanding of the mathematical aspects of the COVID-19 pandemic and its relation to school mathematics education. *Educational Studies in Mathematics, 108*, 201–225.

Hine, C. (2012). Headlice eradication as everyday engagement with science: An analysis of online parenting discussions. *Public Understanding of Science, 23*(5), 574–591.

Horrigan, J. B. (2016). *Lifelong learning and technology*. Pew Research Center, 22. https://www.pewresearch.org/internet/2016/03/22/lifelong-learning-and-technology/

Hurd, P. D. (1958). Science literacy: Its meaning for American schools. *Educational Leadership, 16*(1), 13–16.

Kahan, D. M., Peters, E., Wittlin, M., Slovic, P., Ouellette, L. L., Braman, D., & Mandel, G. (2012). The polarizing impact of science literacy and numeracy on perceived climate change risks. *Nature Climate Change, 2*(10), 732–735.

Kahneman, D. (2003). A perspective on judgment and choice: Mapping bounded rationality. *American Psychologist, 58*(9), 697.

Kaiser, F. G., & Fuhrer, U. (2003). Ecological behavior's dependency on different forms of knowledge. *Applied Psychology, 52*(4), 598–613.

Kolstø, S. D., Bungum, B., Arnesen, E., Isnes, A., Kristensen, T., Mathiassen, K., Mestad, I., Quale, A., Tonning, A. S. V., & Ulvik, M. (2006). Science students' critical examination of scientific information related to socioscientific issues. *Science Education, 90*(4), 632–655. https://doi.org/10.1002/sce.20133

Kunda, Z. (1990). The case for motivated reasoning. *Psychological Bulletin, 108*(3), 480.

Lammers, J., Crusius, J., & Gast, A. (2020). Correcting misperceptions of exponential coronavirus growth increases support for social distancing. *Proceedings of the National Academy of Sciences, 117*(28), 16264–16266.

Laslo, E., & Baram-Tsabari, A. (2021). Expressions of science literacy in online public discussions of animal experimentation. *International Journal of Science Education: Part B*. https://doi.org/10.1080/21548455.2020.1871103

Laslo, E., Baram-Tsabari, A., & Lewenstein, B. V. (2011). A growth medium for the message: Online science journalism affordances for exploring public discourse of science and ethics. *Journalism: Theory, Practice and Criticism, 12*(7), 847–870.

Layton, D., Jenkins, E., Macgill, S., & Davey, A. (1993). *Inarticulate science? Perspectives on the public understanding of science and some implications for science education*. Studies in Education.

Lewenstein, B. V. (2016). Can we understand citizen science? *Journal of Science Communication, 15*(1), E1.

McClune, B., & Jarman, R. (2010). Critical reading of science-based news reports: Establishing a knowledge, skills and attitudes framework. *International Journal of Science Education, 32*(6), 727–752.

Mezirow, J. (1990). How critical reflection triggers transformative learning. In *Fostering critical reflection in adulthood* (Vol. 1, p. 20).

Mezirow, J. (1997). Transformative learning: Theory to practice. *New Directions for Adult and Continuing Education, 1997*(74), 5–12.

Most, T., Ingber, S., & Heled-Ariam, E. (2012). Social competence, sense of loneliness, and speech intelligibility of young children with hearing loss in individual inclusion and group inclusion. *Journal of Deaf Studies and Deaf Education, 17*(2), 259–272.

National Science Board. (2020). Science and technology: Public attitudes, knowledge, and interest. In *Science and engineering indicators 2020*. U.S. Government Printing Office.

Norris, S. P. (1995). Learning to live with scientific expertise: Toward a theory of intellectual communalism for guiding science teaching. *Science Education, 79*(2), 201–217. https://doi.org/10.1002/sce.3730790206

Norris, S. P., & Phillips, L. M. (1994). Interpreting pragmatic meaning when reading popular reports of science. *Journal of Research in Science Teaching, 31*(9), 947–967.

Northern, J. L., & Downs, M. P. (2002). *Hearing in children*. Lippincott Williams & Wilkins.

Organisation for Economic Co-operation Development [OECD]. (2016). *PISA 2015 assessment and analytical framework: Science, reading, mathematic and financial literacy*. OECD Publishing.

Orr, D., & Baram-Tsabari, A. (2018). Science and politics in the polio vaccination debate on Facebook: A mixed-methods approach to public engagement in a science-based dialogue. *Journal of Microbiology & Biology Education, 19*(1). https://doi.org/10.1128/jmbe.v19i1.1500

Osborne, J., & Dillon, J. (2008). *Science education in Europe: Critical reflections*. Nuffield Foundation.

Perkins, D. N., & Salomon, G. (1988). Teaching for transfer. *Educational Leadership, 46*(1), 22–32.

Peters, H. P., Dunwoody, S., Allgaier, J., Lo, Y. Y., & Brossard, D. (2014). Public communication of science 2.0. *EMBO Reports*, e201438979. https://doi.org/10.15252/embr.201438979

Pritchard, D. (2015). Epistemic dependence. *Philosophical Perspectives, 29*(1), 305–324. https://doi.org/10.1111/phpe.12067

Roberts, D. A. (2007). Scientific literacy/science literacy. In S. K. Abell & N. G. Lederman (Eds.), *Handbook of research on science education* (Vol. 1, pp. 120–144). Lawrence Erlbaum Associates.

Roberts, D. A., & Bybee, R. W. (2014). Scientific literacy, science literacy, and science education. In N. G. Lederman & S. K. Abell (Eds.), *Handbook of research on science education*. Routledge.

Romine, W. L., Sadler, T. D., & Kinslow, A. T. (2017). Assessment of scientific literacy: Development and validation of the Quantitative Assessment of Socio-Scientific Reasoning (QuASSR). *Journal of Research in Science Teaching, 54*(2), 274–295.

Rudolph, J. L., & Horibe, S. (2016). What do we mean by science education for civic engagement? *Journal of Research in Science Teaching, 53*(6), 805–820. https://doi.org/10.1002/tea.21303

Ryder, J. (2001). Identifying science understanding for functional scientific literacy. *Studies in Science Education, 36*, 1–44.

Sadler, T. D., & Donnelly, L. A. (2006). Socioscientific argumentation: The effects of content knowledge and morality. *International Journal of Science Education, 28*(12), 1463–1488. https://doi.org/10.1080/09500690600708717

Sandoval, W. A., Sodian, B., Koerber, S., & Wong, J. (2014). Developing children's early competencies to engage with science. *Educational Psychologist, 49*(2), 139–152. https://doi.org/10.1080/00461520.2014.917589

Scharrer, L., Rupieper, Y., Stadtler, M., & Bromme, R. (2016). When science becomes too easy: Science popularization inclines laypeople to underrate their dependence on experts. *Public Understanding of Science*. https://doi.org/10.1177/0963662516680311

Schejter, A. M., & Tirosh, N. (2016). Media policy and theories of justice. In *A justice-based approach for new media policy* (pp. 51–59). Springer.

Sharon, A., & Baram-Tsabari, A. (2020a). Can science literacy help individuals identify misinformation in everyday life? *Science Education, 104*, 873–894. https://doi.org/10.1002/sce.21581

Sharon, A. J., & Baram-Tsabari, A. (2020b). The experts' perspective of "ask-an-expert": An interview-based study of online nutrition and vaccination outreach. *Public Understanding of Science, 29*(3), 252–269.

Shauli, S., & Baram-Tsabari, A. (2019). The usefulness of science knowledge for parents of hearing-impaired children. *Public Understanding of Science, 28*(1), 19–37.

Shea, N. A. (2015). Examining the nexus of science communication and science education: A content analysis of genetics news articles. *Journal of Research in Science Teaching, 52*(3), 397–409. https://doi.org/10.1002/tea.21193

Simonneaux, L. (2008). Argumenttaion in socio-scientific Context. In S. Erduran & M. P. Jiménez-Aleixandre (Eds.), *Argumentation in science education: Perspectives from classroom-based research* (pp. 187–199). Springer.

Snow, C. E., & Dibner, K. A. (Eds.). (2016). *Science literacy: Concepts, contexts, and consequences*. The National Academies Press.

Taragin-Zeller, L., Rozenblum, Y., & Baram-Tsabari, A. (2020). Public engagement with science among religious minorities: Lessons from COVID-19. *Science Communication, 42*(5), 643–678.

Vosoughi, S., Roy, D., & Aral, S. (2018). The spread of true and false news online. *Science, 359*(6380), 1146–1151.

Wolfensberger, W. P. (1977). A multi-component advocacy/protection schema.

Wynne, B. (1996). Misunderstood misunderstanding: Social identities and public uptake of science. In A. Irwin & B. Wynne (Eds.), *Misunderstanding science?: The public reconstruction of science and technology* (pp. 19–46). Cambridge University Press.

Zillien, N., Haake, G., Fröhlich, G., Bense, T., & Souren, D. (2011). Internet use of fertility patients: A systemic review of the literature. *Journal für Reproduktionsmedizin und Endokrinologie-Journal of Reproductive Medicine and Endocrinology, 8*(4), 281–287.

# Chapter 2
# Engaging in Argumentation as Critical Evaluation of the Anti-vaccination Movement

**Blanca Puig and Noa Ageitos**

## 2.1 Introduction

Biology education faces many challenges nowadays, some of them related to teaching health controversies presented by the post-truth era, which demands critical thinking (CT) and social responsibility. A disconnect between scientific consensus and public opinion on topics such as vaccine safety has existed for a number of years (Mian & Khan, 2020); however, the health crisis produced by the coronavirus disease has triggered a fast spread of misinformation and denial movements, such as the anti-vaccination movement, which have been dismissing credible sources of information. The anti-vaccination movement is not a new problem; it has had a demonstrable impact on vaccination policy, and individual and community health (Poland & Jacobson, 2001). However, despite its social impact and potential influence on health teaching, it has not been deeply explored in teacher training programmes in our country (Spain). This chapter argues that teacher education needs to explore learning experiences and ways of instruction that support health literacy through engaging learners in argumentation and CT on socio-scientific issues.

As a construct, health literacy includes the articulation of five core components: theoretical knowledge, practical knowledge, CT, self-awareness, and social responsibility (Paakkari & Paakkari, 2012). This study follows this view and understands the practice of CT as the articulation of cognitive skills through the practice of argumentation (Giri & Paily, 2020). One of the central features in argumentation is the development of epistemic criteria for knowledge evaluation (Jiménez-Aleixandre & Erduran, 2008), which is a necessary skill to be a critical thinker. When debating

B. Puig (✉) · N. Ageitos
Department of Experimental Sciences, University of Santiago de Compostela, Santiago de Compostela, Spain
e-mail: blanca.puig@usc.es; noa.ageitos@edu.xunta.gal



K. Korfiatis, M. Grace (eds.), *Current Research in Biology Education*, Contributions from Biology Education Research,
https://doi.org/10.1007/978-3-030-89480-1_2

a health controversy, participants should consider diverse aspects, including scientific knowledge but also ethical and social aspects. Given the complexity of these issues, people are required to use their CT to challenge established ideas, understand new ideas and evaluate multiple perspectives (Jiménez-Aleixandre & Puig, 2012). In addition to all of these aspects, the advent of the 'post-truth' era has highlighted new challenges to teaching about health.

This chapter presents a case study developed with a group of primary pre-service teachers (PSTs) in the context of teaching health controversies through the use of argumentation and decision-making tasks. The study was carried out at during a significant moment for teachers and educators in our region, when vaccination was established as mandatory for students in kindergarten in all our state schools. Health education is part of the curriculum in Primary education in Spain, and among the contents teachers at this educational level are expected to introduce are the human body and infectious diseases, and ways to prevent and treat them, including vaccines. Therefore, it is important to prepare them in this domain. Furthermore, taking into account the rise in anti-vaccination movements and the importance of CT development from early years, we decided to design a CT activity that engages future teachers in the practice of CT along with knowledge application. The research aims to extend empirical investigation on health controversies and argumentation in PSTs' training. The main goal is to explore PSTs' positions regarding vaccination and their capacity to critically argue about diverse premises supporting the anti-vaccination movement.

## 2.2 Argumentation and Critical Thinking About Health Controversies

It is suggested that argumentation on socio-scientific issues (SSIs) makes biology learning meaningful and promotes CT (Puig & Jiménez-Aleixandre, 2011), since these issues provide a context where students are expected to make argued decisions and take an active role. CT can be characterized as a self-regulating process that judges what to think or do in a particular context (Facione, 1990). Science learning has broadly focused on the transmission of knowledge and has paid less attention to scientific argumentation (Giri & Paily, 2020). Argumentation is framed in this study in the sociocultural perspective, which supports the notion that higher thinking processes originate from socially intermediated activities (Erduran & Jimenez-Aleixandre, 2012). Scientific argumentation, thus, is considered as a scientific practice that involves the evaluation of knowledge in the light of evidence (Jiménez-Aleixandre, 2010). According to Kuhn (2019), we should expand the view of thinking to the construct of CT, which means understanding CT as a dialogic practice that people enact and thereby become predisposed towards practicing. We see CT as reasonable and independent thinking (Jiménez-Aleixandre & Puig, 2012) that entails a set of skills and dispositions, which can be developed through the practice (Facione et al., 2000). Argumentation, understood as the evaluation of claims using evidence (Jiménez-Aleixandre, 2010) is deeply connected with this definition of CT. This has educational implications, for example on the creation of learning environments and learning activities that promote the development of CT. In biology education, socio-scientific contexts such as health controversies could help towards these goals since they engage students in the practice of CT.

CT and health literacy are increasing their importance in biology education in the post-truth era, as the coronavirus pandemic and denial movements related to its origin, prevention and treatment are showing. Critical thinkers of the twenty-first century need to question their place in a world where the biggest problems society faces continue and demand critical actions and independent thinking. Jiménez-Aleixandre et al. (2017) have proposed a characterization of CT that includes two main components: one related to deliberate judgement and the other to citizen literacy. The capacity to develop an independent opinion and challenge social and culturally established ideas, and the ability of criticizing and analyzing discourses that justify inequality (Fairclough, 1995) are embedded in this view of CT. This is a fundamental component of CT, required to be able to make independent and reasoned decisions on health issues in the post-truth condition. The propagation of 'fake news' and disinformation, the use of pseudo-therapies and the increasing number of opponents to vaccination have become worrying, and are issues that should be addressed in teacher training programmes and in biology lessons when looking at the topic of health. Biology education should involve the creation of a dialogic discourse among members of a class, which goes beyond 'what we know' and places equal or even more important focus on the teaching and learning of 'how did we come to know' and 'why we accept that knowledge over alternatives' (Duschl, 2020). The question is: *How we can support our students to be critical thinkers in this context?*

Despite the wealth of theoretical studies on CT, further research is needed on how CT is articulated through the practice of argumentation in biology instruction. The CRITHINKEDU project (Dominguez, 2018) in which international experts from diverse European countries worked together in the development of an educational protocol to introduce CT in higher education, provides some useful guidelines for teachers. This protocol rests on two main claims (Elen et al., 2019): (1) students will develop their CT by explicitly engaging in appropriate learning activities; (2) becoming stronger in CT requires repeated engagement in CT processes. This implies that CT should be practiced continuously in biology education by allowing students to think critically and engage in open-ended activities that invite students to argue and practice CT skills. Hence, according to Vincent-Lacrin et al. (2019), research should focus on the design of learning environments that provide opportunities to practice CT without diminishing subject matter, content, and procedural knowledge. This chapter attends to this view and engages PSTs in the practice of argumentation and critical evaluation in a task on the anti-vaccination movement. A study developed in Spain (Maguregi-González et al., 2017) has revealed that PSTs face diverse problems when dealing with this issue, and experience difficulties applying the model of immunization and putting it into practice in a CT context.

## 2.3 Methodology

We have used a qualitative, case study research methodology (Gerring, 2007), and we draw from discourse analysis (Gee, 2011) to address the research questions as presented below.

### *2.3.1 Study Context and Instructional Design*

The study focuses on the analysis of a task carried out by 39 PSTs taking the 3rd course year in a public university in Spain. The subject, related to biology instruction in primary education, is designed to tackle biology content through the participation in scientific practice and CT. The first author, who was the teacher of the class, led the activity, whereas the second author attended as a passive observant. The study was undertaken in a single 90-min session in which data were collected from the individual students who were training to be primary teachers. The design of the activity draws from a previous one developed and implemented by the authors with a group of secondary PSTs (Ageitos & Puig, 2016). It includes three parts, the first one being the presentation of a hypothetical real-life situation: the case of a 3-year old child named Lía, whose parents were interested in enrolling her in a state kindergarten despite their decision to stop vaccinating her. Students had to consider several pieces of information provided (Lía's vaccination card, the official vaccination card and the regulation that states that vaccination is compulsory in our region to be enrolled in a state kindergarten), to decide whether or not this girl would be able to attend the kindergarten.

In the second part of the task the anti-vaccination movement was introduced. First, we asked students if they were familiar with this movement and what they knew about it. Secondly, they were presented with five premises that support the anti-vaccination movement. The goal was to critically assess and argue about their position regarding each premise.

Thirdly, students were provided with the text below, which shows the reasons Lía's parents decided to stop vaccinating their daughter:

> We decide on matters of our daughter's health. Lía already has a lot of vaccines in place, and it turns out that we read that vaccines can be dangerous. There are boys and girls who have become autistic; it is better for them to pass the disease naturally; as we are using so many vaccines, they are becoming less efficient and are making us weaker (…).

Participants were asked to express their own opinion regarding this situation and then explain whether they agreed or disagreed with it and justify their positions.

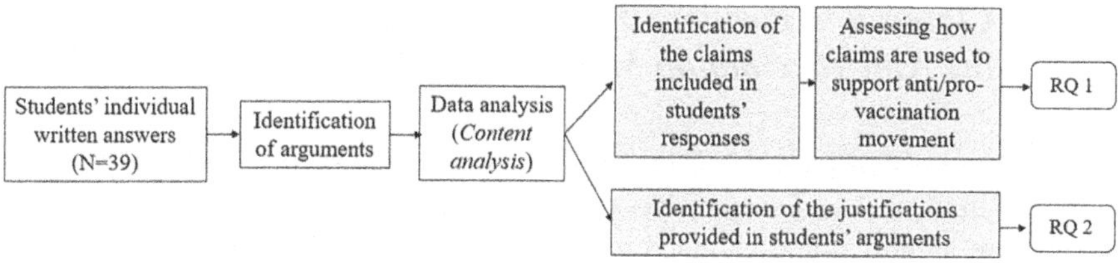


**Fig. 2.1** Data analysis process

Lastly, participants were asked to reflect on their knowledge on vaccination and about the importance of introducing this topic in primary education and teacher training. This article will focus on the analysis of students' written answers to the third part of the task.

### 2.3.2 *Data Collection, Analysis and Coding*

Each student's individual reports (N = 39) were collected and numbered to guarantee their anonymity. Students' written responses were examined based on the arguments identified and then coded by applying content analysis methods (Lincoln & Guba, 1985). Both authors carried out the codification, and discrepancies were discussed to reach an agreement. Figure 2.1 presents the process of analysis which the authors followed to address the two research questions.

The first research question focuses on the identification of the claims provided by Lia's parents and used in students' written responses. The content of the five claims provided by Lía's parents referred to: (a) *free will*; (b) *dangerousness of vaccines*; (c) *low effectiveness and weakening power*; (d) *causing autism*; (e) *go through the disease naturally*. Besides this identification, content analysis was used to determine whether students used these claims to support the anti-vaccination movement, to go against it or whether they were neutral to the movement.

The second research question focused on the justifications used by students to support their own arguments regarding each claim. First, students' answers were segmented into arguments, with the analysis then focusing on the content of the justification provided.

## 2.4 Results

We begin by addressing RQ1, about the claims identified regarding the anti-vaccination movement and students' positions in relation to them. RQ2, relating to the justifications used by students to support their arguments, is then tackled.

### 2.4.1 Claims and Positions Regarding the Anti-vaccination Movement

The analysis of the first research question is summarized in Fig. 2.2. The five claims provided in the task and used by Lía's parents to support the anti-vaccination movement were identified and included in students' answers. Most students (32 out of 39) used at least one of the claims, but none of the students addressed all five claims in their written answers. Figure 2.2 shows in different shades of grey the distribution of the claims according to their position regarding the anti-vaccination movement. Three categories were identified: *pro-vaccines*, *anti-vaccines* and *neutral.* The last category included students' answers that did not show a particular position regarding the anti-vaccination movement.

The analysis shows that the most common claim students' used in their written responses was the '*dangerousness of vaccines*'. It appears in 21 of the responses. The next most frequent were '*vaccines cause autism*' and that it was preferable to '*go through the disease naturally*'. These appear in a total of 15 responses each. As seen in Fig. 2.2, the majority used all the premises to go against the anti-vaccination movement (a total of 5 anti-vaccination accounts against 65 pro-vaccination), although two premises ('*free will*' and '*go through the disease naturally*') were used in favour by a big proportion. Examples of students assessing the five premises showing their position to the vaccination movement (in favour or against) are presented below.

***Free Will*** A total of 12 students referred to this claim in their written answers. Most of them (9) to support the pro-vaccination movement, and three against it.

A26: (…) *They have the right to make decisions about their daughter.*

In the example, student 26 is supporting the anti-vaccination movement by agreeing with the fact that there should be free will.

A7: *To begin with, the statement that "they decide on their daughter's health" is not at all appropriate, first because they lack (I suppose) appropriate knowledge, and second because they could neither morally nor legally make a decision against their daughter's health.*

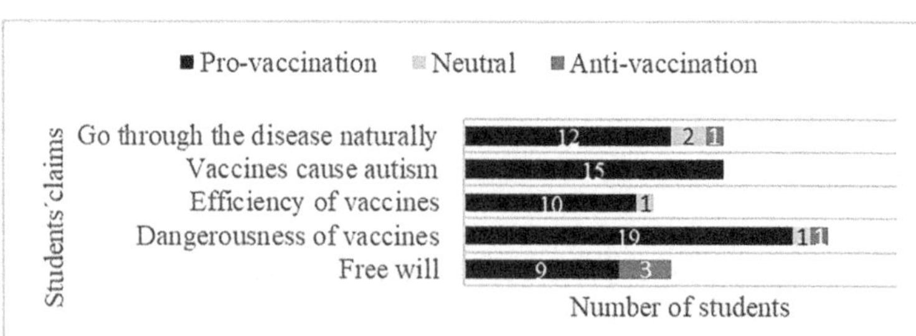


**Fig. 2.2** The most often used claims to argue about the anti-vaccination movement

This student shows a disagreement regarding the free will premise provided by Lía's parents. He provides two reasons against it: one related to knowledge and the other to ethical aspects. This answer can be positioned against the anti-vaccination movement.

***Dangerousness of Vaccines*** This is the most common premise used by students. Most of them (19 out of 21) mentioned it to support the use of vaccines. Only one student was against it and another one showed a neutral position.

A3: *Vaccines are not dangerous; in fact, they pretend that the organism is ready against diseases and issues.*
A1: *There may be some case in which it [a vaccine] has produced some side effect.*

Student A1 acknowledges that side effects are possible when getting a vaccine. He seems to support the premise, despite presenting a mild argument.

***Efficiency of Vaccines*** This is a less common claim used, appearing a total of ten times supporting vaccination and one time not supporting or arguing against it.

A38: *Maybe vaccines are becoming less efficient, I don't know, but I think that if it were true it would be because our immune system is stronger rather than abusing vaccines.*

This student acknowledged his/her lack of knowledge on the subject. However, he/she seems to have a position against the anti-vaccination movement by relating the supposed reduced efficiency to the role of our immune system instead of the overuse of vaccines.

A18: *The last argument refers to the effectiveness of vaccines, again based on beliefs and no kind of verification.*

In this case, the student identified the premise as a social belief and made reference to the lack of evidence behind it.

***Vaccines Cause Autism*** A total of 15 students used this premise in their answers, being the only claim used to support vaccination in all the responses.

A5: *A child cannot "turn into" autistic, a child is born being autistic, it has nothing to do with vaccines.*

This student rejected the premise, pointing to the genetic origin of the disease.

***Go Through the Disease Naturally*** 15 students refer to this claim.

A6: *I would only agree with Lía's parents in going through the disease naturally in cases of mild and short diseases.*

This student agreed with the claim, however, with certain caveats. He seems to be in favour of going through a disease depending on the effects of the disease; this would imply a certain level of support of the anti-vaccination movement related to this issue.

A11: *In addition, in the last lines they say that it is better to pass the disease naturally, but this decision does not only involve her daughter; the fact that Lía is not vaccinated can cause the people around her to become infected.*

Student A11 was opposed to this claim, and argued his/her position on the role of vaccination in providing herd immunity and social protection.

It is important to note that some students used one premise to support the anti-vaccination movement and another to support the pro-vaccination movement. However, as the results in this section show, most students are against the anti-vaccination movement and references agreeing with it were scarce.

### *2.4.2 Justifications Presented by Students to Support Arguments Against or in Favour of Vaccination*

Regarding the second research question, two main categories of justifications emerged from the content analysis applied to students' individual arguments: *information provided* and *usefulness of vaccination*. A description is provided below including examples of students' arguments.

1. **Information provided:** this category includes justifications related to the information and/or data behind the anti-vaccination arguments. It can be divided into two subcategories, as described below.

1a) **Lack of information/data:** this includes justifications that make reference to the lack of data behind Lía's parents' arguments. It is quite frequent, appearing in 18 students' responses. Examples are provided below.

A4: *I understand that they are afraid, but they should inform themselves better about the topic, because they are wrong.*

A5: *I don't agree because they don't talk about the data (…) needed to support themselves.*

1b) **Source of information:** this category includes students' justifications that point to the source of information behind Lia's parents' arguments. It appears a total of 14 times in students' answers.

A10: *If we knew the source, we could assess its reliability. If our arguments were based on sources without scientific rigor, we would not have an objective and reasoned opinion.*

A11: *Lía's parents argue that they read that vaccines can be dangerous, but it would be important to know where they read it, as many times we accept the first information we read and we don't go any further. We don't check the reliability of the source or that the information provided has been contradicted.*

These students pointed to the importance of knowing the source of information in order to assess its validity.

2. **Usefulness of vaccines:** this category embodies students' responses about the benefits of vaccination in the population. It appears a total of 18 times.

A16: *Vaccines are important for child health; non-vaccinated children can get serious diseases.*
A11: *The fact that Lía is not vaccinated can cause people around her to get infected. Thus, this decision would have consequences.*

Student 11 explained how one person failing to get vaccinated could compromise other people in the community.

## 2.5 Discussion and Conclusions

Argumentation is a prominent topic in biology education research on SSIs. However, teacher training and our educational system need to pay further attention to how to develop CT through the practice of argumentation in contexts that are particularly relevant nowadays, such as the anti-vaccination movement. The term 'vaccine hesitancy' is increasingly used to describe the spread of such vaccine reluctance (Peretti-Watel et al., 2015). It is an approach to the issues raised by vaccination which is more acute than the opposition between pro and anti-vaccination (e.g. Johnson et al., 2020; Peretti-Watel et al., 2015). This study follows the view of these authors, who indicate that vaccine hesitancy is a feature attributed to large and heterogeneous groups of people who share varying degrees and reasons of indecision and who hold an intermediate position along a continuum ranging from full support for vaccination to strong opposition to any vaccine. The findings of our study show that this group of PSTs were able to identify different premises related to the anti-vaccination movement and discuss them. As Atwell and Salmon (2014) point out, it is difficult to label a person as pro-vaccination or anti-vaccination. The analysis of the discussion related to each premise revealed that some students showed different positions (against or in favour of vaccines) when they assessed the different premises. For instance, some students argued against vaccination on the basis of free will/choice but were pro-vaccination when they assessed the premise about vaccines causing autism. All in all, the participants in this study are closer to the pro-vaccination movement.

As the results of the second research question show, most participants were able to justify their answers when assessing the premises. The most frequent justifications were related, on the one hand, to the information used, and on the other hand, to the usefulness of vaccines and vaccination. A proportion of students identified the lack of information supporting Lia's parents' premises, but also pointed to the need for reliable information, such as consulting experts. Similar results have been previously reported in other literature in this context, such as Navarro Alonso et al.'s

(2001) research that shows how Spanish families acknowledge pediatricians as experts to be consulted in health topics such as vaccination. This contrasts with a later investigation that shows how one out of four pediatric professionals express doubts about including some vaccines in the official immunization schedule (Picchio et al., 2019). Students not only acknowledge the lack of information, but also, as for instance student A38, their own limitations and lack of information on the subject, which can be related to the CT skill of 'self-regulation' (Facione et al., 2000). This CT skill is defined by these authors as self-consciously monitoring one's cognitive activities. In this case this refers to the knowledge or information required to assess the claims that support the anti-vaccination movement.

We are aware that these results present some limitations. The findings should not be generalized, as this is a case study and cannot be completely replicated elsewhere. Moreover, the academic context may influence students' answers, and the choice of the premises presented in the task may not cover all the information related to the anti-vaccination movement. However, the study has provided evidence regarding how to engage PSTs in CT through the practice of argumentation. This chapter seeks to contribute to debates around how to equip teachers to foster the development of CT in their students. We think that educating good critical thinkers is more than training teachers and students to put into practice a set of cognitive skills, and that it should involve modelling how to critically evaluate socio-scientific information, which has implications in teacher training programmes on argumentation in biology education.

**Acknowledgments** This work was supported by the Spanish Ministry of Science, Education and Universities, partly funded by the European Regional Development Fund (ERDF). Contract grant PGC2018-096581-B-C22. This study was developed in RODA group.

## References

Ageitos Prego, N., & Puig, B. (2016). ¿Debería ser obligatoria la vacunación? *Alambique Didáctica de las Ciencias Experimentales, 83*, 78–79.

Atwell, J. E., & Salmon, D. A. (2014). Pertussis resurgence and vaccine uptake: Implications for reducing vaccine hesitancy. *Pediatrics, 134*, 602–604. https://doi.org/10.1542/peds.2014-1883

Dominguez, C. (coord.) (2018). *A European review on critical thinking educational practices in higher education institutions*. UTAD. ISBN: 978-989-704-258-4. Available at: https://www.researchgate.net/publication/322725947_A_European_review_on_Critical_Thinking_educational_practices_in_Higher_Education_Institutions

Duschl, R. (2020). Practical reasoning and decision making in science: Struggles for truth. *Educational Psychologist, 3*, 187–192. https://doi.org/10.1080/00461520.2020.1784735

Elen, J., Jiang, L., Huyghe, S., Evers, M., Verburgh, A., … Palaigeorgiou, G. (2019). In C. Dominguez & R. Payan-Carreira (Eds.), *Promoting critical thinking in European higher education institutions: Towards an educational protocol*. UTAD.

Erduran, S., & Jimenez-Aleixandre, J. M. (2012). Research on argumentation in science education in Europe. In D. Jorde & J. Dillon (Eds.), *Science education research and practice in Europe: Retrospective and prospective* (pp. 253–289). Sense Publishers.

Facione, P. A. (1990). *Critical thinking: A statement of expert consensus for purposes of educational assessment and instruction. Research findings and recommendations*. The California Academic Press.
Facione, P., Facione, N. C., & Giancarlo, C. A. (2000). The disposition toward critical thinking: Its character, measurement, and relationship to critical thinking skill. *Informal Logic*, (1). https://doi.org/10.22329/il.v20i1.2254
Fairclough, N. (1995). *Critical discourse analysis. The critical study of language*. Longman.
Gee, J. P. (2011). *An introduction to discourse analyses. Theory and method*. Routledge.
Gerring, J. (2007). *Case study research: Principles and practices*. Cambridge University Press.
Giri, V., & Paily, M. U. (2020). Effect of scientific argumentation on the development of critical thinking. *Science & Education, 29*, 673–690.
González, M., Ibarlueza, U., & Etxaburu, B. (2017). Modelización, argumentación y transferencia de conocimiento sobre el sistema inmunológico a partir de una controversia sobre vacunación en futuros docentes. *Enseñanza de las Ciencias, 35*(2), 29–50. https://doi.org/10.5565/rev/ensciencias.2237
Jiménez-Aleixandre, M. P. (2010). *Ideas Clave. Competencias en argumentación y uso de pruebas*. Graó.
Jiménez-Aleixandre, M. P., & Erduran, S. (2008). Argumentation in science education: An overview. In S. Erduran & M. P. J. Aleixandre (Eds.), *Argumentation in science education: Perspectives from classroom-based research* (pp. 3–27). Springer.
Jiménez-Aleixandre, M. P., & Puig, B. (2012). Argumentation, evidence evaluation and critical thinking. In B. Fraser, K. Tobin, & C. McRobbie (Eds.), *Second international handbook of science education*. Springer.
Jiménez-Aleixandre, M. P., Brocos, P., & Puig, B. (2017). Personal or social responsibility? Paper presented at the symposium *challenging social ideas as a component of critical thinking*. EARLI conference, 29 August–2 September 2017, Tampere, Finland.
Johnson, N. F., Velasquez, N., Jonhson Restrepo, N., Leahy, R., Gabriel, N., El Oud, S., Zheng, M., Manrique, P., Wuchty, S., & Lupu, Y. (2020). The online competition between pro-and anti-vaccination views. *Nature, 582*, 230–234.
Kuhn, D. (2019). Critical thinking as discourse. *Human Development, 62*(3), 146–164.
Lincoln, Y., & Guba, E. (1985). *Naturalistic enquiry*. Sage.
Mian, A., & Khan, S. (2020). Coronavirus: The spread of misinformation. *BMC Medicine, 18*, 89. https://doi.org/10.1186/s12916-020-01556-3
Navarro Alonso, J. A., Bernal González, P. J., & Niguez Carbonell, J. C. (2001). Analysis of factors influencing vaccine uptake: Perspective from Spain. *Vaccine, 20*(Suppl 1), S13-5, discussion S1. https://doi.org/10.1016/S0264-410X(01)00300-0
Paakkari, L., & Paakkari, O. (2012). Health literacy as a learning outcome in schools. *Health Education, 112*(2), 133–152. https://doi.org/10.1108/09654281211203411
Peretti-Watel, P., Larsoon, H. J., Ward, J. K., Schulz, W. S., & Verger, P. (2015). Vaccine hesitancy: Clarifying a theoretical framework for an ambiguous notion. *PLOS Currents Outbreaks*. https://doi.org/10.1371/currents.outbreaks.6844c80ff9f5b273f34c91f71b7fc289
Picchio, C. A., Carrasco, M. G., Sagué-Vilavella, M., & Rius, C. (2019). Knowledge, attitudes and beliefs about vaccination in primary healthcare workers involved in the administration of systematic childhood vaccines, Barcelona, 2016/17. *Euro Surveillance, 24*, 18001172019.
Poland, G. A., & Jacobson, R. M. (2001). Understanding those who do not understand: A brief review of the anti-vaccine movement. *Vaccine, 19*(17–19), 2440–2445.
Puig, B., & Jiménez-Aleixandre, M. P. (2011). Different music to the same score: Teaching about genes, environment and human performances. In T. D. Sadler (Ed.), *Socio-scientific issues in the classroom: Teaching, learning and research* (pp. 201–239). Springer.
Vincent-Lacrin, S., González-Sancho, C., Bouckaert, M., de Luca, F., Fernández-Barrera, M., Jacotin, G., Urgel, J., & Vidal, Q. (2019). Fostering students' creativity and critical thinking: What it means in school. In *Educational research and innovation*. OED Publishing. https://doi.org/10.1787/62212c37-en

# Chapter 3
# Theory of Planned Behavior in the Context of Stem Cell Donation

**Julia Holzer and Doris Elster**

## 3.1 Introduction

Leukaemia is diagnosed in approximately 13,000 German citizens every year (Robert Koch Institute, 2016). Stem cell transplantation represents a curative treatment of leukemia, but the demand for new donors is still very high. In about 20% of the cases no suitable match between donor and leukemia patients is found. Therefore, this study tries to raise the awareness of this health issue and inform young people about stem cell donation so that they are able to make a well-founded decision regarding registration as stem cell donors in the future as young adults.

Based on the Theory of Planned Behaviour (Ajzen, 2005) the teaching unit "Wake up – Sensitization of young people for stem cell donation" is set up and conducted with students of the upper secondary level. The overarching goal of this study is to gain a better understanding how different predictive factors, as well as the adolescents' intention for or against stem cell donation, can be influenced by the teaching unit.

## 3.2 Theoretical Frame

The Theory of Planned Behaviour (TPB) by Ajzen (2005) helps to understand variables that determine behaviour and defines *intention* as an indicator of how strongly an individual is willing to carry out a certain behaviour. It is assumed that the

J. Holzer (✉) · D. Elster
Institute for Science Education – Biology Education, University of Bremen, Bremen, Germany
e-mail: doris.elster@uni-bremen.de



K. Korfiatis, M. Grace (eds.), *Current Research in Biology Education*, Contributions from Biology Education Research,
https://doi.org/10.1007/978-3-030-89480-1_3

stronger or more positive the intention is in relation to the behaviour, the more likely it is that the behaviour will actually be carried out. Initially, TPB emerged from the TRA (The Theory of Reasoned Action), where intention was determined by only two factors, namely *attitudes* (ATT) and the *subjective norm* (SN). In the TPB, in addition to the two determinants mentioned above, a third factor is added, that is, *perceived behaviour control* (PBC). ATT shows how positively or negatively a person evaluates a certain behaviour. The SN describes the subjectively perceived social environmental pressure that an individual feels when performing or refraining from an action. PBC shows how easy or how difficult it is for an individual to carry out a given behaviour. In summary, the *intention* to perform a particular behaviour is more likely to be achieved the more positively a person evaluates a behaviour and believes that other people will welcome it if they perform it, as well as if they have a high level of control over their own behavioural performance (Ajzen, 1991).

The three predictors of intention (ATT, SN, PBC) are determined by beliefs, namely *behavioural*, *normative* and *control* beliefs (Fig. 3.1) (Ajzen, 2005). The beliefs represent the cognitive basis of TPB. The behaviour-related beliefs represent likelihoods of occurrence (expectations) of possible behavioural outcomes. The behavioural beliefs are weighted by the ratings of the particular behavioural consequence by multiplying the probability of occurrence of a consequence by its (positive/negative) rating (evaluation). The final attitude towards the questioned behaviour results from the sum of all relevant behavioural beliefs-evaluation products (Ajzen, 1991; Ajzen & Fishbein, 2002). A person who is convinced that the performance of the behaviour is highly likely to lead to positive consequences shows a more positive attitude towards the performance of such behaviour than, for example, a person who is highly likely to associate the negative consequences with the performance of the behaviour (Ajzen, 1991).

Similarly, the subjective norm and perceived behaviour control are derived from beliefs. With regard to normative beliefs, however, it should be noted that these represent a person's perception of the extent to which their relevant reference

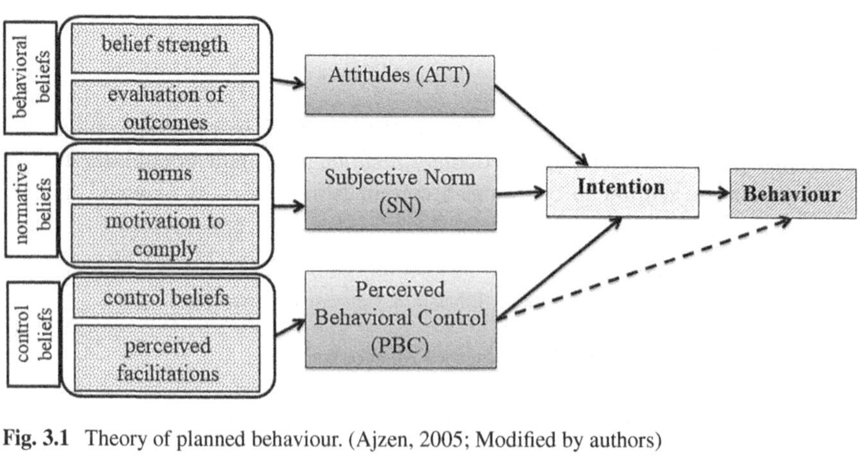


**Fig. 3.1** Theory of planned behaviour. (Ajzen, 2005; Modified by authors)

persons would welcome or reject the performance of the behaviour. In order for normative beliefs to have behavioural effect, the person must also assess the extent to which he or she wishes to comply with the expectations of the relevant reference persons (*motivation to comply*) (Eagly & Chaiken, 1993).

Control beliefs include all the resources (abilities, skills, willpower, information) that a person has for the performance of behaviour, as well as the existing restrictions that prevent the performance of actions (Ajzen, 2008). Many external factors, such as time, opportunity to carry out the action, potential situational obstacles or dependence on other people, which may favour the execution of the behaviour, can have a restrictive effect. All these beliefs determine the perceived behaviour control. In general, the higher the normative beliefs, control beliefs and attitude-related beliefs, the more likely is it that the behaviour in question will actually be carried out by the person (Ajzen, 2008).

Fishbein and Ajzen (1975) initially assumed that all predictors not included in the theory (TRA) only have an indirect effect on behaviour by influencing attitude or subjective norm. Many studies have strongly questioned this based on their research. Finally, in 1991 Ajzen agreed that it can make sense to include other factors in the theory if they contribute to the variance explanation of intention or behaviour independently of the three predictors already included. Research studies demonstrate that different model external factors, such as *content knowledge* (Bednall et al., 2013), *moral obligation* (Armitage & Conner, 2001), *moral reasoning* (Rushton, 1980) and *empathy* (Oliner & Oliner, 1988), may increase the predictive power of the TPB model. Therefore, these factors are inserted in our "extended TPB research model" which is the basis of this study.

According to Ajzen (1991), personal norms express the feeling of responsibility and duty to carry out a behaviour. Schwartz (1977) assumes that the perception of a distress (for example of another person) activates the "moral and value system" of the observer. In this process, people act in concrete situations based on the values and norms they consider appropriate. Personal norms derived in this way ultimately generate a sense of moral obligation (Schwartz & Howard, 1981) that motivates behaviour (Schwartz, 1977; Schwartz & Howard, 1981).

In short, *moral reasoning* summarizes the evaluation and processes in which a certain behaviour is assessed as morally correct. In this process, decisions on action are made on the basis of *moral reasoning* (Richardson, 2018). External moral standards play an important role here, i.e. the persons or their decisions are based on what is expected of them in view of moral standards. *Empathy* expresses the perception and ability to experience the need or suffering of another person. The observer feels an emotional arousal, which is called *empathy* (Schwartz, 1977).

And finally, this study looks at *knowledge*. Kaiser and Frick (2002) distinguish between three forms of knowledge, namely *system knowledge*, *action-related knowledge* and *effectiveness knowledge*. All three types of knowledge or some combinations of them can have a predictive effect on behaviour. Even if the distinction between the three dimensions of knowledge could not always be empirically confirmed (Kaiser & Frick, 2002) – which this study also showed – the differentiation on the theoretical level still seems to make sense: the differentiation of knowledge increases the degree of specificity in knowledge. *System knowledge* is described by Frick et al. (2004) in the context of ecological behaviour as knowledge about the dynamics of (eco)-systems and knowledge about (environmental) problems. *Action-related knowledge* includes knowledge about possible options for action and provides an assessment of whether corresponding actions can be taken and what "costs" they entail (Frick et al., 2004). The *effectiveness knowledge* indicates how effective an option for action can be and whether it is ultimately worthwhile to bear the costs associated with taking action. Therefore, *effectiveness knowledge* describes the potential of a particular action or the relative potential of different actions (Kaiser & Frick, 2002).

## 3.3 Research Questions and Method

The main objective of this study is to examine the impact of the intervention on the components of the extended TPB model. Internal factors, namely ATT, SN and PBC are considered as well as external factors, namely *content knowledge*, *moral obligation* and *moral reasoning*. The following research question was formulated:

> To what extent do the internal and external TPB constructs change after participation in the intervention "Wake up"?

### *3.3.1 Methodology*

A total of 263 students (age ø: 16.78; N = 162♀/N = 101 ♂) participated in the "Wake up" intervention and filled in a questionnaire (pre-post design). [In addition, 68 students who had not participated in the intervention were surveyed as a control group. The results of this survey can be requested from authors.] The scientific investigation at public schools (investigation on students) in the federal state of Bremen (Germany) was examined and approved in accordance with §13 of the Bremen School Data Protection Act (§13 *Bremisches Schuldatenschutzgesetz*). All students and their parents who agreed to participate in the research study had been informed about §13 and had signed a consent form, where, among other aspects, the objectives of the study were explained to the respondents. The collected data were processed in anonymous form in the present study.

For data analysis in SPSS mainly t-tests were performed. Belief-based-attitude is derived from the product of the perceived likelihood of outcomes (outcome beliefs) and the evaluation of those outcomes (evaluations). The same procedure is used to form multiplicative assumptions of belief-based subjective norm as well as for belief-based perceived behavioral control. Finally, we summed up the products of each construct and used the means for further statistical tests.

Based on the results of the factor analysis, this study distinguishes between two types of positive and negative belief-based attitudes. Also the PBC is examined in the form of two factors, i.e. PBC-IF, which reflects the internal factors such as fears of needles and blood, and PBC-EF, which covers the external factors such as awareness of opportunity, time taking and others. Considering the model external factors, the *content knowledge,* which was conducted with 8 questions, and *moral obligation* were developed by the authors, whereby the formulation of *moral obligation* was strongly oriented on the recommendations of Schwarz (1977). For organizational reasons, however, only the *system knowledge* and *action-related knowledge* could be operationalized in this study. The results of the factor analysis failed to

**Table 3.1** TPB-factors and model external factors

| Factor (number of items) | Example | Post-test α |
|---|---|---|
| Attitude[a] (5) | How do you personally feel about actually registering as a stem cell donor in the foreseeable future? | .873 |
| Subjective norm[a] (3) | Most people who are important to me think that I should register as a stem cell donor for leukemia patients within the forseeable future. | .869 |
| Perceived behavioral control[a] (5) | How difficult do you think it is for people your age to register as stem cell donors? | .840 |
| Belief based-(positive) attitude[b] (4) | Saving a life by donating stem cells, I think... | .834 |
| Belief based-(negative) attitude[b] (5) | The danger of contracting leukaemia oneself when donating, I think... | .757 |
| Belief-based subjective norm[b] (4) | If you think of your parents, how important/unimportant is it to you to meet their expectations regarding your registration as a stem cell donor? | .912 |
| Belief-based perceived behavioral control-(IF)[b] (2) | For registration as a stem cell donor for leukaemia patients in the foreseeable future, my fear of needles is… | .862 |
| Belief-based perceived behavioral control-(EF)[b] (5) | For the registration as a stem cell donor for leukaemia patients in the foreseeable future, the availability of an opportunity to do so is... | .762 |
| Intention (2) | I think that I will register as a stem cell donor in the near future. | .967 |
| Moral obligation (4) | I feel morally responsible to support leukemia patients by registration as a potential stem cell donor. | .917 |
| Moral reasoning (5) | I choose a course of action that considers the rights of all people involved. | .840 |
| Empathy (5) | I sometimes find it difficult to see things from the other person's point of view. | .573 |
| Content knowledge (8) | Stem cells are ordinary body cells, that do not pass on their genetic material to the next generation. | .526 |

[a]Direct measurement of attitude, subjective norm and perceived behavioural control; [b]Indirect measurement of attitude, subjective norm and perceived behavioural control; only α –coefficients of the evaluation component are presented

Abbreviation: *IF* internal factors, *EF* external factors

demonstrate two dimensions of knowledge, so that knowledge is considered as a one-dimensional construct here. The items selected for measuring *empathy* and *moral reasoning* come from the standardized Prosocial Personality Battery (Penner, 2002). An overview about scales (factors), item numbers and Cronbach's α values of the questionnaire is given in Table 3.1.

## 3.3.2 Teaching Unit "Wake Up"

Based on the extended TPB model (Fig. 3.1), the 5-h teaching unit "Wake up" was developed. The teaching intervention "Wake up" is divided into three teaching phases (introduction, elaboration [different activities at four stations] and evaluation; see Table 3.2), whereby in each phase not only the topic "Leukemia and stem cell donation" is examined in depth, but also constructs of the research model are promoted and deepened by carrying out different tasks. It addresses many beliefs

**Table 3.2** Teaching unit "wake up"

| Lecture phase | | Topics | Reference to the constructs of the research model |
|---|---|---|---|
| Introductory phase (approx. 30 min) | Interactive introduction: "Stem cells are special cells!" Jonas tells his story | Blood formation<br>Stem cells<br>The role of stem cells in medicine and scientific research | Cognitive level of TPB (beliefs)<br>Knowledge<br>Empathy<br>Moral reasoning |
| Elaboration phase: learning at different stations (approx. 140 min.) | **Station 1:** What is leukemia? | Forms of leukemia, causes & consequences of leukaemia: Analyze and evaluation of genetic aberration | Cognitive level of TPB (beliefs)<br>Knowledge<br>Empathy<br>Moral reasoning<br>Moral obligation |
| | **Station 2:** What the blood tells us? | Analyze and evaluation of blood | |
| | **Station 3:** Stem cell donation – what reveals behind it? | Accessible sources of adult stem cells in humans<br>Types of transplants<br>Side effects and risks | |
| | **Station 4:** From HLA-typing to transplantation | Understanding of HLA-typing<br>Important steps before stem cell transplantation | |
| Evaluation phase (approx. 40 min.) | Working in groups: Evaluation, discussion | Different opinions about stem cell donation | Cognitive level of TPB (beliefs)<br>Knowledge<br>Empathy<br>Moral reasoning<br>Moral obligation |

The teaching intervention has a duration of about 5 h (including pre- and post-tests as well as breaks)

and misconceptions that are linked to the topic of stem cell donation and leukemia, whereby the content knowledge is continuously built up and elaborated. In this way, the reflection processes regarding stem cell donation are stimulated, whereby there is also time for the clarification and discussion of new and old misconceptions regarding the topic.

## 3.4 Findings

Table 3.3 presents significant changes of factors of the TPB model and in model external factors. ATT, SN as well as PBC increase significantly. The items of *belief based-negative-attitude* were "reversed" before the data analysis, so that the increase in this scale can be interpreted as its decrease. *Belief based-negative-attitude* in particular has decreased significantly after the intervention, while *belief based-positive-attitude* shows no significant change. Negative attitudes include fears and mistrust regarding stem cell donation, whereby positive attitudes express positive consequences and evaluations of donation, e.g. stem cell donation is described as a good deed, it can save lives or it describes reciprocity ("*I would appreciate a stem cell donation if I needed one.*"). Also *belief based-perceived behavioural control* (external factors), which measure external resources such as availability of time or support of others, increases significantly after intervention. Furthermore the intention to register as a stem cell donor changes significant from a negative (M = 3.5) to

**Table 3.3** Results of the t-tests (pre-post)

| Factors | M (pre) | SD (pre) | M (post) | SD (post) | N | T | p | Cohen's d |
|---|---|---|---|---|---|---|---|---|
| ATT[a] | 5.18 | .95 | 5.79 | .92 | 263 | −11.479 | .000 | .652 |
| SN[a] | 4.79 | .98 | 4.99 | 1.14 | 263 | −3.993 | .000 | .181 |
| PBC[a] | 4.35 | 1.05 | 4.77 | 1.25 | 263 | −6.947 | .000 | .358 |
| Belief based-(positive) attitude[b] | 42.95 | 6.27 | 42.81 | 7.24 | 262 | .425 | .671 | – |
| Belief based-(negative) attitude[b] | 23.63 | 9.26 | 31.38 | 9.58 | 263 | −13.994 | .000 | .822 |
| Belief based-subjective norm[b] | 20.86 | 8.73 | 21.55 | 9.75 | 262 | −1.822 | .070 | – |
| Belief based-perceived behavioral control-(IF)[b] | 21.19 | 13.71 | 20.56 | 12.95 | 262 | .812 | .417 | – |
| Belief based-perceived behavioral control-(EF)[b] | 25.39 | 9.13 | 27.24 | 10.07 | 263 | −4.197 | .000 | .191 |
| Intention | 3.50 | 1.66 | 4.52 | 1.60 | 263 | −13.926 | .000 | .620 |
| Moral obligation | 4,84 | 1.28 | 5.32 | 1.30 | 263 | −7.888 | .000 | .369 |
| Moral reasoning | 5,07 | .94 | 5.39 | .87 | 263 | −7.213 | .000 | .352 |

[a]Direct measurement; [b]Indirect measurement of attitude, subjective norm and perceived behavioural control

Abbreviation: *IF* internal factors, *EF* external factors

**Table 3.4** T-test results (pre-post) for negative attitude-related beliefs

| Items | | | N | M | SD | T | df | Sig. (2-tailed) | Cohen's d |
|---|---|---|---|---|---|---|---|---|---|
| Leukemia is contagious | A2 | Pre | 263 | 29.98 | 14.85 | −13.9 | 262 | .000 | 1.102 |
| | | Post | | 44.22 | 10.46 | | | | |
| Fear of pain | A4 | Pre | 263 | 19.43 | 13.06 | −4.899 | 262 | .000 | .298 |
| | | Post | | 23.62 | 14.88 | | | | |
| Harm to health | A5 | Pre | 263 | 22.12 | 12.21 | −10.85 | 262 | .000 | .731 |
| | | Post | | 31.95 | 14.46 | | | | |
| Fear of fainting | A6 | Pre | 263 | 21.62 | 12.61 | −3.112 | 262 | .002 | .175 |
| | | Post | | 23.99 | 14.32 | | | | |
| Distrust in medical system | A9 | Pre | 263 | 24.99 | 13.72 | −10.17 | 262 | .000 | .611 |
| | | Post | | 33.41 | 13.84 | | | | |

All items were reversed before product formation. Each belief is derived from a multiplication term [1 item (expectation) × 1 item (evaluation component) = product term of the belief]

a more positive (M = 4.52) intention. Model external factors such as *moral obligation* as well as *moral reasoning* show a significant increase after intervention. Also, the factor *content knowledge* shows a positive and significant change in post-test (Wilcoxon-Test: pre: M = 12.92/SD = 4.69; post M = 21.49/SD = 2.14; Z = −13.792; p = .000; effect size *r* = .602).

This study examined not only summarized belief-based TPB measures, but also single beliefs. Below, beliefs of belief-based (negative) attitude and control beliefs are examined in more detail (Tables 3.4 and 3.5), as their combined constructs also showed significant changes (Table 3.3). All negative attitude-related beliefs are significantly reduced after the intervention (Table 3.4). The belief that leukemia is contagious (M = 29.98pre/44.22post; p = .000; d = 1.102) becomes most positive after the intervention. Also the fear of pain (A4) decreases significantly after the intervention, but only with a small effect size d = .298 (=small effect). The fear of fainting (A6) also decreases significantly after the intervention. Here, the effect size d = .175 according to Cohen (1988) is interpreted as no effect. While the beliefs "harm to health" (A5) (M = 22.12pre/31.95post; p = .000; d = .731) and "distrust in medical system" (A9) (M = 24.99pre/33.41post; p = .000: d = .611) are reduced with a mean effect size after the intervention.

Finally, when control beliefs are examined, it can be summarized that some of these beliefs increase significantly after the teaching intervention. Here, the belief "low level of knowledge" rises the most from 12.34 to 31.07, which also has a large effect size (d = .1600) (Table 3.5). This conviction was also "reversed" before the data analysis, whereby the increase in "low level of knowledge" is interpreted as an increase in knowledge. The conviction PBC4 ("*I would register as a stem cell donor if important people would accompany me to the registration campaign.*") increases (M = 27.07pre/28.84post) slightly in post-test with an effect size d = .134, which is interpreted as no effect according to Cohen (1988). Other control beliefs (PBC6; PBC7; PBC8) increase only with low effect sizes, which according to Cohen (1988)

**Table 3.5** T-test results (pre-post) for control beliefs

| Items | | | N | M | SD | T | df | Sig. (2-tailed) | Cohen's d |
|---|---|---|---|---|---|---|---|---|---|
| Perceived inappropriate state of health | PBC1 | Pre | 263 | 25.94 | 13.92 | 0.299 | 262 | .765 | – |
| | | Post | 263 | 25.64 | 15.74 | | | | |
| Fear of needles | PBC2 | Pre | 262 | 19.56 | 14.56 | 0.89 | 261 | .374 | – |
| | | Post | 262 | 18.82 | 13.63 | | | | |
| Fear of blood | PBC3 | Pre | 263 | 22.92 | 15.05 | 0.769 | 262 | .443 | – |
| | | Post | 263 | 22.25 | 14.27 | | | | |
| Support of the most important persons: Accompany to the registration campaign | PBC4 | Pre | 263 | 27.07 | 13.02 | −2.569 | 262 | .011 | .134 |
| | | Post | 263 | 28.84 | 13.30 | | | | |
| Low level of knowledge | PBC5 | Pre | 263 | 12.34 | 9.89 | −19.133 | 262 | .000 | 1.60 |
| | | Post | 263 | 31.07 | 13.23 | | | | |
| Help despite the effort | PBC6 | Pre | 263 | 25.06 | 11.98 | −3.921 | 262 | .000 | .205 |
| | | Post | 263 | 27.71 | 13.61 | | | | |
| Awareness of opportunity | PBC7 | Pre | 263 | 22.46 | 11.99 | −3.821 | 262 | .000 | .207 |
| | | Post | 263 | 25.06 | 13.12 | | | | |
| Time consuming | PBC8 | Pre | 263 | 21.89 | 9.79 | −2.921 | 262 | .004 | .206 |
| | | Post | 263 | 24.12 | 11.67 | | | | |
| Support of family and friends: Accompany to the registration campaign | PBC9 | Pre | 263 | 30.36 | 12.32 | −0.134 | 262 | .894 | – |
| | | Post | 263 | 30.45 | 13.74 | | | | |

All items were reversed before product formation. Each belief is derived from a multiplication term [1 item (expectation) × 1 item (evaluation component) = product term of the belief]

are interpreted as small effects (Table 3.5). Here also the conviction PBC8 is negatively operationalized, so that also here the increase is to be understood as its reduction. In addition it should be noted that some control beliefs (Table 3.5: PBC1; PBC2; PBC3) become even negative after the five-hour teaching intervention, even if these observations are not significant.

## 3.5 Discussion

The global constructs of intention, namely ATT, SN and PBC, increase significantly after the intervention, even if no effect is attributed to SN according to Cohen (1988) (Table 3.3). The intention also increases significantly in the post-test. This shows that the intervention effectively promotes the TPB constructs. The positive effect of the intervention is also evident in belief-based TPB measures. Overall, individual positive attitude-related beliefs, such as "I want to do a good deed", "saving a life" or "help" are already highly pronounced in the pre-test (Table 3.3), so no significant increases were observed after the intervention. In the study by Glasgow and Bello

(2007), the beliefs "helping another person" and "saving a life" were also cited by respondents as leading reasons for registration as a stem cell donor.

All operationalised negative attitude-related beliefs decrease after the intervention (Table 3.4) that reflects a positive impact of the teaching unit "Wake up". The greatest decrease is observed in the conviction that "leukemia is contagious" and in "harm to health" followed by "distrust in medical system" (Table 3.4). According to this, students, for example, associate the donation of stem cells after the intervention less harmful to their health or that they may develop leukemia as a result of the donation. Overall, their confidence in the bone marrow donation system also increases. After the intervention, the participants also perceive less fear of pain and they associate the stem cell donation less with the fear of feeling faint.

Regarding the control beliefs, it can be summarized that after the intervention the students have a higher level of knowledge about stem cell donation and leukaemia, they perceive the process of stem cell donation as less time consuming, they take more opportunities for a possible donation and they are more willing to become donors despite the effort. The teaching unit can positively influence not only beliefs and direct measured ATT, SN and PBC, but also *moral obligation* and *moral reasoning* as well as *content knowledge*.

In this study, on the one hand, beliefs are considered which reflect structural conditions (PBC4 and PBC9: social support; PBC7 awareness of opportunity; PBC8: time-consuming) and, on the other hand, beliefs which reflect the (negative) consequences of donation (A2: leukemia is contagious; A4: fear of pain; A5: harm to health; A6: fear of fainting; PBC1: perceived inappropriate state of health). In addition, beliefs that reflect fears about the donation process are also examined (PBC2 and PBC3: fear of needles and blood). In particular, beliefs about the consequences of donation were more strongly promoted in the intervention than beliefs that reflect structural conditions. Fears about the donation process remain unchanged after the intervention. For further studies it is recommended, to test different methods that can successfully reduce these fears.

From the literature research in context of blood studies many useful methods can be identified, which can be adopted in context of stem cell donation. Ferguson et al. (2007) suggests that campaigns should focus more on the donors themselves and highlight positive consequences and experiences of donation to counteract negative perceptions and fears. For example, potential donors could be assured that current donors do not feel nervous and tense before or after giving blood. Donor reports can be useful for this purpose. While the fear of blood is difficult to overcome, the fear of needles can be counteracted with active exercises. In order to change "fear of needles", the "pinching method" can be demonstrated to the participants as a desensitisation technique. They can be asked to pinch themselves, appealing to the fact that a pinch of a needle hurts less than a pinch. This method was also tested in the blood donation study by Ortberg et al. (2002). In addition to fears about blood and needles, the literature cites fear of fainting as a major reason why blood donors are more likely not to donate (France et al., 2005; Bednall et al., 2013). From this study it becomes clear that the "fear of fainting" decreases significantly among the respondents, but with a low effect size (Table 3.4). The study by France and France (2019) shows that the "feeling of fainting" or perceived fainting in the context of blood donation is overestimated by respondents and does not correspond to actual reality, calling for education about the low risk of fainting and fainting reactions. France and France (2019) also conclude that the subjects should learn to perceive the feeling of fainting as a physiological and short-term reaction to blood sedimentation and thus as benign rather than something pathological associated with the loss of control. However, for a greater and more lasting change in anxiety and fears, the potential donor may need to develop a greater sense of control or confidence in their own ability to avert and/or deal with such reactions. The teaching of overcoming expected affective or physiological reactions (e.g. through hydration, applied muscle tension techniques) can be helpful in this respect. As France et al. (2008) demonstrated, participants exposed to a booklet with information on how to overcome frequent affective (e.g. anxiety) and physiological (e.g. vasovagal reactions or feeling faint or the actual feeling of powerlessness) reactions showed improvements in attitude, fear, effectiveness and intentions to donate blood. Given the positive correlation between anxiety before donating blood and the assessment of symptoms after donating blood, a reduction in anxiety before donating blood can also increase the likelihood that non-donors who donate blood for the first time will donate blood again in the future (France et al. 2008).

Overall, the consideration of risk perception in connection with negative beliefs is very important and should be given more attention in further studies. The focus on risk perception in the study by Benthin et al. (1993), led to the conclusion that people are more likely to engage in risky behaviour if they understand the risk and feel that they have personal control over the risk than those who do not. Some studies from other contexts have also shown that risk perception varies greatly between the two genders, with male participants showing a lower risk perception (Flynn et al., 1994). All these considerations and methods should be tested in further studies in context of stem cell donation.

## 3.6 Conclusions

To sum up, the positive and significant effects of the intervention "Wake up" are based on significant changes in belief based TPB factors. The teaching unit can positively influence not only beliefs, but also direct measured ATT, SN and PBC as well as *content knowledge*, *moral obligation* and *moral reasoning*. It follows that also student's *intention* regarding stem cell donation is influenced. Nevertheless, further studies should test and compare other methods and approaches to promote stem cell donation, such as the use of information brochures with different coping strategies for the unpleasant consequences of donation (e.g. feeling faint). Based on the findings of this study a stronger focus on risk perception regarding donation is recommended.

## References

Ajzen, I. (1991). The theory of planned behavior. *Organizational Behavior and Human Decision Processes, 50*(2), 179–211.

Ajzen, I. (2005). *Attitudes, personality and behavior* (2nd ed.). Open University Press.

Ajzen, I. (2008). Cosumer attitudes and behavior. In C. P. Haugtvedt, P. M. Herr, & F. R. Kardes (Eds.), *Handbook of consumer psychology* (pp. 525–548). Lawrence Erlbaum Associates.

Ajzen, I., & Fishbein, M. (2002). *Understanding attitudes and predicting social behavior*. Prentice-Hall, Inc.

Armitage, C. J., & Conner, M. (2001). Social cognitive determinants of blood donation. *Journal of Applied Social Psychology, 31*, 1431–1457.

Bednall, C. T., Bove, L. L., Cheetham, A., & Murray, L. A. (2013). A systematic review and meta-analysis of antededents of blood donation behavior and intentions. *Social Science & Medicine, 96*, 86–94.

Benthin, A., Slovic, P., & Severson, H. (1993). A psychometric study of adolescent risk perception. *Journal of Adolescent, 16*, 153–168.

Cohen, J. (1988). *Statistical power analysis for the behavioral sciences* (2nd ed.). Erlbaum.

Eagly, A. H., & Chaiken, S. (1993). *The psychology of attitudes*. Wadsworth Cengage Learning.

Ferguson, E., France, R. C., Abraham, C., Ditto, B., & Sheeran, P. (2007). Improving blood donor recruitment and retention: Integrating theoretical advances from social and behavioral science research agendas. *Transfusion, 47*(11), 1999–2010.

Fishbein, M., & Ajzen, I. (1975). *Belief, attitude, intention, and behavior: An introduction to theory and research*. Addison-Wesley.

Flynn, J., Slovic, P., & Mertz, C. K. (1994). Gender, race, and perception of environmental health risks. *Risk Analysis, 14*, 1101–1108.

France, R. C., & France, L. J. (2019). Estimating the risk of blood donation fainting for self versus others: The moderating effect of fear. *Transfusion, 59*, 2039–2045.

France, C. R., Rader, A., & Carlson, B. (2005). Donors who react may not come back: Analysis of repeat donation as a function of phlebotomist ratings of vasovagal reactions. *Transfusion and Apheresis Science, 33*(2), 99–106.

France, C. R., Montalva, R., France, J. L., & Trost, Z. (2008). Enhancing attitudes and intentions in prospective blood donors: Evaluation of a new donor recruitment brochure. *Transfusion, 48*, 526–530.

Frick, J., Kaiser, F. G., & Wilson, M. (2004). Environmental knowledge and conservation behavior: Exploring prevalence and structure in a representative sample. *Personality and Individual Differences, 37*(8), 1597–1613.

Glasgow, M. E. S., & Bello, G. (2007). Bone marrow donation: Factors influencing intentions in African Americans. *Oncology Nursing Forum, 34*, 369–377.

Kaiser, F. G., & Frick, J. (2002). Entwicklung eines Messinstrumentes zur Erfassung von Umweltwissen auf der Basis des MRCML-Modells [Development of a measurement tool to investigate environmental knowledge based on the MRCML-model]. *Diagnostica, 48*, 181–189.

Oliner, S. P., & Oliner, P. (1988). *The altruistic personality: Rescuers of Jews in Nazi Europe*. Free Press.

Ortberg, J. C. J.-R., Gorsuch, L. R., & Kim, J. G. (2002). Changing attitude and moral obligation: Their independent effects on behavior. *Journal for the Scientific Study of Religion, 40*(3), 489–496.

Penner, L. A. (2002). Dispositional and organizational influences on sustained volunteerism: An interactionist perspective. *Journal of Social Issues, 58*(3), 447–467.

Richardson, H. S. (2018). Moral reasoning. In E. N. Zalta (Ed.), *The Stanford encyclopedia of philosophy*. https://plato.stanford.edu/archives/fall2018/entries/reasoning-moral/. Accessed 10 Oct 2019.

Robert Koch-Institut (Eds.). (2016) *Bericht zum Krebsgeschehen* in Deutschland 2016. Cancer Report in Germany 2016. www.krebsdaten.de. Accessed 10 Jan 2021.
Rushton, J. P. (1980). *Altruism, socialization, and society*. Prentice Hall.
Schwartz, S. H. (1977). Normative influences on altruism. In L. Berkowitz (Ed.), *Advances in experimental social psychological* (pp. 221–279). Academic.
Schwartz, S. H., & Howard, J. A. (1981). A normative decision – Making model of altruism. In J. P. Rushton & R. M. Sorrentino (Eds.), *Altruism and helping behavior* (pp. 189–211). Lawrence Erlbaum.

# Chapter 4
# Prior Knowledge, Epistemic Beliefs and Socio-scientific Topic Context as Predictors of the Diversity of Arguments on Socio-scientific Issues

**Andreani Baytelman, Kalypso Iordanou, and Costas P. Constantinou**

## 4.1 Introduction

Socio-scientific issues (SSIs) are controversially discussed topics with a scientific basis, which always have an important social relevance (Zeidler, 2015). The use of SSIs in science classes aims to improve scientific literacy, hence, instructional approaches try to encourage students to argue more reflectively and to empower them to participate in current and future SSI debates and decision-making (Sadler & Zeidler, 2005; Zeidler, 2015). Studies have produced findings that suggest that curricula which include SSIs support argument skills (Rudsberg et al., 2013; Zeidler et al., 2019).

The factors that facilitate the development of argument skills when people engage in thinking about or debating SSIs remain underdetermined (Asterhan & Schwarz, 2016; Iordanou et al., 2019; Zeidler et al., 2019). Specifically, the factors that enable students to analyze an issue from different perspectives and produce different types of arguments (diversity of arguments) on complex SSIs are still not clear (Newton & Zeidler, 2020). On the other hand, the ability to construct different types of arguments that integrate multiple positions or accounts is essential for SSI

A. Baytelman (✉) · C. P. Constantinou
Learning in Science Group, Department of Education, University of Cyprus, Nicosia, Cyprus
e-mail: baytel@ucy.ac.cy; c.p.constantinou@ucy.ac.cy

K. Iordanou
School of Sciences, University of Central Lancashire UCLan, Pyla, Cyprus
e-mail: KIordaou@uclan.ac.uk



K. Korfiatis, M. Grace (eds.), *Current Research in Biology Education*, Contributions from Biology Education Research,
https://doi.org/10.1007/978-3-030-89480-1_4

debates and a vital component of the development of a functional scientific literacy (Zeidler et al., 2019). This reflects one's depth of reasoning (Nussbaum & Schraw, 2007), as well as the ability to appreciate the complexity and multidimensionality of SSIs, the ability to consider alternative views and appreciate the need to include different evidence and diverse perspectives in the arguments, which are fundamental for decision-making (Herman et al., 2018).

In the present study, we address this gap in the literature, namely whether and to what extent, prior-knowledge, epistemic beliefs and SSI-topic context can predict the diversity of arguments about SSIs constructed by university Education students. By doing this, we hope to gain a better understanding of the development of: (i) the ability to perceive and reason through the complexity inherent to SSIs, (ii) the ability to analyze an issue and potential solutions from the perspectives of different stakeholders, (iii) the ability to determine how scientific knowledge and processes may contribute to the resolution of an SSI; and to recognize dimensions of the issue that cannot be addressed by science (Zeidler et al., 2019).

We investigate the diversity of arguments using three different health SSI topics with social, scientific, ethical, economic and ecological aspects. We used health-SSIs because many SSIs are rooted in scientific advances and technologies and are related to human health (Ratcliffe & Grace, 2003) and biological sciences. The motivation to focus on university education students comes from prior research which has shown that university students demonstrate weak argument skills when dealing with SSIs (Driver et al., 2000; Iordanou & Constantinou, 2014).

In particular, we set out to answer the following research question: *What is the relationship between university Education students' prior-knowledge, epistemic beliefs, SSI-topic context and the diversity of arguments they construct about a controversial health SSI topic?*

### *4.1.1 Socio-scientific Issues and Argument Skills*

The term argument refers to a product, constructed by an individual, consisting of a claim and one or more supporting reasons or evidence that are connected to the claim with warrants (Toulmin, 2003). To gain understanding of how individuals' argument skills develop, science education and psychology researchers have become interested in examining several factors, such as: epistemic beliefs, which refer to individuals' beliefs about the nature of knowledge and the process of knowing (Hofer & Pintrich, 1997), and prior knowledge, values, desires, expectations and experience as potential contributors to the construction of arguments (Rundgren et al., 2016).

However, researchers (Asterhan & Schwarz, 2016; Zeidler et al., 2019) propose more research for the factors that facilitate the development of argument skills, including when people engage in thinking about or debating controversial SSIs.

### 4.1.2 Prior-Knowledge and Argument Skills Relating to Socio-scientific Issues

Discussions on SSIs in the science education literature are accompanied by the assumption that individuals' content-knowledge, potentially, contributes to their reasoning and argumentation in the context of SSIs (Sadler & Fowler, 2006; Sadler & Zeidler, 2005). However, according to empirical studies (Baytelman, 2015; Means & Voss, 1996; Sadler & Fowler, 2006; Sadler & Zeidler, 2005) it seems that previous research has examined the relationship between prior-knowledge and argument skills, focusing on how prior knowledge affects the quantity and quality of arguments. The relationship between prior knowledge and the diversity of different types of arguments on controversial SSIs produced is not clear.

In the present study, prior knowledge is defined as prior domain-specific conceptual content knowledge, and includes the knowledge of concepts, principles, facts and theories of a subject, but also an understanding of how concepts and principles of a subject are organized (Shulman, 1986). The more connections that exist among facts, ideas, and procedures, the better the understanding (Hiebert & Carpenter, 1992). Additionally, according to Novak (2010), concept-maps can provide evidence for a distinction between meaningful and rote learning. Novak explains that rote learning is limited in terms of recall and transfer and subsequent change,

> ...because [concepts learnt by rote] are stored arbitrarily and non-substantively in cognitive structure, [they] soon cannot be recalled and confer interference with new, related learning and recall of related information. (Novak, 2010, p. 69)

So, concept-maps can be a direct method of looking at the organization and structure of an individual's knowledge within a particular domain and at the fluency and efficiency with which the knowledge can be used (Williams, 1998).

### 4.1.3 Epistemic Beliefs and Argument Skills Related to Socio-scientific Issues

In recent decades, students' beliefs about the nature of knowledge and the process of knowing, have received increased attention from researchers (Hofer & Pintrich, 1997; Schommer, 1990). Most of this work has been rooted in Perry's (1970) early research, suggesting that college students start out believing that knowledge consists of simple, unchanging facts, handed down by authority and then progress toward a conception of knowledge as consisting of complex, tentative concepts based on reasoning. While one important line of research on epistemic beliefs has continued Perry's (1970) effort to identify developmental stages in students' epistemic thinking, Schommer (1990) initiated a line of research using a multidimensional approach.

According to the multidimensional approach of epistemic beliefs (Schommer, 1990), epistemic beliefs could be described as a system of more or less independent beliefs (epistemic dimensions), conceptualized as beliefs about the certainty (related with the stability of knowledge), simplicity (related with the structure of knowledge), and source of knowledge, as well as beliefs about the speed and ability of knowledge acquisition. While the dimensions of certainty, simplicity and source in Schommer's conceptualization fall under the more generally accepted definition of epistemic beliefs, known as beliefs about the nature of knowledge (certainty, simplicity) and knowing (source) (Hofer & Pintrich, 1997), the dimensions speed and ability have been controversial because they mainly concern beliefs about learning (speed) and intelligence (ability). Hofer and Pintrich (1997) argued that epistemic beliefs should be defined more purely, with two dimensions concerning the nature of knowledge (what one believes knowledge is) and two dimensions concerning the nature or process of knowing (how one comes to know).

According to Hofer and Pintrich (1997), the two dimensions concerning the nature of knowledge are: (a) Simplicity of knowledge, ranging from the belief that knowledge consists of an accumulation of more or less isolated facts to the belief that knowledge consists of highly interrelated concepts, and (b) Certainty of knowledge, ranging from the belief that knowledge is absolute and unchanging to the belief that knowledge is tentative and evolving.

The two dimensions concerning the nature of knowing are: (a) Source of knowledge, ranging from the conception that knowledge originates outside the self and resides in external authority, from which it may be transmitted, to the conception that knowledge is actively constructed by the person in interaction with others, and (b) Justification for knowing, ranging from justification of knowledge claims through observation and authority, or on the basis of what feels right, to the use of rules of inquiry and the evaluation and integration of different sources (Hofer & Pintrich, 1997). Accordingly, Hofer and Pintrich's model differs from Schommer's by omitting the nature of learning factors and adding another nature of knowing factor – Justification. Additionally, Conley et al. (2004) suggested a new dimension of epistemic beliefs, the Development of knowledge, which is related with the nature of the development of knowledge.

Previous empirical studies (Barzilai & Eshet-Alkalai, 2015; Mason & Scirica, 2006; Sadler & Fowler, 2006) indicated that there is a relation between individuals' epistemic-beliefs and argument skills. However, the possible contribution of epistemic beliefs to the diversity of arguments that individuals produce has not yet been clearly investigated.

### 4.1.4 Socio-scientific Topic Context and Argument Skills

A few studies suggest that SSI topic context (context of a particular scenario of a SSI topic) is a factor that can make a difference in the way students approach SSIs. Specifically, some researchers (Sadler & Zeidler, 2005; Herman et al., 2018) claim that SSI topic context, but not the topic alone, influences the reasoning students employ in multiple ways, engaging with the SSI (including quality, information attended to, and degree of emotional involvement). Additionally, they claim that students prioritize different dimensions of certain issues differently, depending on how that issue is presented to them.

In a study of genetics topics, Sadler & Zeidler (2005) noted:

> ...the context of an issue significantly influenced how individuals responded to that issue. (p. 125)

Yet, Herman et al. (2018) demonstrated that an SSI intervention on Yellowstone national park in an area community, supported the development of more nuanced emotive reasoning (e.g. shifting from a stance that affected parties "deserve their fate" to a more compassionate handling of the issue). In the present work, we aim to gain a deeper understanding about the possible contribution of prior knowledge, epistemic beliefs and SSI topic context to the diversity of arguments that students construct on health SSIs.

## 4.2 Research Design and Method

### *4.2.1 Participants*

In this study, participants were university Education students at a public University in Cyprus (n = 243); 93% were females and 7% were males (the imbalance imposed by the student population of this University). All participants were elementary or early childhood Education majors and were enrolled in a required science education course. Participation in the research was optional and all data were treated anonymously and confidentially.

### *4.2.2 Data Collection*

For this study, we used three different controversial health SSI topics with social, scientific, ethical, economic and ecological aspects. The first SSI topic (SSI 1) was on *Usage versus non-usage of vaccines against the NUEVO flu virus*. The second SSI topic (SSI 2) was on *Consumption of bottled versus tap water*. The third SSI topic (SSI 3) was on *Usage of underground versus overhead high voltage lines in residential areas*.

Our rationale for the choice of these SSI topics was the following:

(a) During the last decade, SSIs such as safety and usage of vaccines, safety and usage of high voltage lines, and consumption of bottled versus tap water have attracted increasing attention in Cyprus, because of the potential impact on human health. Therefore, students might have been more motivated to engage in thinking about these topics;

(b) the participants of this study had already learned about vaccines, drinking water and high voltage lines and their relation with public health in their science classes at secondary school, and they had basic prior knowledge regarding these issues;
(c) all the three health SSI topics of the study have different social, scientific, ethical, economic, and ecological aspects, where valid, yet opposing, arguments can be formed from multiple perspectives.

For each health SSI topic, we developed a text with a scenario and a leaflet. Each text of a scenario presented a SSI dilemma with conflicting information and different views on the SSI, considering a SSI from multiple perspectives and alternative viewpoints. More specifically, the first part of each text was a neutral introduction to the SSI dilemma. The second part of the text introduced one position of the dilemma, whereas the third part presented the opposing position of the dilemma. Both positions were supported with extra evidence (separate sheet/leaflet). Each leaflet of a particular scenario presented conflicting scientific and non-scientific information related to human health, with social, scientific, ethical, economic and ecological aspects from different sources, such as official websites of relevant government agencies, articles by laypeople and non-experts, articles in scientific and professional journals and magazines, newspaper articles, and teaching resources. The sequence of the information of different positions of the leaflets was reviewed by three researchers to make sure that the sequence could not affect the results of the study.

To assess participants' epistemic beliefs, we used the Dimensions of Epistemological Beliefs toward Science (DEBS) Instrument (Baytelman & Constantinou, 2016), which is based on the multidimensional perspective of epistemic beliefs and has been validated in the particular culture in which the research was conducted. The 30-item DEBS Instrument captures three dimensions concerning knowledge (Certainty, Simplicity and Development of knowledge), and two dimensions concerning knowing (Source and Justification of knowledge). Each dimension consisted of six items and the items were rated on a four-point Likert-scale, ranging from strongly disagree (1) to strongly agree (4). High scores on this measure represent more sophisticated beliefs, while low scores represent less sophisticated beliefs.

To assess students' prior knowledge, participants were asked to answer five open-ended questions and to construct a network concept-map for each health SSI. We evaluated prior knowledge at two different levels: (a) shallow (surface) conceptual understanding, which was assessed by open-ended questions that examined distinct basic scientific concepts related to each SSI (Mazur, 1997), and (b) deep conceptual understanding, which was assessed by a concept map requiring appropriate relationships between the relevant concepts. The open-ended questions were scored from 0 to 2 on the basis of their correctness and completeness by the first author and an independent judge (Cohens' $k = .91$). The concept maps were scored counting the number of appropriate concepts and the number and quality of appropriate relationships between concepts (propositions) for each student's concept map (Cohens' $k = .90$).

For the investigation of the effects of the context of the three health SSI topics on the diversity of arguments produced, two dummy variables were created, which were related to the context of the SSI topic scenarios. The SSI 3 *Usage of underground versus overhead high voltage lines in residential areas* was used as the reference category. The first dummy variable was SSI 1(vs. SSI 3), and the second dummy variable was SSI 2 (vs. SSI 3).

Students' individual argument skills were assessed using a written instrument based on published work (Baytelman, 2015). In particular, students were asked to take a position for each topic dilemma and justify it by formulating supportive arguments, counterarguments, and rebuttals, also utilizing information from the leaflet given to them. Students were prompted to construct different types of arguments-social, ethical, economic, scientific and ecological - expressing their opinion by providing a particular space in the instrument for each type of argument. For example, for the topic of whether vaccines are safe and should be used against the NUEVO flu virus, students were asked to answer the following questions:

1. Are you for or against vaccination for the NUEVO flu virus?
2. If you want to convince a friend about your position, what arguments will you put forward to convince them?
3. If somebody holds an opposite position from yours on this issue, what arguments may s/he have?
4. According to the argument you mentioned in Question 3, can you write down your opposite arguments to justify your position?

For each participant, we computed the diversity of arguments. Each supportive argument, counterargument, and rebuttal was considered valid if it involved the presentation of a claim, and the legitimacy of that claim was documented through justification based on evidence. For coding the diversity of argument types, we followed two steps:

(a) We checked the nature of each argument provided under each prompt for argument type in the instrument, giving 1 point for each simple argument type. In this step, it was possible for one argument to be coded as a type other than the argument type in which the participant chose to register it under;
(b) If there were multiple arguments provided under one prompt (e.g. an economic argument with elements of ethical or social considerations, which were not provided again in the space for ethical or social arguments) we coded them separately.

To examine whether prior knowledge, epistemic beliefs and the SSI topic context can predict the diversity of arguments produced, multiple hierarchical regression analyses were carried out with prior knowledge (network concept map and open-ended questions), epistemic beliefs (epistemic dimensions) and SSI topic context as predictors.

### 4.2.3 Procedure

Participants engaged in three sessions. During the first session (20 min), each individual answered the DEBS instrument. In the second session (40 min), the prior knowledge instrument was administered. In the third session (about 60 min), the tasks for the participants were: (a) reading the SSI scenario on one of the three health topics, (b) reading the information related to the SSI scenario, and (c) constructing different types of arguments.

## 4.3 Results

Table 4.1 displays the means, standard deviations, minimum and maximum scores, of all variables of this study.

As seen in Table 4.1, participants' scores on the prior knowledge measure suggested relatively low knowledge on the concept map and higher knowledge on open-ended questions. Participants' scores on the epistemic beliefs measure suggested relatively sophisticated beliefs about Certainty, Justification and Development of knowledge, and less sophisticated beliefs about Simplicity (structure of knowledge) and Source of knowledge. Participants constructed social, scientific, ethical, economic and ecological arguments (supportive arguments, counterarguments, rebuttals) for each SSI topic. Participants' scores on the number of scientific arguments were higher than other types of arguments and ranged from 0 to 15 with mean of 5.72 (*SD* = 2.42). However, participants' scores on the diversity of types of arguments were high, ranging from 2 to 5 with a mean of 4.5 (*SD* = 0.77).

**Table 4.1** Descriptive statistics for all variables of the study

| Variable | M | SD | Min | Max |
|---|---|---|---|---|
| *Prior knowledge* | | | | |
| Network concept-map | 13.60 | 8.40 | 0.00 | 40.00 |
| Open-ended questions | 5.90 | 1.94 | 0.00 | 10.00 |
| *Epistemic beliefs* | | | | |
| Certainty of knowledge | 3.01 | 0.38 | 1.00 | 3.88 |
| Simplicity of knowledge | 2.65 | 0.36 | 1.67 | 3.50 |
| Source of knowledge | 2.62 | 0.42 | 1.50 | 3.83 |
| Justification of knowledge | 3.20 | 0.31 | 2.43 | 4.00 |
| Development of knowledge | 3.33 | 0.35 | 2.29 | 4.00 |
| *Diversity of argument types* | 4.50 | 0.77 | 2.00 | 5.00 |
| Number of social arguments | 4.45 | 2.16 | .00 | 10.00 |
| Number of scientific arguments | 5.72 | 2.42 | 0.00 | 15.00 |
| Number of ethical arguments | 3.16 | 2.07 | 0.00 | 9.00 |
| Number of economic arguments | 4.32 | 1.65 | 0.00 | 7.00 |
| Number of ecological arguments | 2.89 | 2.26 | 0.00 | 8.00 |

**Table 4.2** Results of stepwise regression analyses for variables predicting diversity of argument types

| Predictor variables | Diversity of arguments on health-SSI | |
|---|---|---|
| Step 1 | B(SE) | β |
| Prior-knowledge concept-map | 0.02 (0.01) | 0.21*** |
| Prior-knowledge questions | 0.02 (0.02) | 0.05 |
| Step 2 | | |
| Prior-knowledge concept-map | 0.01 (0.01) | 0.15* |
| Prior-knowledge questions | 0.02 (0.02) | 0.04 |
| Certainty of knowledge | −0.05 (0.14) | −0.03 |
| Simplicity of knowledge | 0.66 (0.13) | 0.32*** |
| Source of knowledge | 0.18 (0.11) | 0.10 |
| Justification of knowledge | −0.19 (0.15) | −0.09 |
| Development of knowledge | −0.12 (0.15) | −0.06 |
| Step 3 | | |
| Prior-knowledge concept-map | 0.02 (0.01) | 0.19** |
| Prior-knowledge questions | 0.01 (0.02) | 0.04 |
| Certainty of knowledge | −0.09 (0.14) | −0.05 |
| Simplicity of knowledge | 0.64 (0.14) | 0.31*** |
| Source of knowledge | 0.20 (0.18) | 0.11 |
| Justification of knowledge | −0.18 (0.15) | −0.08 |
| Development of knowledge | −0.10 (0.15) | −0.01 |
| SSI1 (vs. SSI3) | −0.27 (0.12) | −0.14 |
| SSI2 (vs. SSI3) | −0.01 (0.12) | −0.00 |

Note: ***$p < 0.001$, **$p < 0.01$, two-tailed; *$p < 0.05$, two-tailed
For diversity of argument types $R = 0.22$, $R^2 = 0.05$, Adjusted $R^2 = 0.04$ for Step 1 ($p < 0.01$), $R = 0.41$, $R^2 = 0.17$, Adjusted $R^2 = 0.15$ for Step 2 ($p < 0.001$), $R = 0.43$, $R^2 = 0.19$, Adjusted $R^2 = 0.16$ for Step 3 ($p = 0.08$)

Table 4.2 displays the unstandardized regression coefficients (B) and intercept, the standardized regression coefficients (β), $R^2$, and adjusted $R^2$ after entry of all independent variables (IVs).

After Step 3, with all IVs in the equation $R^2 = .19$, $F(9,231) = 5.96$, $p < .001$. The adjusted $R^2$ value of .16 indicates that 16% of the variability in the diversity of arguments is predicted by PK, EBs, and SSI-context. After Step 1, with PK (concept map and open-ended questions) in the equation, $R^2 = .05$ $F_{inc}(2, 238) = 6.07$, $p < .01$. The adjusted $R^2$ value of .04 indicates that 4% of the variability in the diversity of arguments is predicted by PK. After Step 2, with the addition of EBs (Certainty, Simplicity, Source, Justification, and Development of knowledge) to the prediction of diversity of arguments by PK, $R^2 = .17$, $F_{inc}(5235) = 6.87$, $p < .001$. The adjusted $R^2$ value of .15 indicates that 15% of the variability in the diversity of arguments is predicted by PK and EBs. The addition of EBs to the equation of PK results in a significant increment in $R^2$ by .12. After Step 3, with SSI-context added to prediction of diversity of arguments by PK and EPs, $R^2 = .19$, $F_{inc}(2238) = 2.51$, $p = .08$. The adjusted $R^2$ value of .16 indicates that 16% of the variability in the diversity of arguments is predicted by PK and EBs. The addition of SSI-context to the equation did not significantly improve $R^2$.

The pattern of results suggests that 5% of the variability in the diversity of arguments constructed is predicted by prior knowledge. Epistemic beliefs contribute significantly to that prediction. The addition of SSI topic context to the equation did not significantly improve $R^2$.

## 4.4 Discussion and Conclusions

The present research extends the current literature examining relationships between epistemic beliefs, prior knowledge, context of scenario of SSI-topic and the construction of different types of arguments (diversity of arguments) on complex controversial health SSIs, focusing only on individual perspectives about SSIs. The findings of the present research contribute to the literature on argument skills related to SSIs by demonstrating that prior knowledge and epistemic beliefs predict the diversity of arguments on health SSIs that are constructed by students in personal reasoning. The context of individual scenarios of complex controversial SSI topics has not demonstrated a significant predictive power for the diversity of arguments.

The finding that prior knowledge (as assessed by a concept map and open-ended questions) and epistemic beliefs' dimension of Simplicity of knowledge (beliefs about the structure of knowledge) predict the diversity of arguments, constitutes a novel contribution, suggesting that students need deep topic conceptual understanding and an epistemic understanding - theorizing knowledge as a complex system of organized theoretical principles and ideas (sophisticated Simplicity beliefs) for dealing effectively with complex health SSIs. Robust content knowledge and epistemic understanding enabled students in the present study to appreciate the complexity and multidimensionality of health SSIs and to analyze them from different perspectives, as was reflected in students' ability to produce different types of arguments from multiple perspectives. We argue that robust prior knowledge is necessary for the construction of different types of arguments on a health SSI because students must have a well-developed conceptual schema in order to incorporate content knowledge in their different types of arguments (Sadler & Fowler, 2006; Sadler & Zeidler, 2005).

The finding that sophisticated simplicity beliefs predicted the diversity of arguments produced on health SSIs, above and beyond prior knowledge, is of particular importance, highlighting the role of epistemic beliefs in particular, for constructing different types of arguments about a health SSI. A possible explanation for this finding that epistemic simplicity beliefs predicted the diversity of arguments, is the following: Students who view knowledge as a collection of simple and discrete pieces of factual information (naïve simplicity beliefs), rather than as a complex system of organized theoretical concepts (sophisticated simplicity beliefs), may tend not to:

(a) develop an understanding of the complex controversial nature of a SSI topic;
(b) consider alternative views; and/or

(c) recognize that some aspects of an issue have more and different implications.

Additionally, if students believe that knowledge is simple, they probably do not appreciate the need to include different evidence and diverse perspectives in their arguments, or the need to seek further evidence or to integrate conflicting information from multiple sources. In the present study, students were presented with complex SSI topics for which connections among different pieces of information, perspectives and evidence were necessary, which probably made the simplicity dimension—about the structure of knowledge – more salient than the other epistemic dimensions.

Participants used three different complex controversial health SSI topics. Each particular scenario of each health SSI topic has social, scientific, ethical, economic, ecological aspects, and the students were able to construct different types of arguments on each SSI topic with high diversity, elucidating the complex and multidimensional nature of each SSI-topic.

Our research has important educational implications, showing that relatively sophisticated simplicity epistemic beliefs and robust conceptual understanding are related to an individuals' ability to construct different types of arguments on health SSIs. More research is required to focus on finding the means for supporting the development of university students' conceptual understanding, their epistemic beliefs, particularly their epistemic simplicity beliefs, and their understanding of the complexity and multidimensionality of SSIs. In doing so, we would empower university staff to promote argument skills in relation to controversial health SSIs among their future students, and prepare them to understand the value of analyzing health dilemmas, the value of potential solutions from the perspectives of different stakeholders and the value of disease prevention, public health, and democratic responsibility.

There are some limitations to the current study that may provide impetus to further work in this area. First, our research focuses only on individual perspective about SSIs and our results have applicability only for single persons in personal reasoning. Second, although the health SSIs addressed in the current study are of international applicability, we cannot impute generalizability for our results are based on a relative small sample consisted of 93% females. Third, we used only questionnaires, which does not probe elaborated participants' responses to items as in-depth interviews would do.

## References

Asterhan, C., & Schwarz, B. (2016). Argumentation for learning: Well-trodden paths and unexplored territories. *Educational Psychologist, 51*(2), 164–187.

Barzilai, S., & Eshet-Alkalai, Y. (2015). The role of epistemic perspectives in comprehension of multiple author viewpoints. *Learning and Instruction, 36*, 86–103.

Baytelman, A. (2015). *The effects of epistemological beliefs and prior knowledge on pre-service primary teachers' informal reasoning regarding socio-scientific issues*. University of Cyprus, Faculty of Social Sciences and Education.
Baytelman, A., & Constantinou, C. P. (2016). Development and validation of an instrument to measure student beliefs on the nature of knowledge and learning. *Themes of Science and Technology in Education, 9*(3), 151–172.
Conley, M., Pintrich, P., Vekiri, I., & Harrison, D. (2004). Changes in epistemological beliefs in elementary science students. *Contemporary Educational Psychology, 29*(2), 186–204.
Driver, R., Newton, P., & Osborne, J. (2000). Establishing the norms of scientific argumentation in classrooms. *Science Education, 84*(3), 287–312.
Herman, B. C., Zeidler, D. L., & Newton, M. (2018). Students' emotive reasoning through place-based environmental socioscientific issues. *Research in Science Education, 1*, 1–29.
Hiebert, J., & Carpenter, T. P. (1992). Learning and teaching with understanding. In D. A. Grouws (Ed.), *Handbook of research on mathematics teaching and learning* (pp. 65–97).
Hofer, B. K., & Pintrich, P. R. (1997). The development of epistemological theories: Beliefs about knowledge and knowing their relation to learning. *Review of Educational Research, 67*(2), 88–140.
Iordanou, K., & Constantinou, C. P. (2014). Developing pre-service teachers' evidence-based argumentation skills on socio-scientific issues. *Learning and Instruction, 34*, 42–57.
Iordanou, K., Muis, K. R., & Kendeou, P. (2019). Epistemic perspective and online epistemic processing of evidence: Developmental and domain differences. *The Journal of Experimental Education, 87*(4), 531–551.
Mason, L., & Scirica, F. (2006). Prediction of students' argumentation skills about controversial topics by epistemological understanding. *Learning and Instruction, 16*(5), 492–509.
Mazur, E. (1997). *Peer instruction: A user's manual*. Prentice-Hall.
Means, M., & Voss, J. (1996). Who reasons well? Two studies of informal reasoning among children of different grade, ability, and knowledge levels. *Cognition and Instruction, 14*(2), 139–178.
Newton, M., & Zeidler, D. (2020). Developing socioscientificperspective taking. *International Journal of Science Education, 42*(8), 1302–1319.
Novak, J. D. (2010). *Learning, creating, and using knowledge: Concept maps as facilitative tools in schools and corporations* (2nd ed.). Routledge.
Nussbaum, E. M., & Schraw, G. (2007). Promoting argument-counterargument integration in students' writing. *The Journal of Experimental Education, 76*(1), 59–92.
Perry, W. (1970). *Forms of intellectual and ethical development in the college years: A scheme*. Holt, Rinehart & Winston.
Ratcliffe, M., & Grace, M. (2003). *Science education for citizenship*. Open University Press.
Rudsberg, K., Öhman, J., & Östan, L. (2013). Analyzing students' learning in classroom discussions about socioscientific issues: Students' learning about socioscientific issues. *Science Education, 97*(4), 594–620.
Rundgren, C., Eriksson, M., & Chang Rundgren, S. (2016). Investigating the intertwinement of knowledge, value, and experience of upper secondary students' argumentation concerning socioscientific issues. *Science & Education, 25*(9), 1049–1071.
Sadler, T., & Fowler, S. (2006). A threshold model of content knowledge transfer for socioscientific argumentation. *Science Education, 90*(6), 986–1004.
Sadler, T., & Zeidler, D. L. (2005). The significance of content knowledge for informal reasoning regarding socioscientific issues: Applying genetic knowledge to genetic engineering issues. *Science Education, 89*(1), 71–93.
Schommer, M. (1990). Effects of beliefs about the nature of knowledge on comprehension. *Journal of Educational Psychology, 82*(3), 498–504.
Shulman, L. (1986). Those who understand: Knowledge growth in teaching. *Educational Researcher, 15*, 4–14.

Toulmin, S. E. (2003). *The uses of argument* (Updated ed.). Cambridge University Press (original work published in 1958).

Williams, C. G. (1998). Using concept-maps to assess conceptual knowledge of function. *Journal for Research in Mathematics Education, 29*(4), 414–421.

Zeidler, D. (2015). Socioscientific issues. In R. Gunstone (Ed.), *Encyclopedia of science education* (pp. 998–1003). Springer.

Zeidler, D. L., Herman, B. C., & Sadler, T. (2019). New directions in socioscientific issues research. *Disciplinary and Interdisciplinary Science Education Research, 1*(11), 1–9.

# Chapter 5
# Toward a Questionnaire to Assess Biology Student Teachers' Knowledge of the Nature of Scientific Inquiry (NOSI)

**Corinne Charlotte Wacker, Marius Barth, Christoph Stahl, and Kirsten Schlüter**

## 5.1 Introduction

The nature of scientific inquiry (NOSI) is one critical component of science literacy and is also becoming increasingly important in view of the current corona pandemic. Teaching and learning the characteristics of scientific inquiry processes, through which scientific knowledge is generated and justified, is not only emphasized in the German educational standards set by the standing conference of the ministers of education and cultural affairs (KMK, 2005), but also worldwide as part of science standards (NGSS Lead States, 2013). Thus, students as well as teachers must be able to understand, conduct and critically assess scientific investigations. However, despite the continued science education efforts, research indicates that teachers and student teachers, as well as students of varying ages, typically hold naïve NOSI views (Lederman et al., 2019; Mesci et al., 2020; Zion et al., 2018). Because teachers need to have an elaborated understanding of NOSI in order to be able to discuss it adequately in their lessons, it is important to assess whether (and how well) this educational goal is actually achieved throughout their university education to provide the necessary resources for them.

To appraise the NOSI proficiency of a person, one must first define what it means to be competent in it. This is particularly important because NOSI and nature of science (NOS) are often used as synonymous terms and are frequently combined and

C. C. Wacker (✉) · K. Schlüter
Institute of Biology Education, University of Cologne, Cologne, Germany
e-mail: cwacker1@uni-koeln.de; kirsten.schlueter@uni-koeln.de

M. Barth · C. Stahl
Research Methods and Experimental Psychology, Department of Psychology, University of Cologne, Cologne, Germany
e-mail: marius.barth@uni-koeln.de; christoph.stahl@uni-koeln.de



K. Korfiatis, M. Grace (eds.), *Current Research in Biology Education*, Contributions from Biology Education Research,
https://doi.org/10.1007/978-3-030-89480-1_5

overlap, because they are interdependent (Mayer, 2007). Nevertheless, Schwartz et al. (2008) distinguish between them by stating that NOSI is about "how" the scientific knowledge is generated and validated, i.e. the nature of the *practices* that are most closely related to the *processes* of inquiry; whereas NOS embodies what distinguishes science from other disciplines. Thus, NOS refers to the characteristics of the scientific knowledge, i.e. the *product* of inquiry processes (Schwartz et al., 2012). Schwartz et al. (2008) and Lederman et al. (2014) identified aspects of NOSI *via* literature reviews and science practices studies. According to Schwartz et al. (2008) these NOSI sub-competences include: (*sc*1) scientific investigations all begin with a question and do not necessarily test a hypothesis; (*sc*2) there is no single set of steps followed in all investigations; (*sc*3) scientific questions scientists choose to pursue stem from many sources and can serve many purposes; (*sc*4) scientific data can be interpreted differently; (*sc*5) scientists recognize anomalous data and handle them in a reflective manner; (*sc*6) scientific data are not the same as scientific evidence; and (*sc*7) scientific inquiry is embedded within a researcher's community.

Various NOSI instruments were developed especially in the last 30 years for various stages of education using a variety of response formats (Temiz et al., 2006). However, they primarily focus on pupils up to 10th grade and little is known about natural science student teachers' NOSI views during their university education (Mesci et al., 2020). Furthermore, many of these testing instruments make use of: either a multiple-choice format, which is considered to be time- and administration-economic, but is susceptible to test-wiseness; or an open-ended format with follow-up interviews, which is considered to be time-consuming and vulnerable to discrepancies in interpretation (Temiz et al., 2006; Thoma & Köller, 2018). Another factor that needs to be considered is that the focus of most instruments lies upon experimentation and by this on causal relationships, due to the fact that the experiment is considered to be "the gold standard" of science. Other research methods, such as observations and comparisons, are often seen as preliminary stages or partial aspects of experimentation (Ayyavoo et al., 2002; Wellnitz & Mayer, 2013). It was therefore decided to develop a curriculum-independent NOSI questionnaire that can be used at any point in the academic education of biology student teachers.

The purpose of our study is: (1) to develop a closed-ended questionnaire to assess biology student teachers' NOSI views; and (2) then to validate this instrument's functioning in order to discuss its potential for research and teaching. There is a need to gain insight into the NOSI competence of future science teachers throughout their university education in order to further improve it.

## 5.2 Method

### *5.2.1 Participants*

The NOSI questionnaire was administered to undergraduate biology student teachers in the introductory course "Basics of biology" at the University of Cologne. This was done during the winter semesters of 2018/19 and 2019/20. The sample of the

148 freshman student teachers of biology was comprised of 108 women (73%) and 40 men (27%), and the average age was 20.7 years (SD = 2.9). Data were collected in an online survey using LimeSurvey 3.21; however, during the first survey phase a paper-pencil version of the instrument was also employed due to server issues. Before participants answered the questionnaire, they were briefed, i.e. a brief introduction concerning the voluntary and anonymous participation in this pilot study was given. The response rate amounted to 77.5%.

### 5.2.2 *Questionnaire Design*

A closed-ended NOSI instrument was designed by a group of experts from the field of biology education and psychology. The developed questionnaire was constructed in reference to the seven previously mentioned NOSI aspects of Schwartz et al. (2008) and Lederman et al. (2014) (see Table 5.1). It is important to note that these authors explicitly state that these seven sub-competences are not the only ones, but nevertheless they are indispensable for students (Lederman et al., 2014; Schwartz et al., 2008; Zion et al., 2018). It was decided to extend the testing instrument to include the additional aspect of Questionable Research Practices (QRPs), i.e. either fabrication or falsification of scientific data or results (Fiedler & Schwarz, 2016). Krishna and Peter (2018) stated that approximately 10% of psychology students, who were writing a Bachelor's or Master's thesis on a psychology course at a German public University, practiced some QRPs and that lecturers have an important function in shaping the students' attitude towards them. Consequently, it was

**Table 5.1** Overview and distribution of items in the NOSI testing instrument. Adapted from Schwartz et al. (2008) and Lederman et al. (2014)

| Sub-competence (*sc*) | Items context | Σ |
|---|---|---|
| *sc*1. Scientific questions guide investigations | Investigations do not necessarily need a hypothesis, but a research question is mandatory. | 5 |
| *sc*2. Multiple methods of scientific investigations | Diverse research methods exist (e.g. experiment, correlation study). | 11 |
| *sc*3. Multiple purposes of scientific investigations | (Theoretical versus practical) goals of scientific research. | 9 |
| *sc*4. Justification of scientific knowledge | Data do not stand alone, but can be interpreted differently. | 4 |
| *sc*5. Recognition and handling of anomalous data | Dealing with anomalous information (e.g. looking for information why these data occur). | 2 |
| *sc*6. Distinctions between data and evidence | Evidence is the result of data analysis and interpretation. | 8 |
| *sc*7. Community of practice (CoP) | Scientific inquiry processes and data are checked by a researcher's community. | 2 |
| *sc*8. Questionable Research Practices (QRPs) | Transparency in data collection and reporting. | 5 |

felt that this area, which also touches on the fifth aspect of recognizing and handling of abnormal data, is also essential for the processes of scientific inquiry, i.e. NOSI. With regards to the focus and utility of this NOSI testing instrument, it is important to acknowledge that some areas of NOSI are difficult to distinguish from each other and these sub-competences also tend to overlap with some areas of NOS. In the preliminary phase of the instrument development, content validity was therefore assessed by repeated discussion sessions by the authors, who examined the wording of the statements as well as whether each item fits in its allocated NOSI aspect.

Adjustments were made to the preliminary item pool of 85 items based on a first test survey with university students, so that items that were too easy to answer or too similar to other items, or that could not be clearly assigned to an aspect, were either modified accordingly or eliminated. Finally, the developed NOSI questionnaire consisted of 46 closed items, whereby respondents had to agree/disagree first with statements (true or false), and then subsequently rate their answer in a confidence rating (How confident are you that your answer is correct (as a percentage)? Answers: 0 = guessing, 20, 40, 60, 80, and 100 = absolutely certain). The development of this testing instrument with two response formats could be an adequate trade-off between economic reasons, i.e. in a time-effective manner, and detailed participants' NOSI views that can also take into account the participants' test intelligence. Moreover, in accordance with the second sub-competence that there is no single scientific method that all (biology) scientists follow, the testing instrument neither concentrated on a specific research method nor on a specific curriculum. The complete 46-item NOSI questionnaire is available at https://osf.io/u9gdz/.

### 5.2.3 *Data Analysis*

In order to create a combined multi-response index for each item, a multiplicative weight for each item based on both response formats (dichotomous: true/false and the post-decision confidence rating: 0%, 20%, 40%, 60%, 80%, and 100%) was calculated. The dichotomous responses were coded either −1 (incorrect) or +1 (correct) and the confidence rating responses were expressed as relative probability, i.e. coded as 0, .2, .4, .6, .8 or 1, respectively. The next step was to multiply both values to calculate a multiplicative weight for each item and each case, i.e. $x_{dichotomous} \cdot x_{confidence\ rating}$ (see Table 5.2). For example, if a test person answered one item incorrectly and was 40% sure about his/her answer, the result would be $-1 \times .4 = -.4$.

Subsequently, item analysis was assessed for selecting items for the NOSI questionnaire. This is the average score of the combined multi-response index instead of item difficulty that uses scaling from −1 to +1, Measure of Sampling Adequacy (*MSA*), and Cronbach's α. In addition, further descriptive data analyses were made to better illustrate student teachers' overall NOSI understanding. This was done by calculating a NOSI total score for every biology student teacher *via* the arithmetic mean of the combined multi-response index of all items. In order to identify

**Table 5.2** Combined multi-response index of both formats

| Dichotomous response was...: | ... incorrect. | | | | | ... guessed. | ... correct. | | | | |
|---|---|---|---|---|---|---|---|---|---|---|---|
| Confidence rating response [%]: | 100 | 80 | 60 | 40 | 20 | 0 | 20 | 40 | 60 | 80 | 100 |
| Multi-response index: | **−1** | **−.8** | **−.6** | **−.4** | **−.2** | **0** | **.2** | **.4** | **.6** | **.8** | **1** |

potential scientific inquiry misconceptions as well as items that could be answered with test intelligence, we examined each NOSI item in detail according to the given responses by the respondents. A Maximum Likelihood (ML) exploratory factor analysis (EFA) followed by oblique rotation was also conducted in order to establish the underlying structure of factors.

## 5.3 Results

All analyses were performed using R 4.0.3 (R Core Team, 2020) with the package *psych* (Revelle, 2020).

### *5.3.1 Item Analysis*

Figure 5.1a shows the distribution of the average score of the combined multi-response index for each item across all participants. It indicates that most items have an average score in the positive scale range, this means that most items were chiefly correctly answered. Nevertheless, all five items of the NOSI sub-competence "Scientific questions guide investigations" and two items (8.1 & 8.2) of the NOSI sub-competence "QRPs" were most difficult for the student teachers to answer due to their negative average scores ranging from –.046 to –.367. However, there are no items that were overall too difficult or too easy according to the histogram borderline areas of $-.8 < score_i < .8$. Measure of Sampling Adequacy (*MSA*) was employed to determine the extent to which an item was suitable for factor analysis, i.e. its discriminatory power. If the items are not at all or only weakly correlated with all other items, it is unlikely that factors can be found by which the multiplicity of the variables can be reduced on a smaller number of dimensions (Ludwig-Mayerhofer, 2017). The authors decided to use a rather low cut-off score of .4 to avoid prematurely eliminating items of the NOSI questionnaire. After excluding five items (2.5, 3.3, 3.4, 6.5, 8.3) from the instrument, which are the five points below the *MSA* limit at around .39 in Fig. 5.1b, with the two points of items 3.3 and 3.4 overlapping each other, 41 items remained in the reduced NOSI item pool within the defined boundaries (see Fig. 5.1c). For the reduced item pool, the overall *MSA* = .65 can be considered useful and the reliability can be considered acceptable with Cronbach's $\alpha = .69$, Guttman's $\lambda_6 = .84$. It was also decided to report Guttman's $\lambda_6$, because Cronbach's $\alpha$ tends to underestimate reliability in tests with strong heterogeneity, such as comprising eight components/aspects (Osburn, 2000).

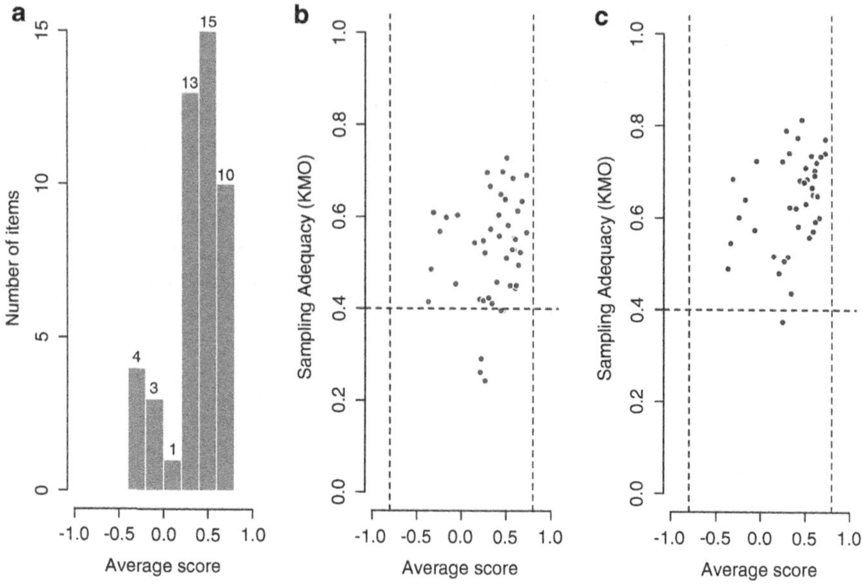


**Fig. 5.1** Average score of the combined multi-response index and *MSA* for each NOSI item of the complete list of 46 items (**a**, **b**) and the reduced list of 41 items (**c**). The limits are depicted as dashed lines in (**b**) and (**c**). The numbers above the bars depict the total number of items in (**a**). Because *MSA* calculates the correlations of an item to all other items, the points in the scatterplots change from Figure (**b**) to (**c**) after eliminating five items

## 5.3.2 Biology Student Teachers' NOSI Competences

The arithmetic mean of the responses to all NOSI items of the test, i.e. the participant ability score, was calculated for each of the 148 freshman biology student teachers (see Fig. 5.2). The positive range of NOSI competence for all participants is between .02 and .65 with an average of $M = .36 \pm .13$.

Despite the overall positive range of NOSI understanding by the freshman biology student teachers, two interesting response patterns were identified within the combined multi-response index by looking at the items in detail. A few items received incorrect responses that were given with high confidence ranging from 80% to 100% (= absolutely certain), and some items were answered correctly, but with a low confidence rating ranging from 0% to 20% (= guessing) by the respondents. The authors focused especially on the items of the testing instrument where more than 10% of the biology student teachers showed these response patterns (see Fig. 5.3). For eight items 14–30% of the participants were certain, i.e. 80–100% confidence, that their answers were correct, although this was not the case. The top item within this group, which is called *NOSI misconceptions*, was: 'A scientific investigation always checks a hypothesis' (item 1.3). Conversely,

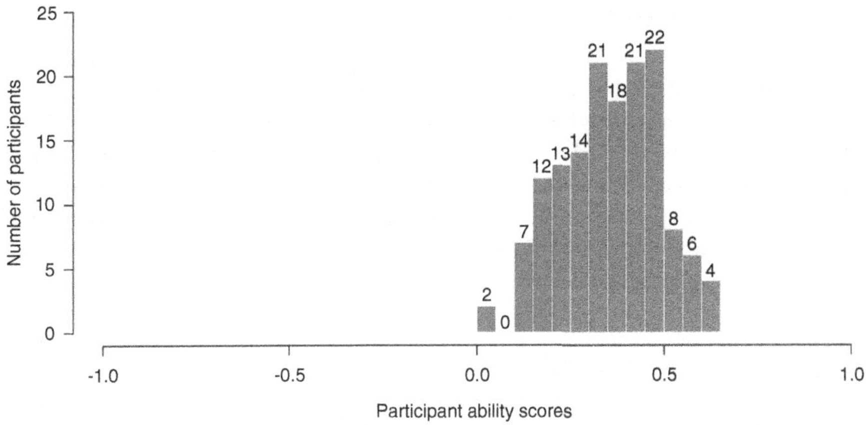


**Fig. 5.2** Biology student teacher's overall NOSI score. The numbers above the bars depict the total number of participants

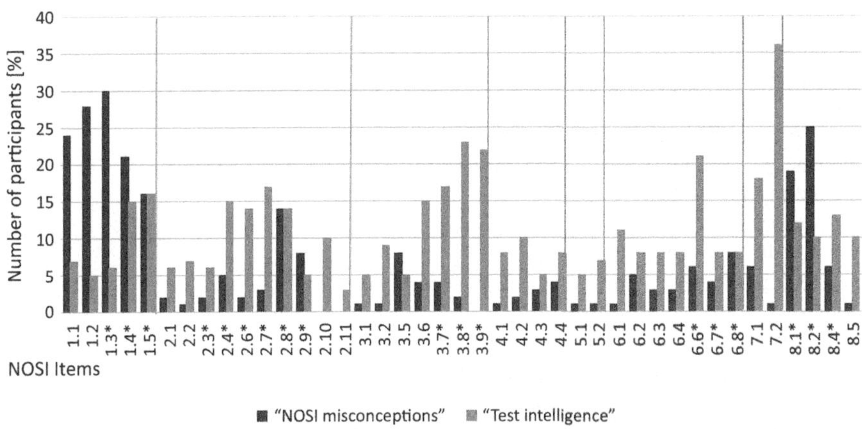


**Fig. 5.3** Proportion of all "high-confidence + incorrect" (80–100% confidence and incorrectly answered) and "low-confidence + correct" (0–20% confidence and correctly answered) responses. Negatively formulated questions are marked with an asterisk. The vertical lines separate the eight NOSI-sub-competences from each other

11–36% of the biology student teachers answered 17 questions correctly, although they felt rather uncertain, i.e. 0–20% confidence, that their responses were right. The top item within this group, which is labelled *test intelligence*, was: 'Scientists organize themselves in professional societies to set standards for scientific work' (item 7.2). In addition, four items showed even both these answer patterns by more than 10% of the participants (1.4, 1.5, 2.8, and 8.1).

### 5.3.3 *Factor Extraction Results*

We conducted a parallel analysis of the remaining 41 items, which suggested keeping five factors for an exploratory factor analysis. Only 17 items with loadings greater than ±.40 were used to characterize the 5-factor construct in Table 5.3. The other 24 items aren't shown here because of their very low communalities. Furthermore, only one or two items loaded on the factors 1, 3, and 4, whereas eight items loaded on factor 2, with loadings ranging from .59 to .40. Moreover, there were four items that loaded on factor 5 with loadings ranging from .59 to .44.

**Table 5.3** Correlation matrix (EFA with oblique rotation). Negatively formulated questions are marked with an asterisk. The communalities are depicted in the last column as $h^2$ indicating each item's variance that can be explained by the corresponding model

| | Factor | | | | | |
|---|---|---|---|---|---|---|
| Item | 1 | 2 | 3 | 4 | 5 | $h^2$ |
| 5.2 Unexpected results can make it necessary to change a scientific theory. | **.99** | | | | | 1.00 |
| 2.2 Depending on the research question, different methods are applied. | | **.43** | | | | .25 |
| 2.10 Different methods can lead to contradictory results. | | **.56** | | | | .37 |
| 2.11 When interpreting scientific data, the methods used should always be considered. | | **.51** | | | | .30 |
| 4.1 Two scientists interpreting the same data can justifiably come to different conclusions. | | **.59** | | | | .33 |
| 4.2 Several investigations using the same procedures may have different results. | | **.40** | | | .21 | .27 |
| 4.4 Scientific research data can be interpreted differently. | | **.58** | | | | .39 |
| 5.1 Unexpected results may cause the data to be reinterpreted. | | **.58** | | | | .39 |
| 6.4 Scientific evidence is based on the newly collected data of an investigation, but also on the previous state of research. | | **.49** | | | | .26 |
| 2.4* For a given research question there is also a given method to answer it. | | | **.46** | .24 | .26 | .41 |
| 6.6* The evidence of a scientific study is not directly related to the research question. | | | **.73** | | | .55 |
| 1.3* A scientific investigation always checks a hypothesis. | | | | **.60** | | .38 |
| 1.4* Before a research question is formulated, hypotheses about possible results of an investigation must have been derived. | | | .23 | **.61** | | .46 |
| 3.1 Scientific investigations can serve to develop technologies. | | | | | **.44** | .29 |
| 3.7* Mechanisms that cannot be directly observed cannot be scientifically investigated. | | | | .24 | **.49** | .43 |
| 6.3 Data are the observations made during a scientific investigation. | | | | −.22 | **.59** | .40 |
| 7.2 Scientists organize themselves in professional societies to set standards for scientific work | | | | | **.54** | .35 |

## 5.4 Discussion

The present study found that the NOSI questionnaire with a combined multi-response index has an acceptable instrument reliability before and after the elimination of five of its items. The average scores of the combined multi-response index (each referring to a single item) in Fig. 5.1a indicate a wide answering range between –.37 and .73, and the overall mean for all 41 items of $M = .36 \pm .13$ indicates an already existing moderate NOSI understanding of all biology student teachers. Moreover, there is no single item with a negative average combined multi-response index score below –.37, which might indicate that all not participants have internalized the same naïve NOSI views in their diverse school biology education. It is important to note that the arithmetic mean of the participant ability scores of all 41 items is based on an unequal number of items in each of the eight sub-competences, which was a result of the item selection process (originally starting from 85 items).

### *5.4.1 NOSI Misconceptions and Test Intelligence*

Two thought-provoking response patterns could be identified by looking at the combined multi-response index of the NOSI questionnaire in more detail (see Fig. 5.3). One group of respondents featured a "false certainty" because they answered some items incorrectly, but were nevertheless sure that their answers were correct. These items could point towards NOSI misunderstandings, which may have been acquired through schooling. In particular, the top negatively coded item in this group, 'A scientific investigation always checks a hypothesis' (item 1.3), hints on the supposed 'general procedure' of 'the Scientific Method'. Almost any inquiry assignment in school science curricula seems to start with generating a hypothesis for an experiment and it seems that a study is only deemed a success if the results serve to confirm this hypothesis (Bencze, 1996). This circumstance could therefore easily lead to a widespread NOSI misconception. In addition, the other four items of the same sub-competence 'Scientific questions guide investigations' (*sc*1), were also answered by more than 10% of student teachers in a similar pattern. Thus, this sub-competence seems to include popular misconceptions about the role of hypothesis *versus* research question. Another sub-competence with two "false certainty" response items refers to 'Questionable Research Practices (QRPs)' (*sc*8). The items are 'On the basis of the data collected, the hypotheses of the study should be adapted.' (8.1) and 'If an expected effect is not yet statistically significant, data collection should be continued so that the effect can become significant.' (8.2). They both indicate that student teachers' NOSI understanding in the areas of HARKing ("Hypothesizing After the Results are Known") and "optional stopping" need to be improved.

There is also a larger group of 17 NOSI items within six sub-competences of the questionnaire, where more than 10% of participants felt rather uncertain that their correct response was right. For instance, the most often "truly guessed" item was: 'Scientists organize themselves in professional societies to set standards for scientific work' (item 7.2). One could argue that 36% of the participants could have deduced from own as well as from heard experiences of others that the existence of CoPs seems highly likely, maybe because meeting with fellow students and (work) colleagues for exchanging experiences is part of everyday life.

Four of the 41 items in the NOSI questionnaire even show both response patterns ("false certainty" and "truly guessed"). In some cases, this could be a hint that there are difficulties in interpreting an item, e.g. item 2.7: 'Chance should not play a role in research'. Both ratings (true as well as false) regarding the correctness of the item have their legitimacy, depending on a student teachers' way of thinking. On the one hand the item is wrong, because chance sometimes plays an important role in science (e.g. discovery of the antibiotic Penicillin by Alexander Fleming in 1928) (Copeland, 2019). On the other hand, the item is correct, because when experimenting, all variables except for the one under investigation must be kept constant. Chance should be excluded in this case as far as possible, because otherwise one would not get any reliable and interpretable results. Therefore, this item cannot be interpreted on its own and must either be reformulated or interpreted in the context of other items. In general, there is a need to continue to improve student teachers' NOSI views so that they have a better understanding about scientific inquiry processes, and by this can fulfill their important role as future teachers in shaping students' NOSI views.

### 5.4.2 Exploratory Factor Analysis (EFA)

The correlation matrix of the EFA of the NOSI questionnaire with 41 items shows no obviously recognizable matrix structure. Although a 5-factor construct can be identified in Table 5.3, the factors 1, 3, and 4 each contain less than three item loadings. Moreover, items of the NOSI aspects 'Justification of scientific knowledge' (*sc*4) and 'Recognition and handling of anomalous data' (*sc*5) are not included in the pattern matrix. A possible explanation for the inconclusive EFA matrix structure – besides the two reliability limitations concerning the number of items and sample size, which are discussed in detail in the next chapter – could be that the respondents' NOSI abilities differ from the authors' theoretical construct. This could be indicated by the fact that the factors 2 and 5 include items from more than three different NOSI sub-competences. Thus, at this point the factor structure of the questionnaire is unclear and future research is required to better understand the underlying constructs and their relations.

### 5.4.3 Limitations

Despite the fact that the NOSI test instrument provides a first insight into the NOSI understanding of biology student teachers and their potential scientific inquiry misconceptions at the start of their academic education, there are obvious limitations to this pilot study. Even though a sequential cross-section research design over more than 1 year was used and the response rate was high (77.5%), the sample of $N = 148$ for a 46-item questionnaire was more than half too small compared to the required size of $N = 400$ (Eid et al., 2017). Therefore, the validation study should be continued or even extended by the participation of, for example, student teachers of other science subjects in order to reach the required sample size. Secondly, the two NOSI sub-competences 'Recognition and handling of anomalous data' (*sc*5) and 'Community of practice' (*sc*6) had only two items instead of the recommended minimum of three items per aspect/factor, which is essential when conducting an EFA of a multidimensional construct such as NOSI (Raubenheimer, 2004). This is due to the fact that in the first NOSI questionnaire test survey, the other items of these sub-competences were eliminated because of their easy item difficulties. Because of these two limitations, the results of the EFA should only be interpreted carefully and a final selection of items for the NOSI questionnaire is therefore premature at this stage. Nevertheless, the identification of the two significant item response patterns within the combined multi-response index, i.e. potential NOSI misconceptions and test intelligence, allows the authors to have an additional decision criterion for the final item selection. One may consider eliminating NOSI items that indicate test intelligence, while retaining items that may point to NOSI misunderstandings. In the target group of freshman student teachers of biology, eight potential scientific misunderstandings could be found (see Fig. 5.3). However, their origin cannot be determined, although they are presumably due to biology or other scientific school lessons. To this end, a mixed-methods study with interviews could be conducted to learn more about the reasons and sources of misconceptions (as well as about the causes for possible test intelligence phenomena). Further studies are currently planned to explore more deeply the informative value of the newly developed NOSI test instrument by applying it to biology (and maybe even other natural sciences) student teachers in other years and phases of their academic education, such as in the Master's programme. This will on the one hand further validate the instrument's functioning and on the other hand help to identify and correct latent NOSI misconceptions that may have been created or propagated throughout former school and university education.

## References

Ayyavoo, G., Bencze, J., Corry, A., & van Oostveen, R. (2002). Correlational studies in school science: Beyond experimentation. *OISE Papers in STSE Education, 3*, 221–230.

Bencze, J. L. (1996). Correlational studies in school science: Breaking the science-experiment-certainty connection. *School Science Review, 78*(282), 95–101.

Copeland, S. (2019). On serendipity in science: Discovery at the intersection of chance and wisdom. *Synthese, 196*(6), 2385–2406.

Eid, M., Gollwitzer, M., & Schmitt, M. (2017). *Statistik und Forschungsmethoden*. Beltz Verlagsgruppe.

Fiedler, K., & Schwarz, N. (2016). Questionable research practices revisited. *Social Psychological and Personality Science, 7*(1), 45–52.

KMK. (2005). *Beschlüsse der Kultusministerkonferenz – Bildungsstandards im Fach Biologie für den Mittleren Schulabschluss*. Luchterhand München.

Krishna, A., & Peter, S. M. (2018). Questionable research practices in student final theses – Prevalence, attitudes, and the role of the supervisor's perceived attitudes. *PLoS One, 13*(8), e0203470.

Lederman, J. S., Lederman, N. G., Bartos, S. A., Bartels, S. L., Meyer, A. A., & Schwartz, R. S. (2014). Meaningful assessment of learners' understandings about scientific inquiry-the views about scientific inquiry (VASI) questionnaire. *Journal of Research in Science Teaching, 51*(1), 65–83.

Lederman, J., Lederman, N., Bartels, S., Jimenez, J., Akubo, M., Aly, S., et al. (2019). An international collaborative investigation of beginning seventh grade students' understandings of scientific inquiry: Establishing a baseline. *Journal of Research in Science Teaching, 56*(4), 486–515.

Ludwig-Mayerhofer, W. (2017). *ILMES – Internet-Lexikon der Methoden der empirischen Sozialforschung: Faktorenanalyse*. Retrieved 5th October, 2020, from http://wlm.userweb.mwn.de/Ilmes/ilm_f3.htm

Mayer, J. (2007). Erkenntnisgewinnung als wissenschaftliches Problemlösen. In *Theorien in der biologiedidaktischen Forschung* (pp. 177–186). Springer.

Mesci, G., Çavuş-Güngören, S., & Yesildag-Hasancebi, F. (2020). Investigating the development of pre-service science teachers' NOSI views and related teaching practices. *International Journal of Science Education, 42*(1), 50–69.

NGSS Lead States. (2013). *Next generation science standards: For states, by states*. Retrieved 1 October 2020, from https://www.nextgenscience.org/

Osburn, H. G. (2000). Coefficient alpha and related internal consistency reliability coefficients. *Psychological methods, 5*(3), 343.

R Core Team. (2020). *R: A language and environment for statistical computing*. R Foundation for Statistical Computing, Retrieved 30 October 2020, from http://www.r-project.org/index.html

Raubenheimer, J. (2004). An item selection procedure to maximise scale reliability and validity. *SA Journal of Industrial Psychology, 30*(4), 59–64.

Revelle, W. R. (2020). *psych: Procedures for personality and psychological research*.

Schwartz, R. S., Lederman, N. G., & Lederman, J. S. (2008). *An instrument to assess views of scientific inquiry: The VOSI questionnaire*. Paper presented at the National Association for Research in Science Teaching (NARST), Baltimore, MD.

Schwartz, R. S., Lederman, N. G., & Abd-el-Khalick, F. (2012). A series of misrepresentations: A response to Allchin's whole approach to assessing nature of science understandings. *Science Education, 96*(4), 685–692.

Temiz, B. K., Taşar, M. F., & Tan, M. (2006). Development and validation of a multiple format test of science process skills. *International Education Journal, 7*(7), 1007–1027.

Thoma, G.-B., & Köller, O. (2018). Test-wiseness: ein unterschätztes Konstrukt? *Zeitschrift für Bildungsforschung, 8*(1), 63–80.

Wellnitz, N., & Mayer, J. (2013). Erkenntnismethoden in der Biologie – Entwicklung und Evaluation eines Kompetenzmodells. *Zeitschrift für Didaktik der Naturwissenschaften, 19*, 315–345.

Zion, M., Schwartz, R. S., Rimerman-Shmueli, E., & Adler, I. (2018). Supporting teachers' understanding of nature of science and inquiry through personal experience and perception of inquiry as a dynamic process. *Research in Science Education, 50*(4), 1281–1304.

# Chapter 6
# Introducing Primary School Students to Aspects of the Nature of Scientific Knowledge

**Marida Ergazaki and Aggeliki Laourdeki**

## 6.1 Introduction

Science advances very fast nowadays and affects society in multiple ways. People need to face complex socio-scientific issues like DNA editing or climate change, and so they are required to be literate about science (Yacoubian, 2018). Scientific literacy is a major educational goal as it represents the things that every citizen should know about science (Durant, 1993). It involves understanding essential scientific concepts, as well as appreciating the nature, aims and limitations of science (Jenkins, 1994). Being an important part of scientific literacy, the nature of science (NOS) is a central theme in science education research and biology education research in particular. Although there is not a unique definition of it, NOS often refers to the nature of the knowledge that scientists produce (NOSK). In fact, NOS and NOSK are considered as synonymous and are used interchangeably (Lederman, 2019). More specifically, NOSK has to do with key features of scientific knowledge, such as for instance its empirical basis or its tentativeness. These features are derived from the way scientists work in order to produce scientific knowledge; in other words, they are derived from the way scientific inquiry is performed (Lederman & Lederman, 2019).

Teaching and learning about NOS appears to be a demanding task (McComas & Kampourakis, 2015). Considering the challenge, the idea of breaking down this complex notion to a number of 'pieces' that when put together they might help students build a basic version of the 'NOS-puzzle', seems rather appealing. In fact, this is the core idea of a popular approach in NOS teaching and learning, the so-called

M. Ergazaki (✉) · A. Laourdeki
Department of Educational Sciences and Early Childhood Education, University of Patras, Patras, Greece
e-mail: ergazaki@upatras.gr



K. Korfiatis, M. Grace (eds.), *Current Research in Biology Education*, Contributions from Biology Education Research,
https://doi.org/10.1007/978-3-030-89480-1_6

'general aspects conceptualization of NOS' (Kampourakis, 2016). NOS (or NOSK) aspects are not strictly defined and thus there are several 'aspects lists', which can serve different teaching/learning goals in different educational contexts, or even be combined if necessary. Despite strong criticisms according to which this NOS conceptualization fails in highlighting the differences between science disciplines, or between science and non-scientific endeavors (Duschl & Grandy, 2013; Irzik & Nola, 2011), the 'aspects lists' seem to be powerful educational tools (Kampourakis, 2016).

The potential of such tools may become even more evident when it comes to younger children, who need to be facilitated the most in their first, challenging attempts to conceptualize NOSK. In fact, there is a growing interest in introducing young children to NOSK aspects as early as the first grades of primary school, or even kindergarten (Bell & St. Clair, 2015). According to Akerson et al. (2010), we should start with more concrete aspects such as the difference between observation and inference, and gradually move to more abstract aspects, such as the influence of the socio-cultural context on scientific knowledge. However, Quigley et al. (2010) suggest that young children may struggle with differentiating between observation and inference as well. According to Akerson et al. (2011), aspects such as the tentativeness of scientific knowledge and the role of creativity or reasoning in its construction, seem to be more suitable for young children.

NOSK aspects may be introduced either implicitly or explicitly. In explicit reflective instruction, which can be quite effective, NOSK aspects constitute explicit learning objectives, whereas students reflect on what they do and what they need to learn (Khishfe & Abd-El-Khalick, 2002). Moreover, in explicit reflective instruction, NOSK aspects can be introduced in connection with scientific concepts and be an integral part of lab inquiries or lessons based on contemporary or historical cases of scientific work (Allchin et al., 2014). Drawing on the history of science to explicitly introduce NOSK aspects, appears to be a quite interesting strand in NOS literature (McComas & Kampourakis, 2015). For younger children, scientists' biographies may be more appropriate than historical scientific controversies, since they are easier to follow; they may provide children with a view of science as an amazing human adventure, and a view of scientists as normal people with strengths, weaknesses or ideas that may be wrong and eventually rejected by their fellow scientists (Hadzigeorgiou, 2018).

In fact, scientists' biographies are considered to have remarkable potential in helping children broaden their views on scientists and NOSK (Kelly, 2018), especially if they emphasize how the scientists in question produced scientific knowledge (Dagher & Ford, 2005). According to Korkmaz (2011), exploring the life stories of several historical scientists (e.g. Lamarck, Mendel) and modern scientists (e.g. Franklin, McClintock) with the aid of story maps, helped eighth graders who participated in her study to improve their views about the profile of bioscientists and about NOSK's empirical and collaborative aspects.

Heisey and Kucan (2010), on the other hand, worked with first and second graders who were engaged in listening to read-aloud life stories of different scientists (Bentley, Anning, Audubon), and discussed them with the aid of 'during-story' discussion protocols. The selected life stories shed light on the hands-on and minds-on activities the three scientists performed during their scientific inquiries. According to Heisey's and Kucan's (2010) findings, the discussion-mediated interaction of children with the biographical texts helped them understand that scientists ask questions about nature, make observations, make experiments, consult books, use tools, think and finally discover how nature works.

Considering the above, we decided to design a learning environment that aims at introducing second grade students to NOSK aspects, as well as to specific scientific notions. Our design was guided by the idea of explicit, reflective instruction of NOSK within the history of science. What we did was using the biographies of two major scientists, Jan Baptist van Helmont and Charles Darwin, to create contexts for pursuing our NOSK-related and concepts-related teaching/learning objectives. The former were the empirical, creative, inferential and tentative aspect of scientific knowledge, whereas the latter do not concern us here. The research question we address in this chapter is:

*Can second graders' understanding about a series of NOSK aspects be enhanced within a learning environment that uses scientists' biographies as a context for introducing them?*

## 6.2 Methods

### 6.2.1 Overview of the Study

This was a mixed-model case study that included:

(i) performing individual, semi-structured pre-interviews with primary school students, in order to highlight their initial understanding about (a) the target NOSK aspects, as well as (b) the target biological notions (plant nutrition and natural selection – which are, however, beyond the remit of this article);
(ii) designing an eight-session learning environment by using the biographies of two major scientists, Jan Baptist van Helmont and Charles Darwin, as contexts for integrating the target NOSK aspects (and the target biological notions);
(iii) implementing the learning environment with peer-group work and whole class discussions;
(iv) performing individual, semi-structured post-interviews; and
(v) assessing the impact of the learning environment through qualitative and quantitative analysis of children's pre- and post-interviews.

### 6.2.2 Participants

The participants of the study were twenty-one, conveniently selected students of a second grade class. More specifically, they were twelve boys and nine girls (age 7–8) who were attending a public primary school, situated in a medium socioeconomic status neighborhood of a small town in Greece. According to the class teacher, the children's school performance ranged from low to high, whereas the nature of science had never been discussed in the classroom up to that point. The parents were informed about the study so that they could give consent for their children's participation. Moreover, in order to protect children from unnecessary psychological pressure, we visited their school, explained to them why and how we needed to work together for some time, and got their own assent for participating in our work.

### 6.2.3 *Learning Environment*

The learning environment aimed at bringing together biology and nature of science in ways that could make sense for young primary school students. In order to create meaningful contexts for the introduction of our target ideas, we used scientists' biographies. A significant reason for choosing the biographies of Jan Baptist van Helmont and Charles Darwin was that they could provide children with complementary insights to how scientific knowledge can be built, and thus help them broaden their views about its nature. The first part of the learning environment was based on the biography of van Helmont (sessions 1–2), the second on the biography of Darwin (sessions 3–7), and the third on both (session 8). For the teaching-learning activities, we took into account the age of children and the need for their active participation, as well as the real classroom conditions. Children's acquaintance with the two scientists was performed through illustrated powerpoint presentations that triggered both peer-group work and whole-class discussions. Children took part in activities that required analyzing data, spotting changes, recording routes, creating timelines, comparing with diagrams. Moreover, they took part in game-like activities requiring from them card matching, word decoding, clue finding, or 'puppet show-induced discussions'. The learning environment is summarized in Table 6.1.

### 6.2.4 *Interview Protocol*

The protocol of the pre- and post-interviews was organized in four parts. Parts A and B was concerned with scientists and NOSK, whereas parts C and D was concerned with biological notions (which are not reported on here). In part A, children were first asked a yes/no question (QA1) about whether they had heard of the word 'scientist' before, and in the case they had, they were asked an open-ended question about what scientists do (QA2). In part B, children were asked two-tier two-choice questions about NOSK aspects. Children were first provided: (a) with a card depicting just the outline of a human being (no visible traits), who was supposedly a

**Table 6.1** An overview of the learning environment

| Sessions | | Overview | Target ideas (Sci work, NOSK) |
|---|---|---|---|
| 1 | Van Helmont and his willow tree experiment | Children watched van Helmont reviewing bibliography, posing a research question, designing and performing his experiment, reporting his work and posing his next research question. Children analyzed his data, went through his reasoning and reached his conclusion. | Scientists' lab work NOSK aspects: science knowledge (SK) is empirical, creative, inferential, tentative |
| 2 | To be continued | Children discussed a timeline of scientists who studied plants after van Helmont. They also took part in an interactive puppet show about van Helmont's work. | Scientists' work NOSK aspects: SK is empirical, creative, inferential, tentative |
| 3 | Darwin: Childhood and studies | Children watched Darwin as a curious boy and young man. They were engaged in making connections of his life and van Helmont's. | Scientists' work NOSK aspects: SK is empirical |
| 4 | Darwin's trip on Beagle: Making observations and collecting data | Children watched Darwin's trip and recorded it on a floor map. They watched him making observations, collecting data, reasoning and communicating with his fellow scientists. Children were also engaged in following his reasoning. | Scientists' field work NOSK aspects: SK is empirical, creative, inferential |
| 5 | Darwin's return to UK: Do living things change over time? | Children watched Darwin studying the fossils of long-dead pilosas collected on the trip. They were also engaged in comparing the fossils with living pilosas, going through Darwin's reasoning and reaching his conclusion. | Scientists' work NOSK aspects: SK is empirical, creative, inferential |
| 6 | How do they change? | Children were introduced to a precursor model of natural selection *(bio only)*. | – |
| 7 | Darwin's book and the years that followed | Children watched Darwin's hesitation to publish his book, followed him through his later years and discussed key qualities like his constant enthusiasm for the study of nature and his constant wish to learn. | Scientists' work & the socio-cultural context NOSK aspects: SK is tentative |
| 8 | Travel in time: Van Helmont & Darwin meet | Children reflected on the two life stories and the NOSK aspects in game-like activities. | Consolidation of the S1-7 target ideas |

scientist; and (b) with two more cards, each in a shape of a 'speech balloon'. In every question, the two speech balloon-like cards presented two optional statements about a specific NOSK aspect. For instance, two optional statements about (i) the empirical, (ii) the creative, (iii) the inferential, and (iv) the tentative aspect of scientific knowledge. Children were asked to choose the speech balloon with the statement that the scientist on the card would actually say, connect it with the scientist and finally justify their choice. The optional statements can be found in the results' section. The order of the part B questions was counterbalanced.

### 6.2.5 Data Analysis

The audiotaped pre–/post-interviews were transcribed and analyzed with 'NVivo'. We coded children's responses to the open-ended question according to their accuracy and fullness; and those to the two-tier two-choice questions, according to the appropriateness of the selected option and the level of children's justification for it. The coding was performed by the authors and cases of disagreement were discussed until consensus was reached.

More specifically, the responses to the open-ended question about scientists' work were coded as follows.

- 'Unaware', when irrelevant or explicitly stating unawareness; e.g. *'I don't know'*.
- 'Naïve', when vague/not to the point; e.g. *'They go to the office and look at things'*.
- 'Transitional', when right but incomplete; e.g. *'They see new things in a lab. They have stuff and they do many things in the lab'*.
- 'Informed', when right and more complete, with adequate references to points made in the learning environment; e.g. *'They do experiments to see if something is true. They go out to observe plants and animals very carefully. They speak with other scientists and tell what they found'*.

Similarly, the responses to the two-tier two-choice questions were coded as follows.

- 'Unjustified', when including either the right or the wrong statement (RS or WS) with no justification (no J); e.g. *'No imagination.* I *don't know why'* (WS/no J).
- 'Naïve', when including either the right or the wrong statement (RS or WS) accompanied by a problematic justification (Naïve); e.g. *'He cannot use both his logic and his imagination. He will get confused. Only logic'* (WS/Naïve).
- 'Transitional', when including always the right statement (RS) accompanied by a right but incomplete justification (trans); e.g. *'They use imagination. Van Helmont had an idea about plants through his imagination'* (RS/trans).
- 'Informed', when including always the right statement (RS) accompanied by a more complete justification, with adequate references to points made in the learning environment) (RS/informed); e.g. *'Like van Helmont. When he was eating his lunch, he saw that the food in the dish was decreasing and an idea came to his mind… as the food in the dish, the soil in the pot must also decrease'* (RS/informed).

We also developed a scoring grid by assigning a score to each category of responses. In the case of QA2, for instance, we assigned the scores 0, 1, 2 and 3 to irrelevant, naïve, transitional and informed responses, respectively. In order to test whether the differences in children's pre–/post-responses were statistically significant or not, we ran Wilcoxon tests for each question.

## 6.3 Findings

Starting with the yes/no question about whether children had heard of the word 'scientist' before (QA1), we just note that even in the pre-interviews all but one children (20/21) claimed that they had; obviously in the post-interviews this was the case for everyone. Moreover, children's ideas about what scientists do were improved. As shown in Fig. 6.1, most of the pre-responses to this open-ended question (QA2) were coded as naive (9/21), whereas most of the post-responses were coded as transitional (17/21). In fact, in the post-interviews, the number of transitional responses was almost tripled (17/21 vs. 6/21), the number of both unaware and naïve responses went all the way down to zero (unaware: 0/21 vs. 6/21; naïve: 0/21 vs. 9/21), and informed responses appeared for the first time (4/21 vs. 0/21). According to the results of the Wilcoxon test, children's ideas about scientists' work were improved in a statistically significant manner ($p = .000$).

In children's own words:

- *'A scientist goes to his work.* [And what does he do?] *He helps other people*' (naïve pre-response); *'They take exams, they study. Like my cousin. He is a scientist. They go to other countries to study and take exams'* (naïve pre-response);
- *'They discover things they didn't know'* (transitional pre-response); *'They study plants and nature and they take notes'* (transitional post-response);
- *'They do experiments and they travel to find new things and do investigations.* [What kind of investigations?] *They investigate strange animals, plants, places*' (informed post-response); '*They investigate.* [What do you mean?] *Like the first scientist. The others said that the plants eat soil and he wanted to see if this was true and he did the experiment* [anything else?] *They read books. They observe out in the nature. Like the second. They write their findings and other scientists read them'* (informed post-response).

Children seemed to do quite well with the empirical aspect of NOSK; i.e. with the idea of using empirical data ('evidence') in the construction of scientific knowledge (QB1). As shown in Fig. 6.2, in the post-interviews, all responses included the right statement *(RS) ('In order to give answers to my questions, I have to gather evidence. I cannot just tell my opinion');* almost half were coded as informed (12/21)

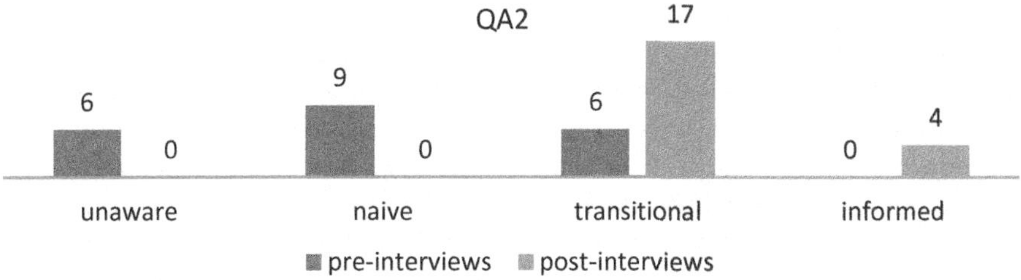


**Fig. 6.1** Students' pre/post-responses about what scientists do

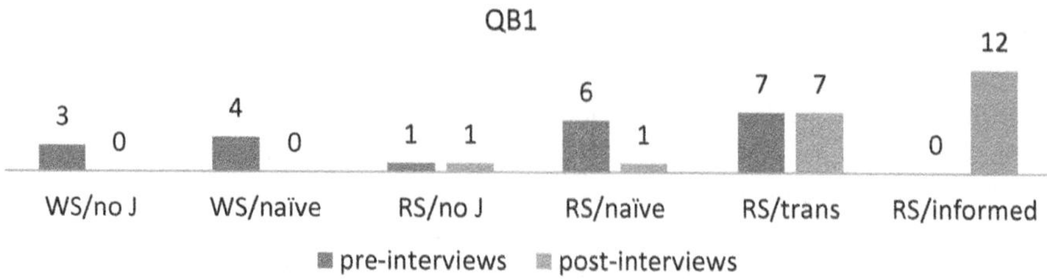


**Fig. 6.2** Students' pre/post-responses about the use of empirical data

and one third as transitional (7/21). Contrastingly, in the pre-interviews, one third of children's responses (7/21) included the wrong statement (WS) (*'In order to have the answers to my questions, I don't have to gather evidence. I can just tell my opinion')*. The remaining two thirds that included the right statement were coded as unjustified (1/21), naïve (6/21) and transitional (7/21). The pre/post differences in children's ideas about the empirical aspect of NOSK were statistically significant (p = .000).

In children's own words:

- *'I choose this one because he is a scientist and he can say his opinion. He does not need anything else.'* (WS/naive pre-response); *'He would say 'I must find evidence'. Because this is the right thing. He must do it'* (RS/naïve post-response);
- *'Like the lawyer in the court. When he says something, he must have evidence. The scientist, if he goes somewhere to show his inventions, they may not believe he is right, so he must have evidence'* (transitional pre-response); *'He needs evidence. He must prove it. Van Helmont proved it* (transitional post-response);
- *'When you give an answer, first you must have evidence. One scientist found that plants don't eat soil, because the soil in the pot of his plant was always the same.* [What was the evidence?] *He had notes. He was writing the weight of the soil and the plant')* (informed post-response); *'He must have evidence like Darwin.* [What evidence?] *He was gathering those with the long nose and the short nose. He was looking at their nose and writing about it in a book and then he showed his notes to other scientists to believe him'* (informed post-response).

Regarding children's ideas about the creative aspect of NOSK, i.e. the use of imagination/creativity in the construction of scientific knowledge (QB2), we note the following. According to Fig. 6.3, in the post-interviews, almost all responses (19/21) included the right statement (RS) *('I'm a little bit of an artist as well. Imagination is used in my own work too'),* and except for some of them which left it unjustified (4/21), all the others (two thirds), were coded as transitional (7/21) and informed (8/21). Contrariwise, in the pre-interviews half of the responses (12/21) included the wrong statement (WS) *('Imagination is for artists. I only use logic in my work')*. The remaining (9/21) that included the right statement were mostly coded as unjustified and naïve (6/9), and sometimes as transitional (3/9). According to the results of the Wilcoxon signed-rank test, children's ideas about the creative aspect of NOSK were improved in a statistically significant manner (p = .000).

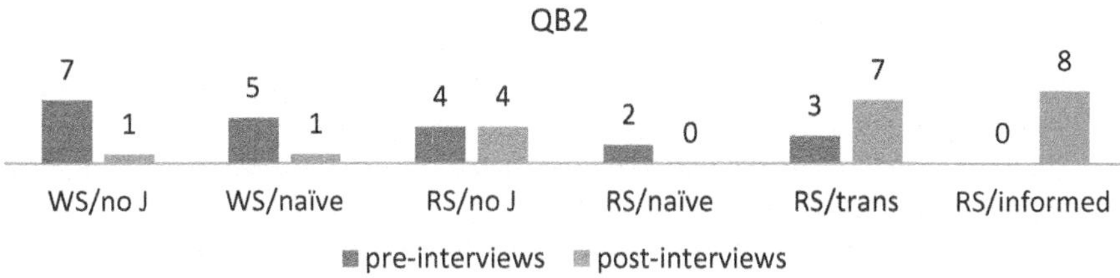


**Fig. 6.3** Students' pre/post-responses about the use of imagination/creativity

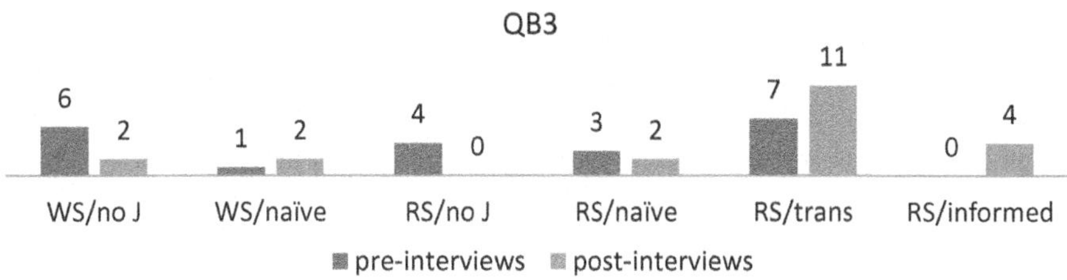


**Fig. 6.4** Students' pre/post-responses about the use of reasoning

In children's own words:

- *'Because scientists find everything with their brain and the computers!'* (WS/naïve pre-response);
- *'It is imagination as well. They have to imagine how the rocket they are building will look like'* (transitional pre-response); *'Van Helmont used his imagination and figured out how to do the experiment'* (transitional post-response);
- *'He can use his imagination. One day two sailors were playing fencing on the ship and there was a narrow carpet underneath their feet and it was... and it started rising up a little bit... they were pushing...* [So, how did he use his imagination?] *He had an idea that the mountain was made like the little mountain on the carpet'* (informed post-response); *'Van Helmont, when eating, he saw that his food was becoming less and less and he thought the same about the soil'* (informed post-response).

Children's ideas about the inferential aspect of NOSK, i.e. the use of reasoning in the construction of scientific knowledge (QB3), were also improved. As shown in Fig. 6.4, in the post-interviews, the number of responses with the wrong statement (WS) (*'I gathered evidence and I'm ok; the information that evidence gives me, are my conclusions')* decreased (4/21 vs. 7/21), whereas the number of those with the right statement (RS) (*'I gathered evidence but this is not enough; I have to think the information evidence gives me and how I can combine them in order to have my conclusions')* became quite high (17/21 vs. 14/21). And more importantly, except for very few post-responses which included a naïve justification for the right statement (2/21), all the others were coded as transitional (11/21 vs. 7/21) and less frequently as informed (4/21 vs. 0/21) (Fig. 6.4). The pre–/post-differences were statistically significant ($p < .05$).

In children's own words:

- *'It is not enough, because he must collect more evidence. He must have the right evidence and not make mistakes'* (RS/naïve pre-response);
- *'He must collect evidence and combine them.* [So, he must do what?] *He must put them all together and look at them one by one*. [Why do you think he needs to do this?] '*To combine them*' (transitional pre-response); '*This one, because except for the evidence, he must think.* [Can you explain it to me?] *Van Helmont studied the plant for five years. He was taking it out of the pot and he had a weight scale to see the weight of the plant and the soil. He did this for five years and then he told about it to the others'* (transitional post-response);
- *'He must combine them. The second scientist... he first went to the trip and he made observations of the animals and he found lots of things... evidence... and he kept all these in his notebook and then we helped him to combine them* [reference to an activity in session 5] *and find his conclusion'* (informed post-response).

Children's ideas about the tentative aspect of NOSK, i.e. the tentativeness of scientific knowledge (QB4), did not show as much progress as their ideas about the other NOSK aspects. More specifically, as shown in Fig. 6.5, in the post-interviews, the number of responses with the wrong statement (WS) (*'My findings will not change in the future')* decreased (5/21 vs. 8/21), whereas the number of those with the right statement (RS) *('My findings may change even very soon')* became quite high (16/21 vs. 13/21). However, the post-responses in which the right statement was grounded on transitional (5/21) or informed justifications (4/21) were slightly more (9/21 vs. 7/21) than those in which the right statement was grounded on naïve justifications (3/21) or left unjustified (4/21). According to the results of the Wilcoxon test, children's ideas about the tentativeness of scientific knowledge were improved in a statistically significant manner ($p < .05$).

In children's own words:

- *'What he found can change because he made a mistake. If he didn't make a mistake, it won't change'* (RS/naïve pre-response);
- *'Because if you think something, it may be wrong and your idea will change.* [What about if it is right?] *It can change. Scientists can change their minds'* (transitional post-response);
- *'Like the first. He read in the book that plants eat soil, but when he investigated himself, they didn't. The thing he read, changed'* (informed post-response).

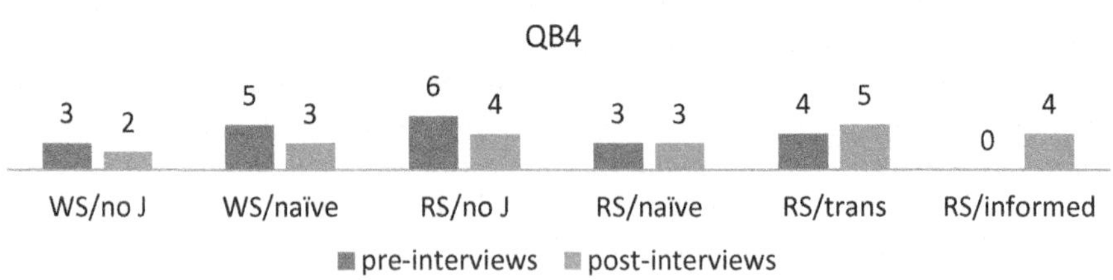


**Fig. 6.5** Students' pre/post-responses about the tentativeness of scientific knowledge

## 6.4 Discussion

According to our findings, the learning environment we designed based on the biographies of van Helmont and Darwin in order to introduce young children to a series of NOSK aspects and biological notions that emerge from these scientists' work, was rather effective with regard to the NOSK aspects in particular. In the two-tier two-choice questions about the target NOSK aspects, most post-responses included right statements with more complete justifications, which actually resulted in their coding at higher level categories. The differences in children's pre−/post-responses to each question were found to be statistically significant.

More specifically, children made remarkable progress with the idea that the construction of scientific knowledge involves empirical data (QB1). After their participation in the learning environment, *all* children were able to recognize that scientists need to collect empirical data, and almost all came up with either informed or transitional justifications for the correct statement. In fact, the number of informed post-responses was higher than in any other question. This might have to do with the fact that in the learning environment children were actually required to 'help' scientists analyze the data they gathered in order to answer their questions. Children's progress with the aspect of imagination/creativity (QB2) was remarkable as well, especially if we take into account that half of the children had initially expressed the typical misconception that imagination is something that concerns only artists, while scientists rely exclusively on their logic. In the post-interviews very few children kept thinking this way, whereas almost all chose the right statement and most justified their choice in either informed or transitional ways. These findings seem to be in line with Akerson et al. (2011), who suggested that creativity is a NOSK aspect that works quite well with young children.

Moving on to the idea of reasoning (QB3), we should note that although children's understanding has also improved, the number of informed post-responses was rather low since half of the children gave transitional ones. Moreover, the number of post-responses with the wrong statement, despite of not being high, was higher than in the previous question. Probably the idea that data and conclusions are respectively the input and output of reasoning, rather than the same thing, was a little bit more demanding for the children. This seems to be in line with Quigley et al. (2010), but at odds with Akerson et al. (2010) who consider this NOSK aspect rather easy for young children. Similarly, after their participation in the learning environment, children appeared to be more familiar with the idea that scientific knowledge is prone to change (QB4). They did improve their understanding, but still only a few post-responses were coded as informed or transitional. Moreover, this time the number of post-responses with the wrong statement was higher than in any previous question. The tentativeness of scientific knowledge can be controversial to children, so there was probably a need for more emphasis in the learning environment.

Children had also the opportunity to build a better view about what scientists do, as shown by their responses to the open-ended question at the beginning of the protocol (QA2). In the post-interviews, they totally shifted to the higher level categories (transitional and informed), leaving the lower level categories empty. This is in line with Heisey and Kucan (2010) who reported that their young participants were able to improve their view about scientists by recognizing that they discover new things as they wonder about nature, observe, make experiments, use tools, study and think. Moreover, it is worth noting that after their participation in the learning environment, the children seemed to have a more realistic view of scientists as people who are very happy with their work, but they also have friends and families and live normal lives like anyone else. In sum, these second graders appeared ready to get familiar with scientists' work and the nature of knowledge they produce. The 'general aspects conceptualization of NOS' and the idea of implementing it in explicit, reflective ways within historical contexts, created with the aid of scientists' biographies which emphasize how scientists work to produce scientific knowledge, seem to be useful for teaching NOS to the younger students of primary school and start investing in their scientific literacy.

## References

Akerson, V. L., Weiland, I., Pongsanon, K., & Nargund, V. (2010). Evidence-based strategies for teaching nature of science to young children. *Journal of Kırsehir Education, 11*, 61–78.

Akerson, V. L., Buck, G. A., Donelly, L. A., Nargund-Joshi, V., & Weiland, I. S. (2011). The importance of teaching and learning nature of science in the early childhood years. *Journal Science Education and Technology, 20*, 537–549.

Allchin, D., Andersen, H. M., & Nielsen, K. (2014). Complementary approaches to teaching nature of science: Integrating student inquiry, historical cases, and contemporary cases in classroom practice. *Science Education, 98*(3), 461–486.

Bell, R. L., & St. Clair, T. L. (2015). Too little, too late: Addressing nature of science in early childhood education. In K. C. Trundle & M. Sackes (Eds.), *Research in early childhood science education* (pp. 125–141). Springer.

Dagher, Z. R., & Ford, D. J. (2005). How are scientists portrayed in children's science biographies? *Science & Education, 14*(3–5), 377–393.

Durant, J. R. (1993). What is scientific literacy? In J. R. Durant & J. Gregory (Eds.), *Science and culture in Europe* (pp. 129–137). Science Museum.

Duschl, R. A., & Grandy, R. (2013). Two views about explicitly teaching nature of science. *Science & Education, 22*, 2109–2139.

Hadzigeorgiou, Y. (2018). Teaching the nature of science through storytelling: Some empirical evidence from a grade 9 classroom. *SFU Educational Review, 10*(2).

Heisey, N., & Kucan, L. (2010). Introducing science concepts to primary students through read-alouds: Interactions and multiple texts make the difference. *Reading Teacher, 63*(8), 666–676.

Jenkins, E. W. (1994). Scientific literacy. In T. Husen & T. N. Postlethwaite (Eds.), *The international encyclopedia of education* (Vol. 9, pp. 5345–5350). Pergamon Press.

Irzik, G., & Nola, R. (2011). A family resemblance approach to the nature of science for science education. *Science & Education, 20*(7–8), 591–607.

Kampourakis, K. (2016). The "general aspects" conceptualization as a pragmatic and effective means to introducing students to nature of science. *Journal of Research in Science Teaching, 53*(5), 667–682.

Kelly, L. B. (2018). An analysis of award-winning science trade books for children: Who are the scientists, and what is science? *Journal of Research in Science Teaching., 55*(8), 1188–1210.

Khishfe, R., & Abd-El-Khalick, F. (2002). Influence of explicit and reflective versus implicit inquiry oriented instruction on sixth-graders' views of nature of science. *Journal of Research in Science Teaching, 39*, 551–578.

Korkmaz, H. (2011). The contribution of science stories accompanied by story mapping to students' images of biological science and scientists. *Electronic Journal of Science Education, 15*(1), 1–41.

Lederman, N. G. (2019). Contextualizing the relationship between nature of scientific knowledge and scientific inquiry. *Science & Education, 28*, 249–267.

Lederman, N. G., & Lederman, J. S. (2019). Teaching and learning nature of scientific knowledge: Is it Déjà vu all over again? *Disciplinary and Interdisciplinary Science Education Research, 1*(6), 1–9.

McComas, W. F., & Kampourakis, K. (2015). Using the history of biology, chemistry, geology, and physics to illustrate general aspects of nature of science. *Review of Science, Mathematics and ICT Education, 9*(1), 47–76.

Quigley, C., Pongsanon, K., & Akerson, V. L. (2010). If we teach them, they can learn: Young students views of nature of science during an informal science education program. *Journal of Science Teacher Education, 22*(2), 129–149.

Yacoubian, H. A. (2018). Scientific literacy for democratic decision-making. *International Journal of Science Education, 40*(3), 308–327.

# Part II
# Teaching and Learning in Biology

# Chapter 7
# Digital Narratives for Biology Learning

**María Napal Fraile, Isabel Zudaire Ripa, Irantzu Uriz Doray, and Lander Calvelhe**

## 7.1 Introduction

*Storytelling*, or telling stories, is a didactical strategy inspired in the oral transmission of culture via songs and stories. This technique relies on the power of narratives – and the emotions they raise – to transmit their message more efficiently. Although narratives often refer to personal reflective writing – used to launch creative writing and education values – they can also encompass instructive stories to explain concepts or processes. For teachers, storytelling can mean an opportunity to transmit knowledge more effectively, in a more interesting, more attractive or clearer way. For students, it can grant a much more active role in the learning process: providing the opportunity of becoming creators of knowledge, fostering creativity and other competences.

Storytelling is a simple but powerful method to help students to make sense of the very complex and unordered world of experience, by crafting story lines (Sadik, 2008) that help students organize and express their ideas and knowledge in an individual and meaningful way (Robin, 2005). Reflection required to tell stories can be a powerful tool to engage in deep learning that is, in nature, reflexive, integrative and progressive (Barrett, 2006). This is especially true for biological concepts,

M. Napal Fraile (✉) · I. Zudaire Ripa · I. Uriz Doray
Department of Sciences, Universidad Pública de Navarra (UPNA), Campus de Arrosadia, Pamplona, Spain
e-mail: maria.napal@unavarra.es; mariaisabel.zudaire@unavarra.es; iranzu.uriz@unavarra.es

L. Calvelhe
Department of Humanities and Educational Sciences, Universidad Pública de Navarra (UPNA), Campus Arrosadía, Pamplona, Spain
e-mail: lander.calvelhe@unavarra.es



K. Korfiatis, M. Grace (eds.), *Current Research in Biology Education*, Contributions from Biology Education Research,
https://doi.org/10.1007/978-3-030-89480-1_7

which are very often complex and multifaceted, and where causal links are not always easy to disentangle.

In *Digital Storytelling (DS)*, the inclusion of Information and Communications Technology (ICT) and authoring tools, such as video or interactive resources, means the opportunity to craft meaningful and attractive stories at low cost and in a relatively short time. And, given the recent proliferation of intuitive creation tools, it can be done with low and decreasing technical requirements (Robin, 2005). Moreover, ICT adds not only interactivity, but also multimodality, or exploring other modes of elaborating and communicating knowledge. In other words, it promotes students from consumers to collaborative producers.

In this sense, DS is one example of the meaningful introduction of ICT into instruction (Sadik, 2008); one ensuring transformation of teaching to the highest levels of Modification or Redefinition in the SAMR model of classroom technology integration, i.e. tasks that could not be achieved without ICT (Puentedura, 2006). But designing impactful situations with ICT requires Digital Competence on the side of both teachers and learners. Digital Competence includes, according to the DigComp framework (Ferrari, 2013), five areas: Informational literacy, Communication and Collaboration, Security, Content Creation and Problem Solving. Those five areas are prerequisites but can also be enhanced by introducing technology-rich activities in the classroom.

However, there are many potential barriers to the implementation of the above suggested approaches. Indeed, they rely heavily on the interconnectedness of scientific concepts, the ability to propose scenarios of situated learning, and meaningful integration of ICT. Taking into account all the above, we launched a research and transference project that sought to provide specific instruction and to accompany teachers in the implementation of Digital Storytelling for the teaching of science.

## 7.2 Objectives

Our objective was to design, implement and test a training programme, specifically designed to provide teachers with the conceptual and technological knowledge required for developing Digital Narratives about biology in their classes. It was delivered purposefully and built to serve as a scaffold for repeating the work with secondary school students. The addressed research questions are:

- Which scientific concepts were successfully integrated by the teachers and proved most useful for building stories about science?
- Which were the main difficulties experienced by teachers during the training?
- Was the level of Digital Competence of the teachers high enough to integrate ICT in their practice?

Our specific interest in these questions was derived from the understanding that they could be used as an indicator of the affordances and difficulties that students might also face, and thus should be specifically tackled during implementation in class.

## 7.3 Research Design and Method

### 7.3.1 *Context*

The described intervention belongs to the project "From research to development and innovation in Secondary Education: Digital Narratives in science using chromebooks" (Gobierno de Navarra, Dept. of Education, grant n° CENEDUCA 08/2019).

The project included 15 h of face-to-face training (five sessions of 3 h), plus 10 h of autonomous work. As a requisite for the completion of the training programme, attendees were requested to submit a 3–5 min-long history answering a driving question related with biology, incorporating Visual Arts and composed using Scratch or tools for video edition.

Teachers (n = 23) of grades 6 to 12 attended the training. It was an open call, included in the catalogue of courses offered by the Department of Education to all teachers in Primary and Secondary levels. As such, it can be considered a convenience sample.

### 7.3.2 *Description of the Intervention*

(A) **Model Digital Story**

Firstly, the researchers designed a digital story that was used as a model during the training sessions. The process to build the digital story was similar to the cycle suggested by Morra (2013), with a special focus on some specific aspects of science teaching (Table 7.1).

(B) **Training sessions**

The 5 training sessions followed the process described above, interspersed with more detailed explanations on topics specific to science teaching, integration of the arts and the language, and technological tools. Transversely, it also included references to different aspects of Digital Competence (INTEF, 2017). Briefly, contents and activities developed in sessions were as shown below and in Table 7.2:

Session 1: Digital narratives

The first session served to introduce the fundamentals of DS; examples from journalism, science communication and science education were shown. The examples were selected to highlight the relevance of designing DS to capturing the attention of the audience, and making it part of the story, rather than treating them as mere consumers. During the session, teachers were requested to do a task where they analyzed the narrative strategies, technology and interactivity presented in the different examples.

**Table 7.1** Design of an example of digital science narrative

| STEP (Morra, 2013) | Specific steps suggested in the training (*With examples from the model story*) |
|---|---|
| 1. **The idea/topic** | To define a driving-question: *Why do a frog, a mushroom, and a woodlouse need water to live?* (Cañal de León et al., 2016) |
| | To identify Disciplinary Core Ideas (DCI) (NSTA, 2014b)<br>– LS1.A: Structure and function: *How do the structures of organisms relate to life functions?*<br>– LS1.B: Growth and development of organisms: *How do organisms grow and develop?*<br>– LS1.C: Organization for matter and energy flow in organisms: *How do organisms obtain and use the matter and energy they need to live and grow?*<br>– LS2.A: Independent relationships in ecosystems: *How do organisms interact with the living and nonliving environments to obtain matter and energy?* |
| | To consider Crosscutting concepts (NSTA, 2014a)<br>– systems and system models<br>– energy and matter<br>– structure and function<br>– diversity and unity[a] |
| 2. **Research** | Use of Thinking routines for exploring ideas (Harvard Graduate School for Education, 2020):<br>*Example of Know, Want to Know, Learnt routine:*<br>*K :"I know frogs are amphibian thus they need water to reproduce, and to breathe.*<br>*W: "Why do mushrooms need water?", "What kind of animal is a woodlouse?"* |
| | Search for information in our PLE[b] in Science:<br>For example: *PBSlearning, Khan Academy, Encyclopedia of Life, Exploratorium, Proyecto Biosfera.* |
| | Design a concept map. |
| **3. The script** (choose the characters, dialogues and scenarios). | – Tadpole: *"Mum, look what a pretty plant!"*<br>– Mother frog: *"It's not a plant darling, it's a fungus. Fungi feed by decomposing dead organic matter".* |
| **4. Storyboard** | <br>Transform the language of the script into images and graphics. |
| **5. Gather and create images, audio, video** | <br>Carlos de Haes<br>Nijmegen (Holland)<br>Prado Museum<br><br>Designed by the authors (L.Calvelhe) |

(continued)

**Table 7.1** (continued)

| STEP (Morra, 2013) | Specific steps suggested in the training (*With examples from the model story*) |
|---|---|
| **6. Put it all together** | *Openshot* and *Scratch*, to produce a final mp4 video product. |
| **7. Share** | The Digital Story created by researchers was shown during teacher training (https://youtu.be/VJA3O4Ggqtg). |
| **8. Reflection and feedback** | Researchers realized the most complex steps of the DS cycle where teachers and students will need support. |

[a]Crosscutting concept suggested by (Cañal et al., 2016)
[b]PLE: Personal Learning Environment: diverse physical locations, contexts, and cultures in which students learn

**Table 7.2** Content of the training sessions

| Title of the session | Topics | Area of teacher digital competence (INTEF, 2017) |
|---|---|---|
| S1. Digital Narratives. | Concept of DS. | Informational literacy. |
| S2. Elements for creating good histories in biology. | Big ideas and crosscutting concepts (CC) in science.<br>Good questions.<br>Heuristic tools. | Informational literacy. |
| S3. Let's talk biology. | Language and scientific thinking.<br>Tools for telling science stories. | Problem solving. |
| S4. Visual Arts (VA) as a tool for creating DS. | Consuming and creating VA. | Content creation.<br>Informational literacy. |
| S5. Digital tools. | Scratch and other digital edition tools. | Content creation.<br>Problem solving.<br>Security. |

Session 2: Elements for creating good histories in biology

During this session, the big ideas of science education (Harlen, 2010), scientific competence and crosscutting concepts (CC) (NSTA, 2014a) were introduced. Researchers also discussed the importance of starting from powerful questions, and distinguished among essential, driving and research questions (McTighe & Wiggins, 2013; Cañal et al., 2016). To highlight the importance of the interplay between activity and theory, the attendees were prompted to reflect on what they knew and they needed to know, before starting to write the story.

The session focused on the informational literacy dimensions of Digital Competence: where and how to search for information for biology lessons, and also how to organize it. Teachers attending the training performed different tasks during this session:

- To analyze the core and crosscutting concepts included in the DS used as a model (*Why do a frog, a mushroom, and a woodlouse need water to live?*)
- To transform this driving question into research and essential questions.
- To perform a thinking routine about the science topic.
- To organize newly gathered information using a concept map (optional).

Session 3: Let's talk biology

In this third session, researchers discussed the central role of language in the building of scientific knowledge, and how the use of proper grammatical structures and a lexicon is essential to science communication. Attendees exercised interpreting, expressing ideas or shifting between different modes of representation (drawings, graphics, symbols including equations, etc.).

Session 4: Visual Arts as a tool for creating DS

The following step implies transforming the script into an audiovisual artefact. In this fourth session the importance of making a good selection of images, voices and audios was discussed. Selection criteria included not only technical features, but also transversal elements that contribute to the hidden curriculum in the school. Teachers were provided in this session with links to search images with licenses compatible with educational uses, and also software to edit them. Together with these technical aspects, the trainer emphasized the need to make use of visual arts, and consider their intrinsic value in STEAM, which goes beyond illustrating or serving to communicate scientific ideas.

The teachers' task included a reflection about their artistic references, and the selection of a background and a character for their own DS, applying the acquired tools and strategies.

Session 5: Digital tools

During this session, the researchers showed different software for editing video and audio in different platforms. Since there is an ongoing plan in the region to provide all secondary school students with Chromebooks and to implement Google Classroom, the training was focused on free apps available for the Google Education suite and evaluation of its features. As an alternative to videos, *Scratch* authoring tool (https://scratch.mit.edu/) was also introduced to allow for more interactive narrative schemes, while developing other aspects of the Digital Competence.

### 7.3.3 Research Methodology: Evaluation of the Intervention

The evaluation of the teachers' outcomes was performed following an interpretative qualitative approach. All the pieces of work produced by the attendees were collected, including both the productions during the training sessions and the final

**Table 7.3** Analysis criteria for the assessments of the teachers' tasks

| Task | Analysis criteria |
|---|---|
| *During training sessions*<br>• To analyze the crosscutting concepts (CC) included in the DS used as a model.<br>• To identify a driving question, and to transform it into research and essential questions.<br>• To gather information (informational literacy)<br>• To use a thinking routine to organize previous knowledge and the newly gathered information (concept map or other).<br>• To use different modes of representation (drawings, graphics, symbols including equations, etc.) to express science. | • Crosscutting concepts (CC) identified, in comparison with the CC initially identified by researchers.<br>• The transformed questions met the criteria of essential and driving questions.<br>• Types of sources (descriptive).<br>• Present/absent.<br>• The alternative modes of representation convey the same message. |
| *Teachers' digital stories*<br>• Scientific content<br>• Technical and artistic aspects | • Presence of driving question.<br>• DCI, CC, scientific practices or elements of nature of science identifiable in productions.<br>• Dynamism of characters expressions and dialogues, role of visual art, interactivity.<br>• Balance of content/ technical aspects. |

story (3 to 5-min video or Scratch story). Individual productions were examined, and assessed to evaluate compliance with the quality criteria for each of the tasks (Table 7.3). All the evidence was evaluated jointly by the researchers, until full agreement was reached. Summaries of the main findings are provided task by task. The inferences based on the analysis of the submissions were completed with non-formal interactions with the attendees.

## 7.4 Results

### 7.4.1 *Analysis of Teachers' Work During the Training Sessions*

One of the most complex activities for the teachers was to analyze the core disciplinary ideas and crosscutting concepts (CC) included in the model story (*Why do a frog, a mushroom, and a woodlouse need water to live?*). Most of the attendees had never considered CC in their science classes, nor were acquainted with this concept. Indeed, although all the seven concepts defined by the NSTA (NSTA, 2014a) were identified in the model story, only some of them were adequately justified, and the proportions were comparable for CC present and absent (according to the judgement of the researchers) (Table 7.4).

By far the most frequently identified concepts were "cause and effect" and "unity and diversity". Except for "stability and change" and "unity and diversity", the

**Table 7.4** Cross-cutting concepts identified by the participants in the model story

| Crosscutting concept | Cross-cutting concept identified by the researchers in the model story (YES/NO) | Identified | Justified |
|---|---|---|---|
| Cause and effect | N | 17 | 8 |
| Unity and diversity | Y | 16 | 16 |
| Energy and matter | Y | 10 | 5 |
| Patterns | N | 9 | 2 |
| Structure and function | Y | 9 | 6 |
| Stability and change | N | 9 | 7 |
| Systems and system models | Y | 5 | 0 |

attendees had difficulties in justifying their choice with reasoning that was coherent with both the content of the story and the meaning of the concept. For example, 7 out of 9 alluded to taxonomy to justify the choice of "Patterns". Almost one third of the teachers (6/17) who selected "cause and effect" provided justifications that linked with other concepts; namely, "patterns" and "stability and change".

As for the transformation of questions (driving to essential/ research questions), the 19 tasks that were submitted included 22 essential and 29 research questions. 10 of the 22 (45%) complied with the requisites for being considered essential, and resembled the model question. Likewise, 17 (59%) were considered acceptable research questions, all of them related to the lack of water in the scenario of the model story.

The teachers had no major difficulties in interpreting different graphical representations, shifting between modes of representation or expressing meanings using graphic organizers.

However, two activities are remarkable for their poor quality. First, only six teachers submitted a thinking routine reflecting about the scientific concepts that would be included in their own digital story, and only three of them were complete. No complete concept maps were submitted. This could be one of the reasons for the absence of sound science conceptualization in some of the artifacts. Second, the teachers stated limited knowledge about sources of information to prepare their science classes, as main sources of information were Google, YouTube and Wikipedia, with much lower frequency of specialized science or educational sites.

### 7.4.2 Analysis of the Teachers' Final Productions

All the groups submitted a full story complying with the technical and formal requirements of the task (format, length, requisites of delivery). All the stories had an implicit driving question, explicit science contents, and could potentially be linked with one or more crosscutting concepts (Table 7.5).

**Table 7.5** Components of the digital stories submitted at the end of the training

| Title | Driving question/ problematic scenario | Content | Crosscutting concepts |
|---|---|---|---|
| Noe's mission | The "Noe's rocket". Distribute species in the stars of the Milky Way Galaxy | Universe and biodiversity | Structure and function |
| Drop by drop, stop floodings. | Why do floods happen? | Ecosystems; interactions, energy and dynamics. | Cause/effect |
| A drop of water | How is life in a drop of water? | Monera and protista kingdom | Patterns<br>Models and system models |
| "Hidratomik and the cells" | Are we the dust of the stars? | The origin of life<br>[In]organic chemistry | Energy and matter<br>Stability/change |
| A football match in Kepler 10-B | Would it be possible to score a goal from mid field in Kepler 10B? | Force and movement | Cause/effect |
| Model catwalk | The evolution of the atomic model | Atomic models | Models and system models |
| Everybody moving | Will Spiderman and Usain Bolt arrive in time to save the child? | Movement | Energy and matter |

A closer analysis of the content of each of the stories revealed the following aspects of interest:

**Scientific Content vs. Digital Artifacts**

Based on the analysis of the final artifacts (videos or Scratch programs) submitted after the training, it became evident that, in most cases, much more time and effort had been devoted to the elaboration of the product than to the process of construction of knowledge. The dialogues with teachers served to confirm this perception.

Nevertheless, the sample of tasks submitted included some interesting examples, where scientific concepts (science, how is science, how it is done) were at the centre. For example, in the story entitled "A drop of water" the dialogues guide observation of the anatomy of several microorganisms, emphasizing distinguishing features and introducing scientific vocabulary in a natural way. In this video, teachers also included procedural aspects such as the use of a microscope.

The video "Everybody moving" succeeded to exploit multimodality, and to combine graphic representations with mathematical formulae and text captions to enhance comprehension (Fig. 7.1a).

The video "Model catwalk" is also an excellent example to teach the nature of science, showing that science knowledge is limited and subject to change in the light of new evidence.

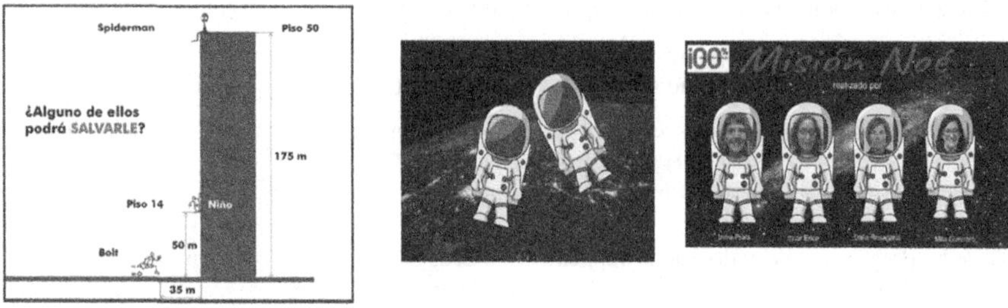


**Fig. 7.1** Remarkable examples in teachers' final productions. (**left**) A graphic representation of the parameters needed to solve the question in *Everyone moving;* (**center**) Slowmotion in Scratch creates the realistic illusion of astronauts floating in *Noe's mission*; (**right**) Final credits and users' licenses

**Technical and Artistic Aspects in the Videos**

The groups made a detailed selection and conscious inclusion of images, some of them famous artworks. As it was shown in the training, they even edited images and fragments of video that enhanced the quality and dynamism of the productions. Although only used by two of the groups, Scratch proved especially useful for introducing movement and a certain degree of interaction, making the scenes and the characters more realistic (especially if they were synchronized with the story), and also to add interactivity with the audience (asking questions, for example) (Fig. 7.1b).

**Digital Competence**

We have no evidence about whether there was any improvement in the informational literacy (increasing the quality or the diversity of the information sources).

In spite of the training specifically addressing the need to consider licenses and author's rights, both when using others' work and when distributing their own work, few groups included credits, specified the license type or referenced their sources (Fig. 7.1c).

The interactions with the teachers made evident their self-consciousness about their technical command of tools, in spite of the training received (content creation), and that this might be limiting their ability to innovate with ICT.

## 7.5 Discussion

The training sessions, scheduled according to Morra's 8 steps for DS (Morra, 2013), and with constant reference to the model story built by the researchers, were aimed to get the attendees familiar with this methodology and facilitate later use in their classrooms. Moreover, the sessions sought to emphasize certain aspects that guarantee the quality of the process and the product, such as the necessity to structure teaching in reference to crosscutting concepts and big ideas, the power of questioning and the utility of considering alternative representation modes (multiliteracy). Lastly, they facilitated teachers' resources and tools with which they are usually not familiar, such as art facilities and coding tools.

The analysis of the productions of the attending teachers during or after the training shed light on some of the dynamics that emerged and the difficulties they encountered during the process. These are developed in the following paragraphs.

As it becomes evident in the final artifacts (videos or short Scratch codes), the technological challenge often attracts all the attention, diverting the focus from the contents. As it is also common in some instances in PBL, the context – the artifact or task – takes over the contents in the control of the real conflict of the project (Domènech, 2019). To prevent this, it could be interesting to use a template to make explicit the objectives of the project (i.e. to elaborate a video to tell a story) and the learning objectives (i.e. science concepts), and to review them at different moments during the project development, to analyse where we are exerting more time and effort (Domènech, 2019).

Likewise, there was a rapid identification and good assimilation of the methodological and technical aspects, such as the incorporation of visual resources or the structure of the process. But, at the same time, we observed a great resistance to changing the way of delivering scientific contents. Indeed, most of the products served to convey a closed fragment of knowledge, far from representing the process of knowledge construction. This could be related to the fact that all but two groups created lineal, non-interactive videos, the main reason being that they did not feel capable of using the Scratch programming interface. The reason they alluded to was their perceived lack of confidence in using new digital tools.

This is a clear example that Digital Competence of teachers is not independent, but may refrain from pedagogical integration. Indeed, only moderate to high levels of technical performance allow for appropriation of ICT, or technology-based innovation (Krumsvik, 2011), while teachers in the sample obtained low to medium levels compatible with adoption or implementation.

In the same vein, concepts tended to be treated as isolated concepts (as it happens in most classrooms), with difficulties to identify and incorporate crosscutting concepts. It became clear that being able to tell stories enables (but does not guarantee) knowledge to be (re)built, and that it is technically feasible to construct narratives without developing deep understanding. This raises awareness about the importance of ensuring this entailment, if we want Digital Storytelling to have real impact on the significant acquisition of concepts.

The gap between the initial expectations and the final result could be attributed to several factors, including deeply held beliefs and powerful inertias, and also limitations of the training programme. It was quite a specific programme, that took for granted several prerequisites pertaining mainly to science teaching. This was obviously an overoptimistic assumption; the training should have been longer, and/ or started from more basic concepts to make it more meaningful and easier to grasp. Other possible limitations were related to the scheduling of the sessions, and of the delivery of the final product with respect to the end of the "lectures". The process of creation of the final artifact was a perfect black box for the researchers, who did not get any insight on the progress of the project until it was delivered. In cases requiring such a profound revision of their own schemes, it is essential that the learners are accompanied during the process.

To conclude, in the literature, Digital Storytelling (Barrett, 2006; Robin, 2005; Robin & Pierson, 2005; Sadik, 2008) is presented as a methodology with positive impact on students' deeper learning of science content, and that can also be a very motivating activity for twenty-first century students, who are used to communicating in a digital environment. In our experience, DS may be also a transversal approach that allows integration of arts and digital competences, fostering multiliteracy. However, this process of creation of knowledge, especially in terms of science learning, is a complex process that requires time and significant supervision during the construction process.

## References

Barrett, H. (2006). Researching and evaluating digital storytelling as a deep learning tool. In *Proceedings of society for information technology and teacher education international conference* (pp. 647–654). AACE.

Cañal de León, P., García Carmona, A., & Cruz-Guzmán Alcalá, M. (2016). *Didáctica de las Ciencias Experimentales en Educación Primaria*. Paraninfo.

Domènech Casal, J. (2019). *Aprendizaje Basado en Proyectos*. Indagación y Controversias.

Ferrari, A. (2013). *DIGCOMP: A framework for developing and understanding digital competence in Europe* (Vol. EUR 26035). Publications Office of the European Union: Institute for Prospective Technological Studies. Joint Research Centre European Commission. https://doi.org/10.2788/52966

Harlen, W. (Ed.). (2010). *Principles and big ideas of science education*. Association for Science Education. http://www.ase.org.uk

Harvard Graduate School for Education. (2020). *Project zero*. http://www.pz.harvard.edu/thinking-routines.

INTEF. (2017). Marco común de competencia digital docente. V.2.0. *Plan de Cultura Digital En La Escuela*, 1–75. https://doi.org/10.2788/52966

Krumsvik, R. J. (2011). Digital competence in Norwegian teacher education and schools. *Högre Utbildung, 1*(1), 39–51.

McTighe, J., & Wiggings, G. (2013). *Essential questions: Opening doors to student's understanding*. ASCD.

Morra, S. (2013). *8 Steps to great digital storytelling*. http://edtechteacher.org/8-steps-to-great-digital-storytelling-from-samantha-on-edudemic/

NSTA. (2014a). *Cross-cutting concepts*. https://ngss.nsta.org/crosscuttingconceptsfull.aspx

NSTA. (2014b). *Disciplinary core ideas*. https://ngss.nsta.org/disciplinarycoreideastop.aspx

Puentedura, R. D. (2006). *Transformation, technology, and education*. http://hippasus.com/resources/tte/

Robin, B. (2005). *Educational uses of digital storytelling. Main directory for the educational uses of digital storytelling*. http://www.coe.uh.edu/digital-storytelling/default.htm

Robin, B., & Pierson, M. (2005). *A multilevel approach to using digital storytelling in the classroom*. http://digitalstorytelling.coe.uh.edu/archive/multilevel-approach.pdf

Sadik, A. (2008). Digital storytelling: A meaningful technology-integrated approach for engaged student learning. *Educational Technology Research and Development, 56*(4), 487–506. https://doi.org/10.1007/s11423-008-9091-8

# Chapter 8
# Digital First? Effects of Digital and Analogue Learning Tools on the Plant Knowledge Acquisition of Future Biology Teachers

**Lars Emmerichs, Meike Mohneke, Sandra Hofhues, and Kirsten Schlüter**

## 8.1 Introduction

Digital tools such as smartphones, laptops and tablets increasingly play an essential role in everyday life of our society as well as in everyday life of pupils (Kuhn et al., 2017). Thus, the use of digital media can be didactically justified relating to everyday life (ibid.).

The *International Computer and Information Literacy Study* (ICILS by European Commission, 2014) shows that Germany is at the lower end regarding the use of digital media in teaching environments. According to current results, the conditions for learning with digital media have changed very little in Germany since the start of the ICILS study in 2013 (Eickelmann et al., 2019).

From the end of the 1990s onwards, the "Schulen ans Netz" initiative took place in Germany, which led to the introduction and increasing use of mobile devices. However, Petko (2012) stated that the integration of digital media was not progressing as hoped. According to more recent data, the average ratio of students to computers at German schools is 11.5 to 1. For this reason, the smartphone appears to be a suitable medium, since it can generally be assumed that every learner in the higher grades has his or her own device (Rathgeb et al., 2020). Based on students being well equipped with smartphones the expression 'Bring your own device (BYOD)'

L. Emmerichs (✉) · M. Mohneke · K. Schlüter
Institute of Biology Education, University of Cologne, Cologne, Germany
e-mail: lars.emmerichs@uni-koeln.de; m.mohneke@uni-koeln.de; kirsten.schlueter@uni-koeln.de

S. Hofhues
Fakultät für Kultur- und Sozialwissenschaften, FernUniversität in Hagen, Hagen, Germany
e-mail: sandra.hofhues@fernuni-hagen.de



K. Korfiatis, M. Grace (eds.), *Current Research in Biology Education*, Contributions from Biology Education Research,
https://doi.org/10.1007/978-3-030-89480-1_8

was established in the research landscape. Therefore, at the time of this study, the smartphone was the medium of choice.

The following section presents concepts that have already been used to examine the effectiveness of digital media in science learning contexts. Lenz (2003) points out that for sustainable learning, action orientation should take precedence over purely cognitive queries. Of course, action orientation is not limited to digital media. Nevertheless, digital media seem to offer more scope due to their more flexible structure. West and Vosloo (2015) stated that due to the technical possibilities and the great familiarity of learners with such devices, the use of digital media will enrich teaching. However, it was found that digital content is often predetermined and not adaptable, as most teachers cannot be assumed to have specific programming skills. This was the starting point for Dellbrügge and Marohn (2017) to develop their concept "Choice$^2$Interact". This concept includes various design criteria for creating digital learning environments in *Prezi* (a platform-independent, but cloud-dependent presentation program). The pupils received a ready-made learning environment that contained the contents of a lesson or even a whole series of lessons. The structure of this learning environment offered their users learning paths that they could follow in a self-determined way (a strategy which had already been suggested by Krause & Eilks, 2014). In addition, Dellbrügge and Marohn (2017) presented mobility as a major advantage in their concept, as the content is theoretically available everywhere. The same advantage was also mentioned by Schulz-Zander (2005), who pointed out that digital tools offer teaching and learning independently of place and time. In another study Hermes and Kuckuck (2016) dealt with a different digital learning tool called *Actionbound*. By embedding playful elements and competitive tasks into this digital learning tool, learning contexts that promote motivation were generated.

However, it is still questionable whether these playful digital elements have a lasting impact on learning success. As an additional advantage of digital learning pathways, the authors stated that no expensive information boards (or in the case of our study: identification guidebooks) needed to be purchased and that digitally available boards can be accessed individually. Weng and Pfeiffer (2016) pointed out another advantage of digital media, which is that learning is most efficient when learners are able to communicate what they are currently learning or have learned to a recipient. Digital media can offer opportunities for extended communication through their technical possibilities (e.g. instant messaging service). This makes it possible to interact with others regardless of location or time (Fauville et al., 2014). Again, this is an aspect that may support non-linear learning strategies. These non-linear learning strategies can particularly favor the development of procedural knowledge (Kuhn et al., 2017; Petko, 2012), which from our point of view, refers to perceiving and adapting one's own learning strategies more consciously. Based on these many expected benefits, biology education research should set a deeper focus on the design of digital learning environments and application of digital media in teaching.

Jäkel and Schaer (2004) pointed out that students' species knowledge, especially regarding native organisms, is deficient. Several studies have proven this (Tunnicliffe

& Reiss, 2010; Gatt et al., 2007; Hesse, 2000; Wandersee and Schussler 1999; Mayer, 1995; Pfligersdorffer, 1991). Wandersee and Schussler (1999) even use the term "plant blindness". They state, that for the untrained eye, plants form an anonymous green mass without individuality (Tunnicliffe, 2001). For these reasons, the subject of botany should receive much more attention in biology education (Wandersee & Schussler, 1999; Lock, 1995). Unfortunately, there has been little work on how species knowledge can be improved through teaching and which methods are particularly suitable for this (Pews-Hocke, 1995). Randier (2002) recommends that student-focused teaching should be combined with teacher-focused teaching to encourage individual work and group work. He states that students learn more actively when working independently. Once a good knowledge base has been generated, Randier (ibid.) considers field trips to be useful to put the acquired knowledge into an ecological context.

## 8.2 Aims and Research Question

The present study aims to contribute towards counteracting the deficit in plant species knowledge by integrating the use of digital media in plant identification lessons. The starting point for this were teacher training courses at the university, as student teachers may later act as multipliers. The aim being addressed was to determine the influence that two different consolidation methods (analogue and digital) might have on acquiring and/or retaining plant species knowledge. Both consolidation methods used should lead to an increased elaboration of the previously covered content (referring to distinguishing features of certain plant families), but for both methods this elaboration process probably happened in different ways. While the intervention group worked on observation tasks under their own guidance by using an app (*Biparcours*), a second group of students (control group) applied a classical identification key for species identification. The research question therefore was:

What influence does the *Biparcours* app have on the acquisition (and retention) of plant knowledge by future biology teachers in comparison to an analogue learning form?

Both treatments seemed to be suitable for the groups to repeat characteristics of the plant families dealt with and to get to know selected representatives of these plant families in more detail. However, we assumed that students using the digital application would show a better performance in a subsequent test addressing plant species knowledge than the students using a traditional identification guide. The reasons for our hypothesis are as follows:

1. Classical identification guides usually apply dichotomous identification keys as opposed to non-linear learning paths. These identification keys are made of many individual questions which must be answered in a specific order. If a person cannot answer one of the many dichotomous questions, he or she will most likely not achieve the given goal. Dichotomous identification keys are therefore

not very tolerant of errors or lack of knowledge. These challenges can partly be overcome by using an app offering independent observations tasks, which do not build on each other.

2. The questions in the app focus on what is specific to a particular plant species and thus put the spotlight on a few preselected features. In contrast, identification guides list many features, whereby the central features can easily be overlooked.
3. A traditional dichotomous identification key consists primarily of text elements and just a very few illustrations. In addition, each identification step covers the same task which is to compare a written identification characteristic with the plant material at hand. Because of this monotony, plant identification can easily be seen as a boring task and might have a negative effect on motivation. Using the app should lead to an advantage in this respect because very different types of tasks can be used which also may include playful elements and may address plant use in everyday life (see Sect. 3.1).
4. Traditional identification lessons still often take place in the classroom (according to the authors' experience). The teacher brings previously collected plants to the course, which are then identified by using a classical identification key. This is time-efficient for the teacher and allows the lesson to run according to the teacher's planning. Of course, it would also be possible to go out into nature with an identification book, but this is rarely done. An app could facilitate this step of looking at plants in their natural habitat, e.g. through other types of tasks that require going out and finding a described plant species in nature. This can be motivating for the students, because plant hunting and moving around in nature may be fun. It is also possible, that plants in nature are perceived as more aesthetic, because they are not lying plucked out on the table. Furthermore, there is the possibility of learning additional information about the habitat and the range of trait variation in a plant species. In addition, Lai et al. (2007) showed that higher order learning outcomes can be achieved when learners use mobile devices outdoors instead of paper and pencil.
5. If, based on the tasks in the app, a person has found places where this plant species naturally occurs, and if this person passes these places more often, then he or she will be reminded of the plant species again. This opportunity for repetition is missing when working with a classical identification book.

## 8.3 Methods

### *8.3.1* **Biparcours** *Application*

*Biparcours* is an app for creating digital learning offers in educational contexts and for fostering the use of extracurricular learning locations. The supplier (Bildungspartner NRW) provides schools and their partners with an interactive, multimedia learning tool free of charge. The app can be used to convey information in different ways and to integrate different types of tasks that need to be solved.

Possible components of the app are (1) quiz elements, (2) recording of photos, videos and audio elements, (3) QR codes, (4) text elements, and (5) coordinates for reaching a certain location. Figure 8.1 shows three examples of such different types of tasks, while Fig. 8.2 depicts examples of students' work results. The app allows users to navigate through the material in a self-determined way. If a person answers a question incorrectly due to the lack of knowledge, he or she can still continue to work in a proper way (which is different with a classic identification key).

### 8.3.2 *Design and Procedure*

Based on previous research and the technical possibilities offered by the app, the working hypothesis of the study was, that learners who use the digital tool to acquire plant species knowledge, achieve correct answers with greater frequency in a subsequent test than when using analogue learning forms.

The study referred to an university seminar called "Identification Exercises in Botany" in the Bachelor's programme for biology student teachers. In 2018, a pilot study was conducted to test the different survey instruments as well as the app as a consolidation and learning method. The pilot study did not have a control group design, because it served just as a preparation for the main study in terms of a design-based research approach. The main study took place from April to June 2019. The course structure contained a theoretical input given by the lecturer as well

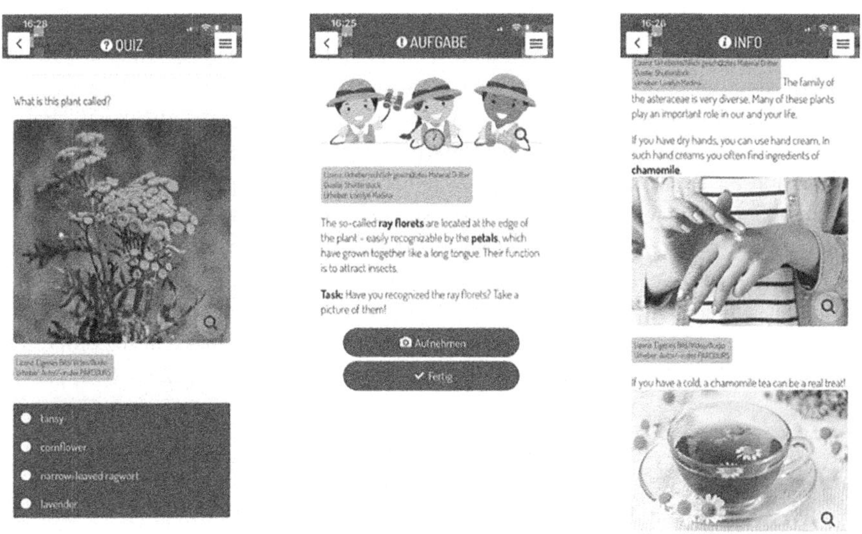


(1) quiz element (2) photo upload (4) contextualization

**Fig. 8.1** Examples from the *Asteraceae* parcours

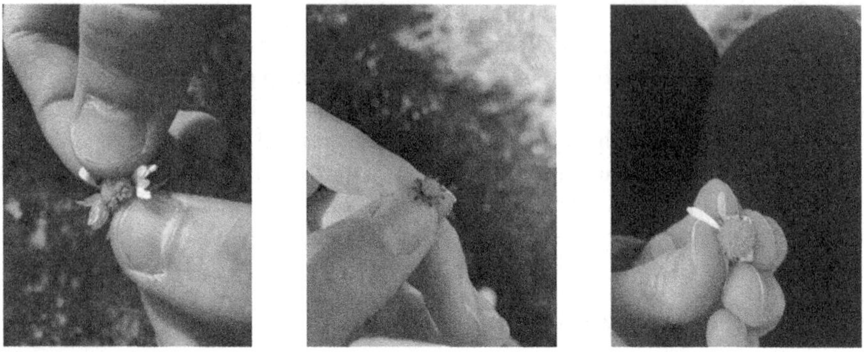

**Fig. 8.2** Examples from the task: Photograph the tubular flowers of a daisy

as learning stations which addressed five different plant families (*Asteraceae, Brassicaceae, Lamiaceae, Rosaceae, Fabaceae*). While this theoretical introductory part was the same in both, the intervention and control group, these groups differed in the subsequent practical part. In the intervention group, the learners used the digital tool *Biparcours* app to deepen their knowledge of plant species outdoors in nature. When using the app, students were asked to look for specific plant species in nature, to describe their distinctive features and to photograph them. The control group instead, studied the same plant species by using an identification book with a simplified, shortened version of a dichotomous identification key and performed the identification tasks inside the classroom. To ensure comparability, the plants to be identified were identical in both groups. The representatives of the different plant families were selected according to their seasonal appearance and to easy recognition of the specific identification features, which had been discussed theoretically in the introductory course part.

For the study, two very different learning scenarios were compared: classical book-based identification exercises in the classroom using just one repetitive task format versus app-based identification exercises in nature using different task formats. Thus, not only one variable was changed, but different variables. The reason for this was to compare two contrary teaching formats as typically found in university courses or school lessons. We hereby wanted to achieve ecological validity according to Bronfenbrenner's definition: "Ecological validity refers to the extent to which the environment experienced by the subjects in a scientific investigation has the properties it is supposed or assumed to have by the investigator" (Bronfenbrenner, 1979, p.29).

### 8.3.3 Sample

In order to analyze students' learning progress, a total of six courses ($N = 125$) were examined. These were parallel courses at the same university taken by student teachers in their fourth bachelor's degree semester in biology. The courses were divided into three control and three experimental groups. An intervention study was done using a pre-post design. Only students who participated in the pretest and posttest were considered in the present study. The intervention group comprised a total of 57 participants (46 female, 9 male, and 2 diverse) and the control group a total of 41 participants (32 female, and 9 male). In the intervention group, the average age of the students was 22.2 ($SD$ = 2.86) years, and the control group had an average age of 22.1 (SD = 1.52) years. All participants were randomly assigned to the two groups before the course started. The test persons were anonymized by a code, and all data were collected in compliance with the data protection regulation (EU-DSGVO).

### *8.3.4 Instrument*

In order to examine students' knowledge gains concerning plant species, a self-developed knowledge test (15 items) was applied (questionnaire available at: https://osf.io/e4zr5/). The adequacy of the items was assessed by two experienced biology lecturers. The items were constructed as single-choice questions with four response options. The answers were dichotomously scored (0 = wrong answer; 1 = correct answer). One point could be given per question. The maximum achievable score was 15 points in total with a high sum score indicating high plant knowledge. Pre- and posttest (April and June 2019), between which there were four seminar dates, contained the same questions but in different orders and with different plant pictures. The participants had no access to their test results. The main focus of the test was on the characteristics of the treated plant families and species, whereby each family was covered by 3 items (with one exemption, where the preference was given to a more general question). Furthermore, item formulation was based on the criteria for good questions, e. g. simple language, clearly formulated, focused on a single issue (Beller, 2016). All data were collected using the survey program 'qualtrics$^{xm}$'. All results were calculated with the statistics software "IBM SPSS Statistics 26". For the items analysis, the limit values given by Moosbrugger and Kelava (2012) were used. According to this, all items with discriminatory power of $r > .3$ and item difficulties between $p = .2$ and $p = .8$ were taken into account for the score formation. Only one item did not meet the selection criteria. It was nevertheless retained in the test because it referred to the basic aspect of knowing different classification levels (taxonomy). For reliability analysis, Cronbach's alpha was calculated to assess the internal consistency of the knowledge tests. The internal consistency of the pretest (15 Items; $\alpha = .31$, $M = 5.90$, $SD = 1.68$) was lower than the posttest (15 Items; $\alpha = .63$, $M = 10.48$, $SD = 1.82$).

## 8.4 Results

A mixed analysis of variance (relevant conditions were checked in advance) was executed to investigate whether there is a difference in students' knowledge growth between pre-test and post-test (effect by 'time') and whether knowledge

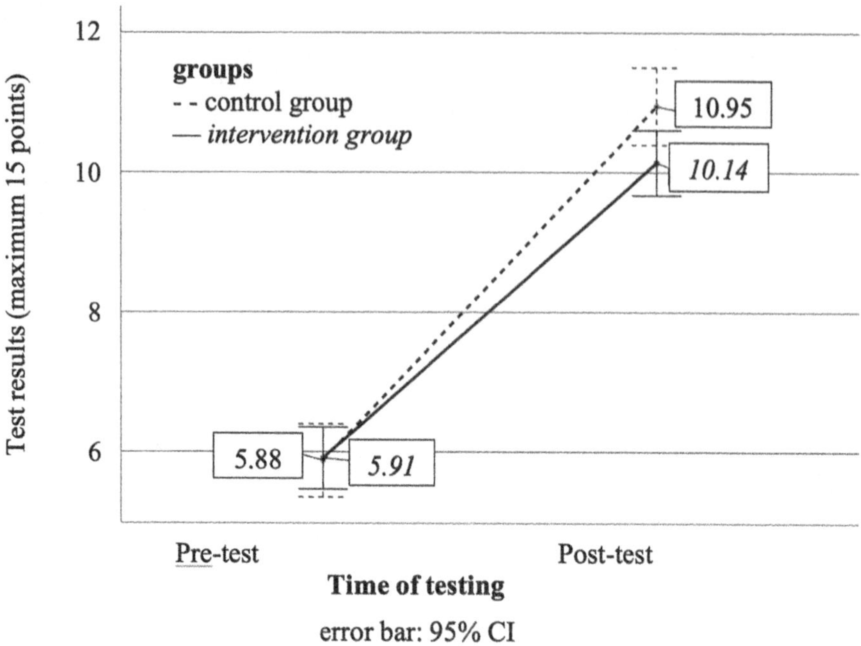


**Fig. 8.3** Test results for the intervention group at the time of pre- and post-test

growth differs between the two groups (interaction effect between 'time' and 'group'). A within-group effect was observed: students' knowledge significantly improved from pre- to post-test, $F(1, 96) = 403.52$, $p < .001$, partial $\eta^2 = .81$. There was no between-group effect, which means that the knowledge level of the intervention group did not differ significantly from that of the control group, $F(1, 96) = 2.065$, $p = .154$, partial $\eta^2 = .02$. In addition, there was no statistically significant difference concerning the interaction effect, $F(1, 96) = 3.331$, $p = .071$, partial $\eta^2 = .34$. These results disprove our hypothesis that our digital learning tool (here: *Biparcours* app) would be more conducive to knowledge growth than an analog learning tool (here: plant identification book) (Fig. 8.3).

## 8.5 Discussion

### *8.5.1 Gain of Knowledge*

The results of the post-test showed a significant increase in knowledge growth for both groups compared to the results of the pre-test. However, there was no significant difference regarding the interaction effect between 'time' and 'group'. Thus, the working hypothesis has to be rejected, because teaching with a digital application did not lead to better plant knowledge than using a traditional identification guide. Both, the digital and the analogue teaching approach, affected learning in a positive way, so that one can speak of an almost equally good increase in performance. There may be different explanations, why the intervention group that worked with the app did not perform any better than the control group:

1. When working with a classical identification book, the depth of information processing being achieved may be even higher than when working with an app. A possible reason for this could be that each identification step in an identification book corresponds to an application task at the same time. When reading a dichotomous identification statement, one has to visualize it mentally and compare it to the plant sample at hand. This hypothesis corresponds to some extent with theoretical statements by Bohnsack (2009, p. 299): "Images are implicated in all signs or systems of meaning. In the terms of semiotics, a specific "signified" which is associated with a specific "signifier" (for instance a word) is not a thing, but a mental image." This would mean that when one hears or reads the word flower, one tends to refer to an abstract image of a flower in one's mind. In addition, with each identification process, one goes through the identification criteria of an identification book anew and thus repeats them permanently.
2. The playful elements of the app, the reference to some everyday use of the plant species dealt with, the search for these plant species in the surrounding area (with the latter tempting one to engage with diverse, non-target explorations) – these may all have been distracting factors that prevented the student teachers from concentrating only on what was essential, i.e. the respective special characteristics of the plants. In this respect, a cognitive overload (Chandler & Sweller, 1991) may have occurred with the app, in that important information was not sufficiently separated from rather extraneous information (although this could be a similar effect as the cognitive overload caused by the many individual comparisons one has to make when working with a classic identification book).
3. For the study group, i.e. student teachers in biology, it could be assumed that they were relatively motivated individuals with an interest in the subject and thus also in knowledge about plants. In this case, the chosen method of learning about plants would have less impact on the learning outcome than if the study was conducted with a different target group, such as pupils, whose interests are more diverse and less focused on biology and plants.

### *8.5.2 Limitations of the Study and Implications for Further Research*

There are some limitations of the study that affect data interpretation and should be considered for the design of subsequent studies.

1. The two teaching approaches might rather have led to different effects, when the practical part of the seminar (teaching with the app versus teaching with analogous material) had covered the whole time of the seminar. For this purpose, however, the basic taught information phase about plant families should be interwoven with the practical part of the seminar, and must thus be integrated into the app as well as into the work with the identification guide. This requires a significant expansion of the app, as it is then used not just primarily for consolidation purposes but also for basic knowledge transfer.
2. It would have been revealing to have a second control group where only the basic information would have been given, without the practical identification phase that followed. In this way, one could have analysed how much knowledge was gained through the practical parts of the seminar or if most of the information was already conveyed through the basic unit. The reason for omitting such a second control group was that otherwise the group sizes would have been too small.
3. In terms of their design, the two teaching approaches were oriented as closely as possible to classroom reality and thus had a high ecological validity. As a result, however, these approaches differed in several aspects and not just in one variable. In order to achieve greater comparability of the two teaching approaches, both the app and the identification book could be worked with outdoors, for example. However, it is questionable whether this alignment of study conditions would really lead to different results. The expectation would rather be that the likelihood for different learning gains to occur would be reduced due to the greater similarity of the study conditions.

A final question that ultimately arises is, what features do apps need to have in order to bring about additional learning gains? Perhaps such an app should combine potential advantages of an identification book (frequent repetition of identifying features, continuous comparisons between the described features and the original plant, stimulating mental visualisation of contrasting features) and potential advantages of a digital learning tool (such as error tolerance through independent tasks, diverse activity options, focus on particularly prominent features, and easy use in the field). However, whether these are really effective design criteria needs to be investigated through separate surveys. These surveys should also refer to interview studies in which participants assess which app and task design is effective for them and for what reasons.

## References

Beller, S. (2016). *Empirisch forschen lernen–Konzepte, Methoden, Fallbeispiele, Tipps, hogrefe, 3*. Auflage.

Bohnsack, R. (2009). The interpretation of pictures and the documentary method. *Historical Social Research/Historische Sozialforschung, 34*(2 (128)), 296–321.

Bronfenbrenner, U. (1979). *The ecology of human development: Experiments by nature and design*. Harvard University Press.

Chandler, P., & Sweller, J. (1991). Cognitive load theory and the format of instruction. *Cognition and Instruction, 8*(4), 293–332.

Dellbrügge, B., & Marohn, A. (2017). *choice2interact–Interaktiv Lernen mit Tablets im Chemieunterricht* (Lernprozesse mit digitalen Werkzeugen unterstützen–Perspektiven aus der Didaktik naturwissenschaftlicher Fächer). Joachim Herz Stiftung.

Eickelmann, B., Bos, W., Gerick, J., Goldhammer, F., Schaumburg, H., Schwippert, K., et al. (2019). *ICILS 2018# Deutschland: Computer-und informationsbezogene Kompetenzen von Schülerinnen und Schülern im zweiten internationalen Vergleich und Kompetenzen im Bereich Computational Thinking*. Waxmann.

Fauville, G., Lantz-Andersson, A., & Säljö, R. (2014). ICT tools in environmental education: Reviewing two newcomers to schools. *Environmental Education Research, 20*(2), 248–283.

Gatt, S., Tunnicliffe, S. D., Borg, K., & Lautier, K. (2007). Young Maltese children's ideas about plants. *Journal of Biological Education, 41*(3), 117–122.

Hermes, A., & Kuckuck, M. (2016). Digitale Lehrpfade selbstständig entwickeln–Die App Actionbound als Medium für den Geographieunterricht zur Erkundung außerschulische Lernorte. *GW Unterricht, 142*, 174–182.

Hesse, M. (2000). Erinnerungen an die Schulzeit—Ein Rückblick auf den erlebten Biologieunterricht junger Erwachsener. *Zeitschrift für Didaktik der Naturwissenschaften, 6*, 187–201.

Jäkel, L., & Schaer, A. (2004). Sind Namen nur Schall und Rauch?: Wie sicher sind Pflanzenkenntnisse von Schülerinnen und Schülern? *Zeitschrift für Didaktik der Biologie (ZDB)-Biologie Lehren und Lernen, 13*, 1–24.

Krause, M., & Eilks, I. (2014). Lernwege mit PREZI modern gestalten–Beispiele zum Teilchenkonzept. *Digitale Medien im naturwissenschaftlichen Unterricht*, 209–215.

Kuhn, J., Ropohl, M., & Groß, J. (2017). *Fachdidaktische Mehrwerte durch Einführung digitaler Werkzeuge. Lernprozesse mit digitalen Werkzeugen unterstützen–Perspektiven aus der Didaktik naturwissenschaftlicher Fächer* (pp. 11–33). Joachim Herz Stiftung Verlag.

Lai, C. H., Yang, J. C., Chen, F. C., Ho, C. W., & Chan, T. W. (2007). Affordances of mobile technologies for experiential learning: The interplay of technology and pedagogical practices. *Journal of Computer Assisted Learning, 23*(4), 326–337.

Lenz, T. (2003). Handlungsorientierung im Geographieunterricht. *Geographie Heute, 4*(24), 2–7.

Lock, R. (1995). Biology and the environment—A changing perspective? Or 'there's wolves in them there woods!'. *Journal of Biological Education, 29*, 3–4.

Mayer, J. (1995). Formenvielfalt als Thema des Biologieunterrichts. *Vielfalt begreifen-Wege zur Formenkunde, S*, 37–60.

Moosbrugger, H., & Kelava, A. (2012). *Testtheorie und Fragebogenkonstruktion*. Springer. https://doi.org/10.1007/978-3-642-20072-4

Petko, D. (2012). Hemmende und förderliche Faktoren des Einsatzes digitaler Medien im Unterricht: Empirische Befunde und forschungsmethodische Probleme. In *Jahrbuch Medienpädagogik 9* (pp. 29–50). Springer.

Pews-Hocke, C. (1995). *Vielfalt der Arten. Eine Einführung in die Grundlagen der Systematik*. Paetec.

Pfligersdorffer, G. (1991). *Die biologisch-ökologische Bildungssituation von Schulabgängern.* Eine empirische Untersuchung über die Kenntnisse von Schülern sowie über die Lehrplangegebenheiten des entsprechenden Unterrichts im weiterführenden Schulwesen der AHS, BHS und BMS Österreichs. ABAKUS-Verlag.

Rathgeb, T., Schmid, T., Feierabend, S., & Reutter, T. (2020). *JIM-Studie 2019. Medienpädagogischer Forschungsverbund Südwest* (LFK, LMK). https://www.mpfs.de/studien/jim-studie/2019/

Randier, C. (2002). Comparing methods of instruction using bird species identification skills as indicators. *Journal of Biological Education, 36*(4), 181–188.

Schulz-Zander, R. (2005). Veränderung der Lernkultur mit digitalen Medien im Unterricht. *Perspektiven der Medienpädagogik in Wissenschaft und Bildungspraxis*, 125–140.

Tunnicliffe, S. D. (2001). Talking about plants-comments of primary school groups looking at plant exhibits in a botanical garden. *Journal of Biological Education, 36*(1), 27–34.
Tunnicliffe, S. D., & Reiss, M. J. (2010). Building a model of the environment: How do children see plants? *Journal of Biological Education, 34*(4), 172–177.
Wandersee, J. H., & Schussler, E. E. (1999). Preventing plant blindness. *The American Biology Teacher, 61*(2), 82–86.
Weng, A., & Pfeiffer, A. (2016). *"Lernen durch Lehren" in der Mathematik: Videotutorials und Apps im Praxistest*. 2016, 16.S. URN. https://doi.org/10.25656/01:12264
West, M., & Vosloo, S. (2015). *Policy guidelines for mobile learning*. UNESCO Publishing.

# Chapter 9
# Comparing Inquiry-Based Learning and Interactive Lectures While Teaching Physiology to Undergraduate Students

**Yvette Samaha and Assaad Yammine**

## 9.1 Introduction

Although inquiry-based learning (IBL) is widely advocated in higher education (The Boyer Commission, 1998; Brew, 2003; Healey, 2005; Spronken-Smith et al., 2007), there is relatively little research on its usefulness at the post-secondary level (Spronken-Smith et al., 2011; Helle et al., 2006). However, recently, there has been an increased interest in this research topic, including detailed case studies comparing different forms of IBL (Spronken-Smith & Walker, 2010), and a study on the impact of IBL on the learning outcomes of social science students (Justice et al., 2009).

We currently have little knowledge, particularly in Lebanon, about the impact of IBL on the learning outcomes of students studying biology at the post-secondary level. To address this issue, this study examines the impact of IBL on the learning outcomes of 57 physical education (PE) students in a general physiology course on muscle fibre function. Our study may provide a starting point to address certain concerns of some university teachers, such as IBL not being suitable for first/second year undergraduate students, or IBL being solely usefully in schools (elementary, or especially intermediate level). Furthermore, this study can help identify the means to foster the link between research and IBL teaching practices, that are specific to a subject area, namely biology (Healey, 2005). Our study also provides key ingredients and procedures for designing a teaching unit for an inquiry-based biology course. This investigative model is applicable and will help faculty members from various disciplines, including biology, to develop an innovative way to engage students in active learning.

Y. Samaha (✉) · A. Yammine
Faculty of Pedagogy-Dean's Office, Lebanese University, Beirut, Lebanon
e-mail: yvette.samaha@ontariotechu.net; assaad.yammine@ul.edu.lb



K. Korfiatis, M. Grace (eds.), *Current Research in Biology Education*, Contributions from Biology Education Research,
https://doi.org/10.1007/978-3-030-89480-1_9

In this chapter, we present a brief review of how IBL has been described in the literature. According to Justice et al. (2007), IBL refers to:

> …a range of instructional practices that promote student learning through student-driven and instructor-guided investigations of student "centered" questions (p.202).

Spronken-Smith et al. (2007) have emphasized that IBL is a student-centered approach in which teachers play the role of facilitators. This approach should be question-driven or problem-driven and based on seeking out and acquiring new knowledge. Similarly, Oliver (2008) described IBL as the set of pedagogical approaches in which "some form of problem or task serves as a catalyst for student engagement and participation" (p.287). According to him learning is a consequence of the information processing that occurs as students explore the problem and seek to solve it.

Although the definitions differ, authors generally view IBL as a learning activity in which students conduct some form of inquiry that may be either question-oriented or problem-oriented. This conception will serve as a basis for our analysis compared to interactive lecture-based teaching (IL), which mainly promotes the exposure of a content or a subject with no emphasis on engaging students in hands-on activities. In the latter approach, students are mostly passive participants.

The Boyer Commission (1998) described IBL as capable of developing in students "a spirit of inquiry". Justice et al. (2007) state that IBL can promote students' ability to "think critically and reflectively about the production of knowledge" (p.203). In the same vein, Spronken-Smith et al. (2007) speak of IBL as a pedagogy which allows "pupils to experience the processes of knowledge creation". IBL is therefore a pedagogical approach including the development of critical thinking and the design of strategies that students develop in a learning context. The implementation of this kind of approach has been extremely successful in increasing student interest and understanding of concepts in the fields of bioinformatics and molecular biology (D'Costa & Shepherd, 2009; Lau & Robinson, 2009).

Students' lack of interest in pursuing scientific studies has been attributed to a range of factors, including unattractive pedagogical approaches and stereotypical academic practices (Boilevin, 2014). Poumay (2014) identified six levers for improving the learning of higher education students. The first lever concerns the concept of constructive alignment; the teacher aims to improve the educational alignment between objectives, methods, and evaluation within the course. The second lever aims to engage students in active learning. The third, fourth and fifth levers concern the factors contributing to students' motivation and autonomy. In particular, the instructor will have to give meaning to the activities by bringing them closer to the students' future professional experiences (third lever), to increase the students' sense of mastery or competence (fourth lever), and to give the students more control over the proposed activities and autonomy in the conduct of his/her learning (fifth lever). The sixth lever concerns the introduction of Information and Communication Technologies in Education.

It is worth mentioning that IBL implementation presents some challenges, including the teacher's readiness to implement it. IBL is influenced by the teacher's beliefs, values, knowledge and skills, as well as student's interest in being involved in the inquiry process (for a review, see Khalaf, 2018). These challenges may be the reason behind lectures still being widely used by university instructors. However, it is well known that students' attention decreases after 20 min of a lecture (Roopa et al., 2013). Fortunately, with a little twist, lecturing can turn into an effective instructional method that favors organization and knowledge transmission (Snell, 1999). In fact, learners better remember the content of a course when it includes brief activities, and when they engage in tasks such as thinking, reading, writing, and talking (Nabors, 2012). In this paper, this instructional method will be referred to as interactive lectures (IL). This method requires less time in terms of preparation and implementation when compared to IBL. Additionally, it does not seem to require curriculum development, training, or guidance of an instructional designer.

To our knowledge, there are no studies on IBL or comparing IBL and IL in the Lebanese higher education context. This study may spark instructors' interest in experimenting with novel teaching strategies.

## 9.2 Research Questions

In the context presented above, this study addresses the following questions:

1. *What are the contributions of IBL implementation in terms of knowledge acquisition in the context of teaching general physiology to first year undergraduate students?*
2. *How efficient is IBL compared to the preferred teaching method in university education (i.e. interactive lectures)?*
3. *Does the efficiency of IBL vary depending on variables such as the educational background of students?*

In this study, we are putting IBL to the test by using a university context and its real-time constraints. Since the goal is to contribute to the academic success of students, we are investigating whether the IBL contributes to the acquisition of learning outcomes.

## 9.3 Research Design and Methods

### *9.3.1 Participants*

The target population in this study is Physical Education (PE) students (N = 57) enrolled in a General Physiology first-year undergraduate course, at the Faculty of Education of the Lebanese University. All students enrolled in this course were initially included in the study. They come from a variety of educational backgrounds and, therefore, do not have the same level of knowledge in biology. There are two sections of this course. In this study, we implement either IBL or IL in each section, and we compared the effect of the methods on students' learning about muscle function. We also checked whether these methods may reduce the gap between students having a strong educational background (i.e. those holding a life sciences high school baccalaureate – LS students), and those with a weaker background in biology (e.g. students holding a socio-economics baccalaureate – SE students). The students belonged initially to two different sections (22 in section A and 35 in section B). Students that were present at all the phases of the study were included in the study (17 in section A and 20 in section B). The PE students hold a variety of high school baccalaureates as shown in Table 9.1.

### 9.3.2 *Experimental Design*

A quasi-experimental non-equivalent group design was used. The intervention was preceded and followed by a test to determine the students' level of understanding of muscle function (See Fig. 9.1). Some students (n = 5), belonging to section B, showed remarkable previous knowledge, and were excluded from the study. The teaching methods were randomly assigned to the sections. Section A was taught using IBL (n = 17) and section B using IL (n = 15) (See Table 9.1 for participants' distribution according to instructional methods and educational background).

### 9.3.3 *Instructional Methods*

The two instructional methods were implemented to achieve the understanding of the following concepts: pathways of ATP production, substrates used, types of muscle fibres, type of activities for which they are recruited. The lesson plans and documents were prepared by the researchers and validated by the course instructor. The documents to be analyzed by students were prepared for the IBL group. Following their validation by the course instructor, the lesson plan and documents were modified to suit the IL instructional method, and in turn, were also validated by the instructor.

**Table 9.1** Participants' distribution according to instructional methods and educational background

| | IBL (Section A, n = 17) | IL (Section B, n = 15) |
|---|---|---|
| Socio-economy | 11 | 5 |
| Life science | 6 | 4 |
| Other | 0 | 6 |

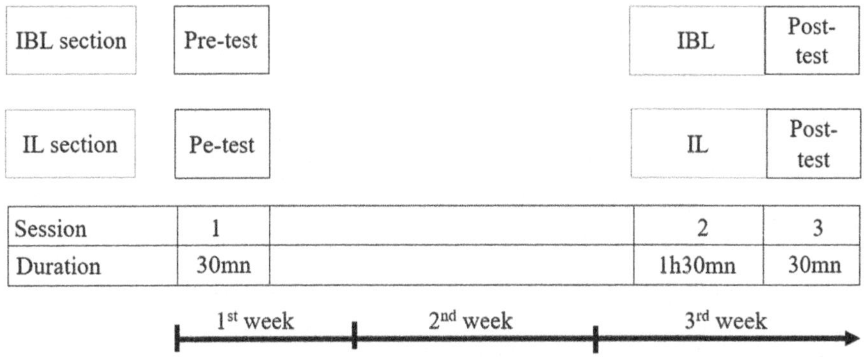


**Fig. 9.1** Diagram summarizing the steps of the study

Although both methods are based on document analysis, they are fundamentally different in the level of learner participation required, the number of learners who can participate in the discussion, the teacher's position in the knowledge guide-transmitter continuum, as well as in terms of time management. Each method has its sets of advantages and challenges. On one hand, IBL requires a more visible participation on the part of the student, the students collaborate in the construction of their knowledge, and the teacher acts as a guide. On the other hand, IBL may require the reduction of the curriculum as IBL implementation is more time consuming than lecturing. Concerning IL, the teacher has greater control over time and class management compared to the IBL. Yet, students are still encouraged to actively engage in the construction of their own knowledge by participation to various activities such as whole-class discussions.

#### 9.3.3.1 The Inquiry-Based Learning Session

We applied IBL with the dual objective of initiating students to the inquiry process and help them understand the functioning of muscle fibres. According to the four levels of IBL by Banchi and Bell (2008), we chose the procedure to be followed by the students during the investigation, as well as the documents to be analyzed and the questions to guide them in their analysis. Students were asked to validate or to refute their hypotheses based on the data they analyzed.

Although some of the knowledge was imparted to them in grade 11, the results of the pre-tests indicated a low level of knowledge. In the following we describe how the IBL session was run. The triggering situation (hook) was the following story:

> Jana decides to participate in the Beirut Marathon. She eats a banana, an orange, and a cereal bar for breakfast. She then joins Jad, Khaled and Joanna at the marathon. After few minutes from the start of the marathon, she feels that her heart is beating faster. She starts breathing heavily. She feels hot and she starts sweating. Twenty minutes later, she begins to feel tired and has a burning sensation in her legs. She attempts a sprint, but after few metres she falls on her knees exhausted and completely out of breath.

This problem situation was chosen because it is supposed to arouse the curiosity of the PE students as many are planning to become personal trainers. More importantly, this situation is reminiscent of various concepts related to the operation of muscle fibres:

- The diet to follow should be adequate to the type of activity as a means of providing the necessary nutrients for muscle fibres;
- The aerobic pathway of ATP production and its signs: acceleration of breathing and heart rate;
- The anaerobic lactic pathway: burning sensation in the muscles;
- The anaerobic alactic route: during the 150 m sprint;
- The different types of fibres and the type of activity for which each is recruited.

The students analyzed and discussed the situation in teams and then with the whole class. The discussion was guided to help in the identification of a problem that could be answered using the documents prepared in advance. The students suggested questions that were discussed and rephrased. For example, the following question "How come she was already tired and managed to sprint?" was rephrased to "How did Jana's muscles allow her to sprint although she was already tired?"

Students discussed this in their respective teams and several hypotheses were generated. The hypotheses were summarized as follow:

- Jana's breakfast was not enough to produce enough ATP (energy reserves assumption);
- There may be different types of muscles; muscles that allow you to run a marathon and others to sprint;
- The amount of ATP needed for a sprint is less than that needed for a marathon;
- There are different means of ATP production.

Students then analyzed the documents and answered the questions in the documents in teams. The goal of this collaborative activity was to guide students into checking their explanatory hypotheses in response to the above question. The students analyzed five documents:

- Document 1: The three ATP production mechanisms and their response time;
- Document 2: Muscle metabolism;
- Document 3: Electronography of muscle fibres;
- Document 4: Characteristics of muscle fibres;
- Document 5: Different types of activity.

We were available to guide students in their analysis when needed. At several points during the section, we discussed the answers as a class to confirm that each document has been properly analyzed by the teams. Students referred to their hypothesis and critiqued it during the whole-class discussions.

#### 9.3.3.2 The Interactive Lecture Session

During IL, students analyzed documents organized in a PowerPoint file, first individually then collectively. These documents were the same as those analyzed during the IBL session. Before the start of the session, we made it clear to them that, for the purpose of the study, they should engage in the whole-class discussion and not engage in pairs or group discussions. The session began with a reminder of the mechanism of muscle contraction explained in the previous session, highlighting the essential role of ATP in muscle contraction.

Then we distributed to the students a paper version of the documents to be analyzed. Two activities were carried out using the information found in the documents analyzed. The first activity consisted of the filling of a table on the characteristics of the three pathways of ATP production. The second activity involved the determination of the type of muscle fibre employed for several activities.

### *9.3.4 Instruments*

A pre-test (Appendix A) and a post-test (Appendix B) were used to determine students' understanding of the following concepts: pathways of ATP production, substrates used, types of muscle fibres, type of activities for which they are recruited. The tests were prepared by the researchers and then were validated by the course instructor as measuring the attainment of the intended learning goals. Firstly, the expected learning outcomes were determined. Then the tests were prepared as follows: a set of exercises targeting the learning outcomes were elaborated (an even number for each outcome), then the exercises were separated into two tests randomly to get both tests measuring the same learning outcomes. Next, the tests were randomly assigned, one as the pretest and the other as the post-test. The pretest was composed of 5 multiple choice questions and 10 short answer questions. The post-test was composed of 4 multiple choice questions, 10 short answer questions, and one fill in the blanks question. The pretest was to check whether students had no remarkable previous knowledge related to muscle function, and whether there was a statistically significant difference between the means on the pretest of the two groups. The tests were based on a 20-point scale. The scores of the pretest and post-test were used to calculate the relative gain of each individual student.

### *9.3.5 Statistical Analysis*

The relative gain (RG) was calculated using the following formula (Duroisin et al., 2011):

(Post score – Prescore)/(Maximum score – Prescore) ×100. Since the sample size was small (less than 50 per group and subgroup), Shapiro-Wilk tests were run to check if the distribution of our data was close to normal and chose the appropriate statistical tests accordingly to compare means. A value of $p < .05$ was considered significant. The results of Shapiro-Wilk tests indicated that the data distributions were close to normal ($p > .05$). Therefore, student t tests (parametric tests) were used to detect significant differences between the means. Paired t tests were used to compare the means of the pretest and post-test scores within a group, and independent t tests were used to compare the means of the pretest or post-test scores between the two groups.

## 9.4 Results

The means of the scores were out of 20. The mean of pre-test scores of IBL (3.02) and IL (3.53) were not significantly different ($p = .34$). The initial level of the two groups was considered similar (Fig. 9.2). The mean of the posttest scores (IL: 9.53 and IBL: 9.63) was significantly greater than the mean of the pretest for both groups ($p = .00$; Fig. 9.2). There was no significant difference between the means of the post-test scores of the two groups ($p = .93$; Fig. 9.2).

This result shows that resorting to IBL in undergraduate education is absolutely possible, and that, the same as with IL, IBL contributes to the acquisition of learning goals in the context of a biology course. The main benefit of IBL is that it represents a concrete application of Poumay's (2014) levers for improving the learning of higher education students. In fact, in view of these levers, an IBL strategy may improve PE undergraduate students' learning. IBL can include activities in which students will have to act as professional exercise scientists (2nd and 3rd levers).

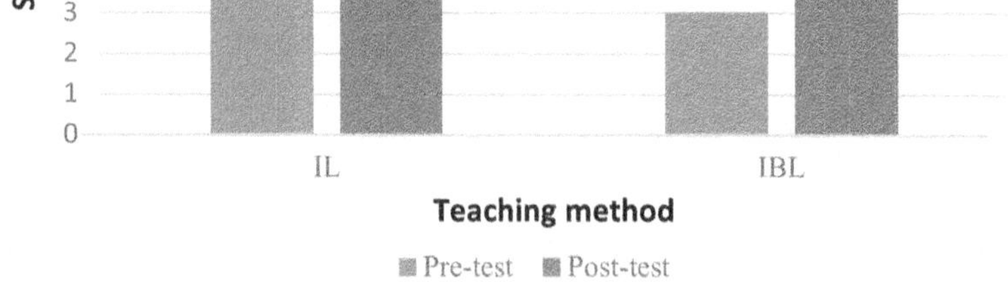


**Fig. 9.2** Bar graph of the means of the scores (/20) obtained on the pretest and the post-test for IBL and IL groups

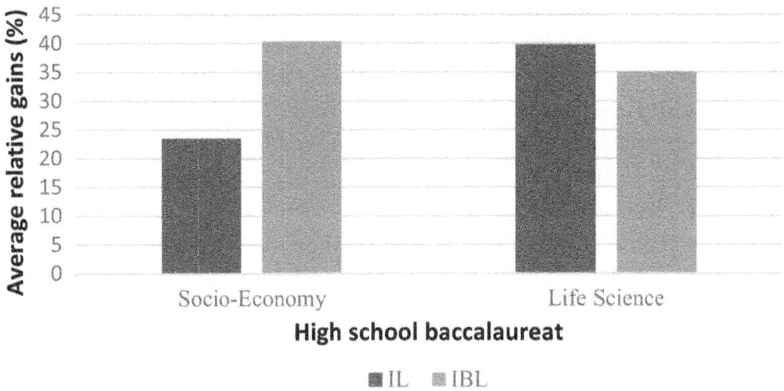


**Fig. 9.3** Bar graph representing the average relative gains (%) of students according to their subgroups (SE and LS)

These activities would also increase students' sense of mastery or competence (4th lever), and allow them to have more autonomy and control over the learning process (5th lever).

The relative gain (RG) allows the calculation of students' progress, regardless of their initial level. An RG greater than 30% indicates a positive learning effect (Temperman et al., 2018). The RG mean of both groups was greater than 30%, showing a positive learning effect of both instructional methods. The mean was greater for the IBL (38.50%) than for the IL group (36.00%), but not significantly ($p = .36$; Fig. 9.3).

The RG mean of socio-economics baccalaureate students (SE students), who were in the IBL group ($n = 11$; mean = 40.36%) was greater than 30%, unlike that of SE students in the IL group ($n = 5$; mean = 23.51%), indicating a positive learning effect of IBL on SE students. Additionally, RG means of SE in the IBL group was about twice that of those in the IL group, although this difference was not statistically significant ($p = .13$; Fig. 9.3).

The mean of the relative gains of life sciences students (LS students) was greater than 30% in both groups. The mean of those in the IBL group ($n = 6$; mean = 35.07%) was lower than that of those in the IL group ($n = 4$; mean = 39.75%). However, this difference was not statistically significantly ($p = .91$; Fig. 9.3).

Thus, IBL efficiency seems to vary depending on students' educational background.

## 9.5 Discussion

There was a significant difference between the scores on the pre-test and the post-test in both groups. This result was expected as most students did not have prior knowledge of the material. However, the objective was to check that the improvement in the scores is not simply due to chance (i.e. p values are <.05). The results of our study are in line with other studies showing the remarkable effect of IBL on achievement (Abdi, 2014; Akpullukçu & Gunay, 2011; Justice et al., 2009; Pandey et al., 2011, Seyhan & Morgil, 2007; Wilson et al., 2010). Our results also confirm those of other studies showing the positive effect of IL on academic performance (Ernest & Colthorpe, 2007; Prakash, 2010). However, our results suggest that IBL and IL lead to a similar acquisition of knowledge. Yet, the results suggest a tendency for IBL to have a more pronounced positive learning effect than IL for students with an SE baccalaureate. The average relative gains of students belonging to the IBL group was approximately twice that of IL. IBL could possibly reduce the gap between students of a heterogeneous group in terms of educational background. However, this difference is not statistically significant. This could be due to the reduced sample size of students in each subgroup. Concerning students with an LS baccalaureate, our results indicate that they benefit equally from an IBL session as from an IL session. A study by Carvalho and West (2011) suggests that voluntary participation in learning activities leads to better test performance than following a traditional course. Students, whose performance improved significantly, in this study, are probably those who were engaged in the activities, regardless of their group or subgroup. The authors could have checked this hypothesis by investigating student engagement either qualitatively, through observation grids, or quantitatively, using a questionnaire. It is also worth mentioning that the results of these studies are not meant to be generalized. Due to the small sample size, short study duration, and lack of instrument reliability tests, future studies with a more robust methodology are needed to compare the benefits of IBL and IL and their divergent effect on the achievement gap between students with strong and weak biology educational backgrounds. Yet, it is worth mentioning that due to the nature of its results, this study responds to the concerns of biology instructors about the usefulness of IBL and fosters realistic expectations of IBL implementation.

## 9.6 Conclusion

The highlight of this paper is that implementing IBL to teach undergraduate students biology is possible. IBL seems to be more efficient than IL for students that have a weak biology educational background (SE students). Thus, it might make the physiological concepts easier to understand for students with a weak background in biology. This study was conducted in one branch of the Faculty of Education. Its results are considered transferable to the other branches of the Faculty. However, more studies with a larger number of participants and longer intervention duration are needed to allow the generalization of these results.

# Appendices

## *Appendix A: Pretest*

### A. Circle the Correct Answer

1. The energy molecule used by myofilaments during contraction is:
   (a) Adenosine triphosphate
   (b) Glycogen
   (c) Myoglobin
   (d) Phosphocreatine
2. The stock of oxygen in the muscles is provided by:
   (a) Adenosine triphosphate
   (b) Glycogen
   (c) Myoglobin
   (d) Phosphocreatine
3. The substrate used to produce ATP through the **an**aerobic **a**lactic pathway is:
   (a) Adenosine triphosphate
   (b) Phosphocreatine
   (c) Glucose
   (d) Oxygen
4. The stock of glucose in the muscles is provided by:
   (a) Adenosine triphosphate
   (b) Glycogen
   (c) Myoglobin
   (d) Phosphocreatine
5. The muscle fibres type I produce ATP particularly through the _____________ because they have___________________:
   (a) Aerobic pathway – a great amount of glycogen
   (b) Aerobic pathway – a large number of mitochondria
   (c) Anaerobic lactic pathway – a great amount of glycogen
   (d) Anaerobic lactic pathway – a large number of mitochondria

### B. Answer the Following Questions

1. What is the role of mitochondria?
2. Name the three pathways (mechanisms) of ATP production.
3. What is the immediate source of ATP production? Phosphocreatine (phosphagens).

4. The glucose is a substrate used by which pathway(s) of ATP production?
5. There are three types of muscle fibres (I, IIA and IIB). Which type of fibres is also called fast twitch oxidative fibres?
6. What is the difference between slow twitch oxidative fibres and fast twitch glycolytic fibres (endurance, color, number of mitochondria, number of glycogen)?
7. The muscle fibre type that is preferably used during a physical activity, depends on the intensity and duration needed. Give one example of physical activity for each type of fibre.
8. To improve his performance at a sprint of very short duration (less than 10″), Jawad decides to drink sugar water just before the sprint. Will his performance improve? Why?
9. Mohammad filled his bottle with a mixture of water and sugar to drink it while he ran the marathon. Is it a good idea? Why?
10. Hassan and Rabih took part in a scientific study. The researcher took a sample of their biceps to observe them under the microscope. The observation of Hassan's sample show that more than 50% of his muscle fibres are red and have a large number of large-sized mitochondria. The observation of Rabih's sample show that more than 50% of his fibres are white and contain glycogen in high quantity. Who will be more able to provide an effort of weak intensity, and for a long period of time, that involves the biceps? Why?

## *Appendix B: Post-test*

### A. Circle the Correct Answer

1. ______[blank]______ is responsible for the production of ATP in presence of oxygen

   (a) Glycogen
   (b) Mitochondria
   (c) Myoglobin
   (d) Phosphocreatine

2. The immediate source of ATP production is ____________________, but the stock is limited and is rapidly depleted.

   (a) Adenosine triphosphate
   (b) Glycogen
   (c) Myoglobin
   (d) Phosphocreatine

3. The substrate used to produce ATP through the **an**aerobic lactic pathway is:

   (a) Adenosine triphosphate
   (b) Phosphocreatine

   (c) Glucose
   (d) Oxygen

4. The other name for type I muscle fibre is:
   (a) Slow twitch oxidative
   (b) Fast twitch oxidative
   (c) Slow twitch glycolytic
   (d) Fast twitch glycolytic

### B. Answer the Following Questions

1. Give the complete name of the energetic molecule produced by the mitochondria and used during muscle contraction.
2. Glucose is a substrate used by muscle fibres to produce ATP. Name the two sources of
3. All cells use oxygen to produce ATP. Name the two sources of oxygen of the muscles.
4. What fibre type is also called fast twitch glycolytic type.
5. What are the differences between slow twitch oxidative fibres and fast twitch oxidative fibres (endurance, color, number of mitochondria, quantity of glycogen, quantity of myoglobin)?
6. Which muscle fibre type produce ATP, the best, through anaerobic lactic pathway? Why?
7. Which type of fibres is used to hold the weight of your head and to maintain your posture?
8. Which muscle fibre type is mostly used during a martial art competition?
9. Two days before a marathon, Lara went for her usual run and then she ate a large quantity of pasta. Is it a good idea? What is the point of doing that?
10. Ali is not used to sprint. However, he decided to give it a try and ran with Jad. At some point, a muscle in his calf stiffened. Explain to him in simple words what happened at the level of his muscle fibres.

### C. Fill in the Blanks with the Fibre Type Recruited

You are helping your friend to move his furniture to his new apartment, and he lives on the 3rd floor of a building with no elevator. You were asked to move a couch from the truck to the living room. While you walk to the truck, only the ______**[blank]**_____ muscle fibres are recruited. This movement required no strength and no speed. The moment at which you lift the couch and walk back to the building, the ______**[blank]**_____ muscle fibres of your legs are recruited in addition to the ______**[blank]**_____ muscle fibres, because this effort requires more strength. When you go up the stairs, the effort is at its maximum. It requires the activation of the ______**[blank]**_____ muscle fibres.

## References

Abdi, A. (2014). The effect of inquiry-based learning method on students' academic achievement in science course. *Universal Journal of Educational Research, 2*(1), 37–41.

Akpullukçu, S., & Gunay, F. (2011). The effect of inquiry based learning environment in science and technology course on the student's academic achievements. *Western Anatolia Journal of Educational Sciences (WAJES), Dokuz Eylul University Institute, Izmir*. Retrieved from https://acikerisim.deu.edu.tr/xmlui/bitstream/handle/20.500.12397/5200/417-422.pdf?sequence=1&isAllowed=y

Banchi, H., & Bell, R. (2008). The many levels of inquiry. *Science and Children, 46*(2), 26–29.

Boilevin, J. (2014). Désaffection pour les études scientifiques et recherche en éducation scientifique. *Review of science, mathematics and ICT Education, 8*(2), 5–23.

Brew, A. (2003). Teaching and research: New relationships and their implications for inquiry-based teaching and learning in higher education. *Higher Education Research & Development, 22*(1), 3–18.

Carvalho, H., & West, C. A. (2011). Voluntary participation in an active learning exercise leads to a better understanding of physiology. *Advances in Physiology Education, 35*(1), 53–58.

D'Costa, A., & Shepherd, I. T. (2009). Zebrafish development and genetics: Introducing undergraduates to developmental biology and genetics in a large introductory laboratory class. *Zebrafish, 6*(2), 169–177.

Duroisin, N., Temperman, G., & de Lièvre, B. (2011). Effets de deux modalités d'usage du tableau blanc interactif sur la dynamique d'apprentissage et la progression des apprenants. *Environnements Informatiques pour l'Apprentissage Humain, Conférence EIAH'2011*, 257–269.

Ernest, H., & Colthorpe, K. (2007). The efficacy of interactive lecturing for students with diverse science backgrounds. *Advances in Physiology Education, 31*(1), 41–44.

Healey, M. (2005). Linking research and teaching to benefit student learning. *Journal of Geography in Higher Education, 29*(2), 183–201.

Helle, L., Tynjälä, P., & Olkinuora, E. (2006). Project-based learning in post-secondary education–theory, practice and rubber sling shots. *Higher Education, 51*(2), 287–314.

Justice, C., Rice, J., & Warry, W. (2009). Developing useful and transferable skills: Course design to prepare students for a life of learning. *International Journal for the scholarship of Teaching and Learning, 3*(2), 1–19.

Justice, C., Rice, J., Warry, W., Inglis, S., Miller, S., & Sammon, S. (2007). Inquiry in higher education: Reflections and directions on course design and teaching methods. *Innovative Higher Education, 31*(4), 201–214.

Khalaf, B. K. (2018). Traditional and inquiry-based learning pedagogy: A systematic critical review. *International Journal of Instruction, 11*(4), 545–564.

Lau, J. M., & Robinson, D. L. (2009). Effectiveness of a cloning and sequencing exercise on student learning with subsequent publication in the National Center for biotechnology information GenBank. *CBE—Life Sciences Education, 8*(4), 326–337.

Nabors, K. L. (2012). *Active learning strategies in classroom teaching: Practices of associate degree nurse educators in a southern state*. Doctoral dissertation, University of Alabama Libraries.

Oliver, R. (2008). Engaging first year students using a web-supported inquiry-based learning setting. *Higher Education, 55*(3), 285–301.

Pandey, A., Nanda, G. K., & Ranjan, V. (2011). Effectiveness of inquiry training model over conventional teaching method on academic achievement of science students in India. *Journal of Innovative Research in Education, 1*(1), 7–20

Poumay, M. (2014). Six leviers pour améliorer l'apprentissage des étudiants du supérieur. *Revue internationale de pédagogie de l'enseignement supérieur, 1*(30). https://journals.openedition.org/ripes/778

Prakash, E. (2010). Explicit constructivism: A missing link in ineffective lectures? *Advances in Physiology Education, 34*(2), 93–96.

Roopa, S., Bagavad Geetha, M., Rani, A., & Chacko, T. (2013). What type of lectures students want? – A reaction evaluation of dental students. *Journal of Clinical and Diagnostic Research: JCDR, 7*(10), 2244.

Seyhan, H., & Morgil, I. (2007). The effect of 5E learning model on teaching of acid-base topic in chemistry education. *Journal of Science Education, 8*(2), 120.

Snell, Y. L. (1999). Interactive lecturing: Strategies for increasing participation in large group presentations. *Medical Teacher, 21*(1), 37–42.

Spronken-Smith, R., & Walker, R. (2010). Can inquiry-based learning strengthen the links between teaching and disciplinary research? *Studies in Higher Education, 35*(6), 723–740.

Spronken-Smith, R., Angelo, T., Matthews, H., O'Steen, B., & Robertson, J. (2007). How effective is inquiry-based learning in linking teaching and research. In *An international colloquium on international policies and practices for academic enquiry* (pp. 19–21). Marwell.

Spronken-Smith, R., Walker, R., Batchelor, J., O'Steen, B., & Angelo, T. (2011). Enablers and constraints to the use of inquiry-based learning in undergraduate education. *Teaching in Higher Education, 16*(1), 15–28.

Temperman, G., Montagne, S., Lièvre, B. D., & Boumazguida, K. (2018). Analyse des usages d'aides par QR codes et de leurs effets dans un manuel scolaire augmenté. *Sciences et Technologies de l'Information et de la Communication pour l'Éducation et la FormFormation, 25*(1), 151–168. https://www.persee.fr/docAsPDF/stice_1764-7223_2018_num_25_1_1759.pdf

The Boyer Commission. (1998). *On educating undergraduates in the research university "reinventing undergraduate education: A blueprint for America's research universities"*. Stomy Brook.

Wilson, C. D., Taylor, J. A., Kowalski, S. M., & Carlson, J. (2010). The relative effects and equity of inquiry-based and commonplace science teaching on students' knowledge, reasoning, and argumentation. *Journal of Research in Science Teaching: The Official Journal of the National Association for Research in Science Teaching, 47*(3), 276–301.

# Chapter 10
# Train the Trainer in the Jigsaw Puzzle of Biology Education: Effects of Mentor Training on Teaching Quality

**Emanuel Nestler, Carolin Retzlaff-Fürst, and Jorge Groß**

## 10.1 Introduction

Mentoring of pre-service teachers is a common part of biology teacher training. Mentors (teachers as teacher educators) usually coach pre-service teachers (student teachers) during teacher training placements (Jasman, 2003). An important but challenging area in the field of mentoring is the training of mentors (e.g. Koballa et al., 2010).

Biology mentor training affects biology teacher training and ultimately biology teaching in the classroom. Biology education, including the interrelationship of mentor training, teacher training and classroom teaching is comparable to an unfinished jigsaw puzzle (Fig. 10.1).

There have been some attempts to connect two or three pieces of the puzzle (e.g. Elster, 2008; Kreis 2012), but there is no overview in the literature of the scientific field of these three connected parts of biology education. With the jigsaw puzzle in mind, we designed a training programme for biology mentors. Afterwards, we evaluated the impact of the programme on mentoring, focusing on the ultimate quality of teaching. This paper is an overview of some pieces of the biology education interrelationship jigsaw puzzle.

E. Nestler (✉) · C. Retzlaff-Fürst
Department Biologiedidaktik, University of Rostock, Rostock, Germany
e-mail: emanuel.nestler@uni-rostock.de; carolin.retzlaff-fuerst@uni-rostock.de

J. Groß
Department Science Education, Otto-Friedrich-University Bamberg, Bamberg, Germany
e-mail: jorge.gross@uni-bamberg.de



K. Korfiatis, M. Grace (eds.), *Current Research in Biology Education*, Contributions from Biology Education Research,
https://doi.org/10.1007/978-3-030-89480-1_10

**Fig. 10.1** Some pieces of the 'biology education' interrelationship jigsaw puzzle

## 10.2 Theoretical Background

The biology education jigsaw puzzle consists of three levels: mentor training, teacher training and classroom teaching (Fig. 10.2). In each level there are general (white) and biology specific (grey) pieces in the puzzle. For this paper it would be redundant to repeat studies about the general research on teaching biology. For a current overview on the research see Gericke and Grace (2018) or Tal and Yarden (2016).

Research has tended to focus on single pieces of the jigsaw puzzle rather than putting them together. For example, Roesler et al. (2018) focused on teaching biology and the interest and motivation of students, while Crasborn et al. (2011) reported the mentor role during mentoring dialogues with pre-service teachers. It is important to note that both examples only focus on one of the three levels mentioned. Förtsch et al. (2016) reported an indirect link between pedagogical content knowledge of biology teachers and student achievement, which is another way to organize the jigsaw puzzle.

We use the metaphor of the "jigsaw puzzle" because we will show in this article that we must pay less attention to the individual components in training processes and more to the interrelationships. Every individual part has meaning but only from connecting all the parts can an overall picture of the quality of the training be determined. But the metaphor also has its limits; in a puzzle, only the individual parts have to fit together. However, within the mentor training in biology education we are dealing with a process: each trainer passes knowledge and skills on to the next level. To make this concept of "training the trainer" clearer, we will avoid the metaphor of the puzzle in what follows.

In the last ERIDOB book (Gericke & Grace, 2018) only two papers (Baytelman et al., 2018; Brauer & Hößle, 2018) out of 24 focused on biology teacher training. Nevertheless, in the last few years several studies have been performed on biology teacher training. For example, Barnett and Friedrichsen (2015) prioritized the pre-service teachers' pedagogical content knowledge, while Bruckermann et al. (2017)

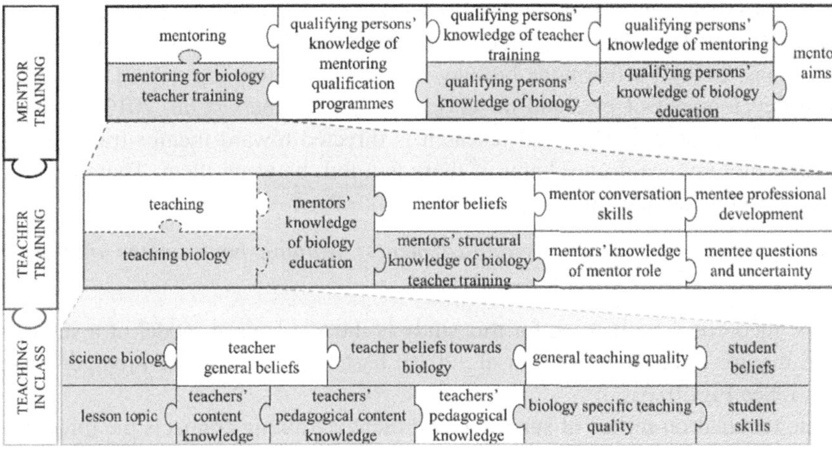


**Fig. 10.2** The biology education jigsaw puzzle with general (white) and biology specific (grey) pieces

focused on experimentation competencies, and Walan (2020) explored different instructional strategies.

Many studies have been published on mentoring in teacher training (Aspfors & Fransson, 2015; Clarke et al., 2014), conversational techniques (Kreis, 2012; Carroll, 2005) and mentor role (Ambrosetti & Dekkers, 2010; Crasborn et al., 2011). In their review of 70 studies, Ellis et al. (2020) identified seven dimensions of a quality pre-service mentor. Reynolds et al. (2015) concluded that it is not the time of professional placement in school but the quality of support that is linked 'with pre-service teachers' perceptions of their ability to apply subject content and teaching' (Reynolds et al., 2015). But only a few studies have looked at mentoring in biology and science teacher education (e.g. Barnett & Friedrichsen, 2015; Hudson, 2005; Koballa et al., 2010). Hudson (2005), in particular, underlines the need for subject specific mentoring.

While there is a need (Hudson, 2013), there are few descriptions of mentor training in the literature (Beutel & Spooner-Lane, 2009; Malmberg et al., 2020; Kreis, 2012; Hall et al., 2017). The subject-related mentor training is a blind spot in biology teacher training. Elster (2008) describes one of the few examples of biology-related mentor training studies. Unfortunately, her study does not include research in biology-related mentor training. So, there are still many unanswered questions about biology teacher training.

If mentoring is important for pre-service teachers (Reynolds et al., 2015) and the importance of subject-related teacher training has been demonstrated, then we need to ask why only a few biology-related mentor training studies have been published. This question becomes even more pressing in the light of the appeal for larger, more comprehensive studies on the subject of biology education, and the need to form a link between the research community and practitioners of biology education that was addressed in McComas's 'Grand challenges in biology education research' (McComas, 2018).

This link between researchers and practitioners helps to face other challenges such as the development of an effective organizational plan for biology instruction (McComas, 2018). Furthermore, biology-related mentor training can aid the professional development of pre- and in-service biology teachers (Chu, 2019; Hudson, 2013). On the other hand, if no research is directed toward mentor training and teaching biology, we do not know if there is a link between them. Therefore, we raise the following question:

*What influence does a biology-related mentor training have on the quality of teaching?*

The theoretical framework for this study is the tetrahedron-model of a subject-based mentor training (Nestler et al., 2020) based on the work of Prediger et al. (2017). See Fig. 10.3.

The tetrahedron-model of subject-based mentor training connects the three levels: teaching, teacher training and mentor training. Every tetrahedron consists of four corners: learner (e.g. student, pre-service teacher, mentor), teacher (e.g. teacher, mentor & mentor qualifier), teaching media and subject. While the subject of teacher training is the whole teaching tetrahedron, the subject of mentor training comprises both the teaching and the teacher training tetrahedron. This theoretical framework provides the ability to describe chains of effect. In this study, we examine whether

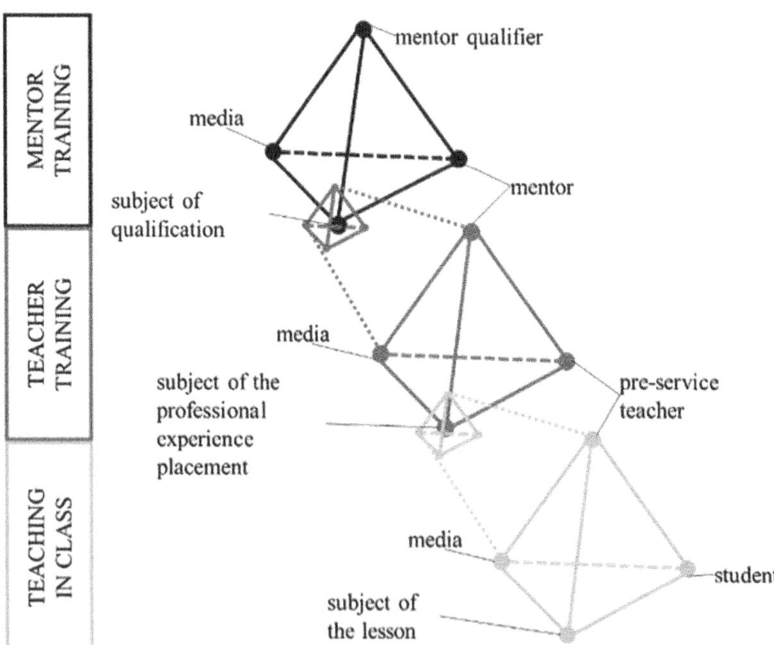


**Fig. 10.3** The tetrahedron-model of subject-based mentor training. (Modified from Prediger et al., 2017)

there is a chain of effects between the biology-related mentor training and teaching quality. This does not include the exact description of the chain of effects. For this description, a mixed-method approach with qualitative elements would be needed. Therefore, the qualitative approach is in the focus of further articles, which are in progress.

## 10.3 Earlier Research and Hypotheses

The field of biology-related mentor training is very unclear due to the lack of recognition of the interrelationship of the three levels: mentor training, teacher training and teaching in class. In addition, *general* and *biology-related* research need to be included. In summary, there are *general* studies regarding:

- teacher and teaching quality (e.g. Helmke, 2015);
- mentoring and pre-service teachers (e.g. Reynolds et al., 2015; Aspfors & Fransson, 2015; Clarke et al., 2014; Ambrosetti & Dekkers, 2010; Crasborn et al., 2011); and
- mentor training and mentoring (e.g. Kreis, 2012; Hall et al., 2017; Malmberg et al., 2020).

There are *biology-related* studies regarding:

- biology teacher and teaching quality (e.g. Förtsch et al., 2016); and
- biology-related mentoring and pre-service teacher and their professional development (e.g. Elster, 2008).

To the best of our knowledge, there appear to be no studies in *biology-related* research that combine biology-related mentor training with mentoring or teaching quality. So, a serious weakness in biology-related teacher training is the lack of knowledge regarding the impact of biology-related mentor training. Therefore, we need to observe mentoring and teaching quality before and after a programme of biology-related mentor training. Based on the theoretical framework, our first hypothesis was: *The perceived teaching quality increases after biology-related mentor training.*

Several studies have been conducted on the impact of the assessor on teaching quality. Tillema (2009) describes the differences in the perceived purpose of appraisal as well as incongruence between the criteria used for assessing teaching quality in mentors, pre-service teachers, and university-based supervisors. Similar results were presented by Harnischmacher and Hofbauer (2011) regarding the assessment of teaching in music; there were significant differences in the levels used by pre-service teachers, in-service teachers, and students. As is underlined by Korver and Tillema (2014), the different perceptions of the feedback provided can impact the acceptance of the assessment and by extension what the pre-service teacher may learn from it. To identify the differences between the observers (mentor, pre-service teacher responsible for this lesson and observing pre-service teacher) we stated our second hypothesis: *There is a difference in the observers' assessments of teaching quality.*

On the other hand, Kreis (2012) described differences in content between the preliminary meeting and the debriefing. To identify these differences in relation to teaching quality, our third hypothesis was: *The perceived teaching quality changes between the preliminary discussion of the lesson and the debriefing after the teaching.*

## 10.4 Research Design and Method

To analyse the chain of effects and evaluate the possible impact of biology-related mentor training on teaching quality, we designed a biology-based mentor training programme. The mentors completed a general qualification and half a year later we applied a biology-based qualification (Nestler & Retzlaff-Fürst, 2020). This qualification involved an overall workload of 90 h (45 h general, 45 h biology-based) for biology mentors ($N = 7$, age: $M = 42.3$). The key topic was to teach scientific inquiry (see Paul et al., 2016). This qualification was co-created with the professor of genetics who introduced the mentors to new findings in genetics (e.g CRISPR/Cas9). Afterwards, the biology mentor carried out various experiments such as in a session of a laboratory course for pre-service teachers. The additional content of the biology-based mentor training (Fig. 10.4) includes the findings of Niggli (2005), Kreis (2012), Helmke (2014) and Mayer (2018). Additionally, there were units focused on an introduction to didactics of biology and the structure of biology teacher training. A more detailed description can be found in the paper of Nestler &

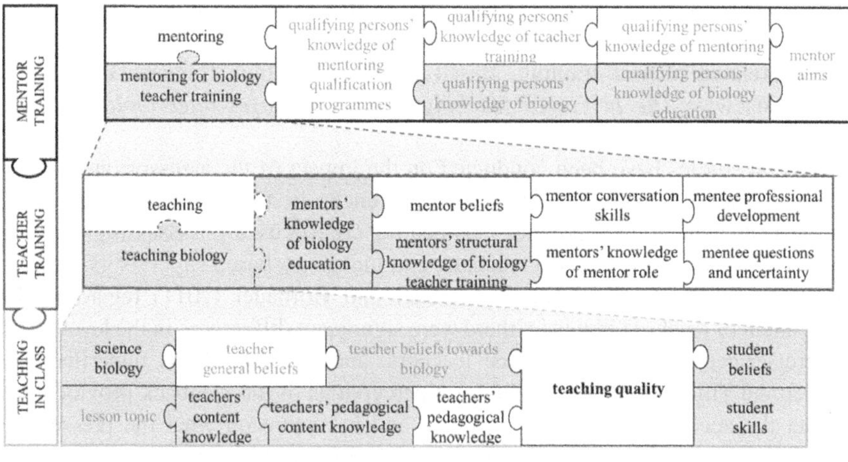


**Fig. 10.4** Contents of the biology-based mentor training (black)

Retzlaff-Fürst (2020). This mentor training can easily be adapted for other biology-related requirements.

The professional experience placement for the pre-service biology teachers was based on weekly teaching during one semester. A group of five pre-service biology teachers took turns teaching one class in secondary education. Every student taught two biology lessons and observed about nine lessons (one of the mentor's). The biology mentor coached the mentee who taught the lesson in a preliminary meeting. After observing the lesson, the group of pre-service biology teachers and the mentor discussed the teaching quality and achievement of lesson goals.

To evaluate the impact of the biology-related mentor training on teaching quality a quantitative approach was chosen. This approach allows more information regarding the impact of the biology-related mentor training and teaching quality to be obtained. Therefore, the effects of single mentors, pre-service biology teachers or the specific situation can be better excluded. While there are qualitative studies developing different methods for mentor development (see Beutel & Spooner-Lane, 2009; Langdon & Ward, 2015; Meetze-Hall, 2018), it is important to investigate if those methods produce consistent improvements in mentoring ability and teaching for mentors in general. Additionally, the large amount of time required for mentor training and mentoring, necessitates short research instruments, as we did not want to put an unreasonable burden on the mentors.

The teaching quality was assessed, by applying the teaching quality paper-pencil-questionnaire of Helmke (2014). The instrument was adapted to fit the time frame of mentoring sessions. Eight items connected to didactical aspects were chosen, such as lesson goals (*'The lesson goals can be found in this lesson and the tasks.')* or orientation on pupils' interests *(The lessons contain examples from the pupils' everyday life/world or are linked to the pupils' interests.'*). The items are rated on a 6-point-scale, ranging from 1 ("don't agree") to 6 ("agree"). The scale has a high internal consistency of $\alpha = 0.87$ in the present study (Cronbach, 1951).

To test our hypotheses, we designed an intervention study. Initially, the mentors supported 25 pre-service teachers (11 female, 10 male, 4 no information; age $M =$ 25.6) in a professional experience setting during the winter semester 2017/2018. Subsequently, the aforementioned mentor training occurred. In the following summer semester, the mentors supported 28 pre-service teachers (16 female, 10 male, 2 no information, age $M = 23.4$). There was no control group in regard to the structural conditions (e.g. 7 mentors, 53 pre-service teachers per semester). The pool of mentors who are responsible for the professional experience setting at our university is largely unvarying, so it is not possible to repeat the study.

## 10.5 Findings

The seven biology mentors and 53 pre-service biology teachers completed 633 questionnaires. Completing the questionnaire after every preliminary discussion and debriefing was their responsibility, and although the participants did not

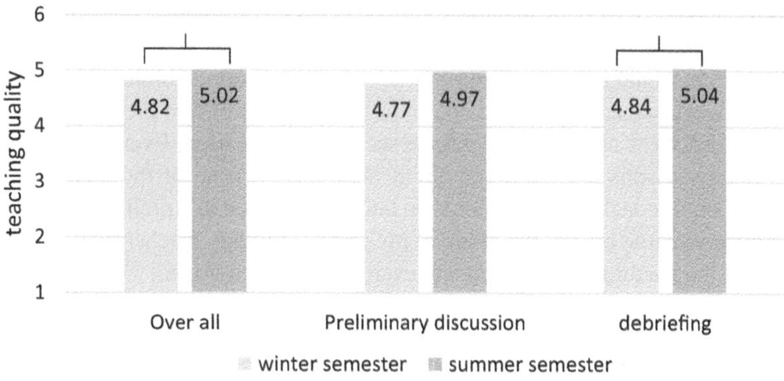


**Fig. 10.5** Teaching quality before and after the mentor training

complete every single questionnaire, the overall response rate was good. The participants estimated their teaching quality as good (4.51 and higher). In general, there was good quality teaching in the professional experience setting. The results were not approximately normally distributed for the quality of teaching, as assessed by the Shapiro-Wilk-Test ($p < .05$). Additionally, the homogeneity of variances was asserted using Levene's Test which showed that equal variances could not be assumed ($p < 0.05$). We report the results for teaching quality based on draft lessons in preliminary discussions and based on actually performed lesson in debriefing separately.

The first hypothesis was: *The perceived teaching quality increases after the biology-related mentor training.*

The tests revealed that teaching quality increased by 0.2 units after the mentor training (Fig. 10.5). There was a significant result over all evaluations (winter semester $n = 330$, $M = 4.85$; summer semester $n = 303$, $M = 5.02$; $z = -2.387$, $p = 0.017$, $r = 0.095$) and for the debriefing (winter semester $n = 252$, $M = 4.84$; summer semester $n = 227$, $M = 5.04$; $z = -2.376$, $p = 0.018$, $r = 0.109$). No significant result was found for teaching quality in preliminary discussions using the Mann-Whitney-U-test (winter semester $n = 78$, $M = 4.77$; summer semester $n = 76$, $M = 4.97$; $z = -0.726$, $p = 0.468$). The mean score for teaching quality in preliminary discussions increases as much (0.2) as in the debriefing but with a smaller group of observers. Teaching quality increases roughly 0.2 points of the mean score in the summer semester for every observer group in preliminary discussion and debriefing (Table 10.1). Overall, we accept this hypothesis.

The second hypotheses states: *There is a difference in the observer's assessment of teaching quality.*

The mean scores of pre-service teachers are higher than the mean scores of the mentors. Mentors take a more critical view on teaching quality than pre-service teachers. Mann-Whitney-U-test results show significant differences in the assessment for debriefings in the winter semester ($z = -2.040$, $p = 0.041$, $r = 0.231$) and summer semester ($z = -2.836$, $p = 0.005$, $r = 0.325$). In both semesters pre-service

**Table 10.1** Sample size, means and standard deviations for the assessment of teaching quality

| | Preliminary discussion before the lesson | | | | Debriefing after the lesson | | | | | |
|---|---|---|---|---|---|---|---|---|---|---|
| | mentor | | Pre-service teacher responsible for this lesson | | Mentor | | Pre-service teacher responsible for this lesson | | Pre-service teacher observing | |
| | *n* | *M (SD)* | *n* | *M (SD)* | *n* | *M (SD)* | *n* | *M (SD)* | *n* | *M (SD)* |
| Winter semester 2017/18 | 32 | 4.51 (1.07) | 46 | 4.94 (0.77) | 48 | 4.64 (0.90) | 60 | 4.79 (0.72) | 144 | 4.93 (0.72) |
| Summer semester 2018 | 28 | 4.72 (0.56) | 48 | 5.11 (0.44) | 41 | 4.87 (0.58) | 59 | 5.01 (0.55) | 127 | 5.11 (0.55) |

teachers observing the lessons had the highest mean scores in their rating of teaching quality. After significant Kruskal-Wallis-tests ($p$ = 0.042) Dunn-Bonferroni-tests reveal that differences in assessment are based on differences between mentor and observing pre-service teacher ($z = -2.477$ $p = 0.040$ Dunn-Bonferroni-tests, $r =$ 0.191) in the summer semester. Overall, we can accept this hypothesis.

The third hypothesis states: *The perceived teaching quality changes between the preliminary discussion of the draft and the debriefing of the teaching.*

Further analysis of the mean scores showed that mentors are more satisfied in the debriefing than in the preliminary discussions. The data shows that pre-service teachers who designed the lesson are less satisfied after the lesson. It appears that pre-service teachers are disappointed by the reality. The drop of the mean score is not as strong after the mentor training (−0.10) as before (−0.15). This drop of mean score is not statistically significant.

It may be a coincidence, but the mean score of teaching quality of pre-service teachers observing the lesson in debriefing is as high as the mean score of teaching quality of pre-service teachers responsible for this lesson in preliminary discussions in both semesters. Overall, we cannot accept hypothesis three because there are no significant differences shown by the Mann-Whitney-U-tests.

## 10.6 Discussion

Our results for teaching quality and mentor training are the first results in this field. Therefore, we cannot put the pieces of the jigsaw puzzle together definitively, but we can lay them side by side. It is possible that biology-related mentor training has an impact on the improvement of teaching quality. We bear in mind that the mentors participated in general mentor training before the winter semester.

Further research needs to replicate and expand upon the empirical evidence to link more pieces of the jigsaw puzzle, i.e. how the different levels of qualification are linked together. One option might be to include biology-based mentor training in the discussion of larger studies in biology education aiming at teaching quality and students' beliefs.

After the biology pre-service teachers gave their lesson, the assessment from the mentor improved. Since we can assume that the pre-service teachers responsible for the lesson conducted their lesson mostly as they described in the preliminary discussion, our results can be optimistically interpreted as a demonstration that pre-service teachers convince the mentors about their planned lesson. In addition, the pre-service teachers responsible for this lesson are less satisfied with the teaching quality after their lesson. Overall, the differences in mean scores of teaching quality start to converge in the debriefing. This supports the conceptual change of the mentor role from experienced senior person to coach or learning partner (Ellis et al., 2020; Heikkinen et al., 2012; Kreis, 2012; Burley & Pomphrey, 2011). Mentors can focus on key aspects of teaching quality during the debriefing because pre-service teachers will have a more realistic view of teaching quality after their lesson. On the other hand, mentors re-evaluate their view after the lesson, so they need to trust pre-service teachers' teaching quality. The high scores for teaching quality observed by the pre-service teachers not involved in lesson planning may speak for including them in preliminary discussions. We recommend that more objective methods are included so we know whether the mentors or the pre-service teacher are closer to an objective measure of teaching quality.

The research in teaching quality increased during and after this study (Heinitz & Nehring, 2020; Joyce et al., 2018; Steffensky & Neuhaus, 2018; Dorfner et al., 2017). It appears that analysis of a videotaped class is a good method to develop a deeper understanding of teaching quality (Dorfner et al., 2017). Despite this, videotape analysis is not a practical application for research in mentor training as part of biology education as it is too time consuming. Further work needs to be conducted to establish convenient methods to evaluate teaching quality. Combining small questionnaires with selected analysis of videotaped classes may be a good opportunity to efficiently evaluate teaching quality.

Overall, there are several possible sources of weakness in the field of research in biology education including mentor training linked to teacher training and teaching. The first challenges are the general and biology-related *theoretical constructs* such as mentoring, teacher professionalisation and teaching quality. We need a broad understanding of the field but need to sharpen our focus for satisfying research. We therefore have to look at the methodical challenges such as the scaling of the levels of mentor training, teacher training and teaching. In this study seven mentors and 53 pre-service teachers were included. If we want to obtain a better and more representative picture of mentoring, we need to involve more mentors. In this case we have to include more pre-service teachers because there needs to be a satisfactory number of mentoring situations before and after the mentor training. All the mentoring consists of three learning situations that we have to evaluate: preliminary discussions, teaching in class and debriefing. Overall, scaling is a major challenge which has led to this unsatisfactory situation of a lack of research in biology-related mentor training.

*The absence of possibilities of real interventional designs of mentor trainings* is another methodical challenge. Teachers do not want to wait one year (the time necessary to build up a control group) to be part of a mentor training cohort. The individual professionalization of pre-service teachers can also vary widely. To address this challenge, there has to be a larger number of mentor and pre-service teacher tandems, which can cause the aforementioned problems with scaling.

Finally, *different educational systems* challenge mentor trainings. There are differences between schools and universities and between different countries. General mentor training on conversational techniques or mentor role cope well with this, but biology-related mentor training depends on educational goals, teacher training curricula and teaching and research in biology education, which focuses teaching as we have seen before.

In conclusion, while acknowledging the limitations and challenges of our research, we have taken the first steps to understand the impact of biology-related mentor training on teaching quality, which may have parallels for other disciplines.

**Acknowledgment** Thank you to Richard Fry for his expertise and support in language editing.

## References

Ambrosetti, A., & Dekkers, J. (2010). The interconnectedness of the roles of mentors and mentees in pre-service teacher education mentoring relationships. *Australian Journal of Teacher Education, 35*, 42–55. https://doi.org/10.14221/ajte.2010v35n6.3

Aspfors, J., & Fransson, G. (2015). Research on mentor training for mentors of newly qualified teachers: A qualitative meta-synthesis. *Teaching and Teacher Education, 48*, 75–86.

Barnett, E., & Friedrichsen, P. J. (2015). Educative mentoring: How a mentor supported a preservice biology teacher's pedagogical content knowledge development. *Journal of Science Teacher Education, 26*, 647–668.

Baytelman, A., Iordanou, K., & Constantinou, C. P. (2018). Challenges in biology education research. In N. Gericke & M. Grace (Eds.), *A selection of papers presented at the XIth conference of European Researchers in Didactics of Biology (ERIDOB)* (pp. 152–169). Karlstad University Printing Office.

Beutel, D., & Spooner-Lane, R. (2009). Building mentoring capacities in experienced teachers. *International Journal of Learning, 16*, 1–10.

Brauer, L., & Hößle, C. (2018). Acquiring diagnostic skills in the field of experimentation in the Wadden Sea's Teaching & Learning Laboratory. In N. Gericke & M. Grace (Eds.), *Challenges in biology education research. A selection of papers presented at the XIth conference of European Researchers in Didactics of Biology (ERIDOB)* (pp. 256–272). Karlstad University Printing Office.

Bruckermann, T., Aschermann, E., Bresges, A., & Schlüter, K. (2017). Metacognitive and multimedia support of experiments in inquiry learning for science teacher preparation. *International Journal of Science Education, 39*, 701–722.

Burley, S., & Pomphrey, C. (2011). *Mentoring and coaching in schools: Professional learning through collaborative inquiry*. Taylor & Francis.

Carroll, D. (2005). Learning through interactive talk: A school-based mentor study group as a context for professional learning. *Teaching and Teacher Education, 21*, 457–473.

Chu, Y. (2019). Mentor professional identity development in a year-long teacher residency. *Mentoring & Tutoring: Partnership in Learning, 27*, 251–271. https://doi.org/10.1080/13611267.2019.1630991

Clarke, A., Triggs, V., & Nielsen, W. (2014). Cooperating teacher participation in teacher education: A review of the literature. *Review of Educational Research, 84*, 163–202. https://doi.org/10.3102/2F0034654313499618

Crasborn, F., Hennissen, P., Brouwer, N., Korthagen, F., & Bergen, T. (2011). Exploring a two-dimensional model of mentor roles in mentoring dialogues. *Teaching and Teacher education, 27*, 320–331.

Cronbach, L. J. (1951). Coefficient alpha and the internal structure of tests. *psychometrika, 16*(3), 297–334.

Dorfner, T., Förtsch, C., & Neuhaus, B. J. (2017). Die methodische und inhaltliche Ausrichtung quantitativer Videostudien zur Unterrichtsqualität im mathematisch-naturwissenschaftlichen Unterricht. *Zeitschrift für Didaktik der Naturwissenschaften, 23*, 261–285. https://doi.org/10.1007/s40573-017-0058-3

Ellis, N. J., Alonzo, D., & Nguyen, H. T. M. (2020). Elements of a quality pre-service teacher mentor: A literature review. *Teaching and Teacher Education, 92*, Article 103072. https://doi.org/10.1016/j.tate.2020.103072

Elster, D. (2008). *Subject-related mentoring in biology teacher education*. Paper presented at "Impact of science education research on public policy" NARST annual international conference (national association for research in science teaching) 29th March-2nd April 2008. https://doi.org/10.13140/2.1.3341.1366

Förtsch, C., Werner, S., von Kotzebue, L., & Neuhaus, B. J. (2016). Effects of biology teachers' professional knowledge and cognitive activation on students' achievement. *International Journal of Science Education, 38*, 2642–2666. https://doi.org/10.1080/09500693.2016.1257170

Gericke, N., & Grace, M. (Eds.). (2018). *Challenges in biology education research. A selection of papers presented at the XIth conference of European Researchers in Didactics of Biology (ERIDOB)*. Karlstad University Printing Office.

Hall, D. M., Hughes, M. A., & Thelk, A. D. (2017). Developing mentorship skills in clinical faculty: A best practices approach to supporting beginning teachers. *Teacher Educators' Journal, 10*, 77–98.

Harnischmacher, C., & Hofbauer, V. C. (2011). Wahrnehmungsdimensionen des Musikunterrichts-Eine explorative Studie zur Unterrichtsbeobachtung von Schülern, Studenten. *Lehramtsanwärtern und Lehrern. Beiträge empirischer Musikpädagogik, 2*(2), 2–14.

Heikkinen, H., Jokinen, H., & Tynjälä, P. (2012). Teacher education and development as lifelong and lifewide learning. In H. Heikkinen, H. Jokinen, & P. Tynjälä (Eds.), *Peer-group mentoring for teacher development* (pp. 3–30). Routledge.

Heinitz, B., & Nehring, A. (2020). Kriterien naturwissenschaftsdidaktischer Unterrichtsqualität – ein systematisches Review videobasierter Unterrichtsforschung. *Unterrichtswissenschaft, 48*, 319–360. https://doi.org/10.1007/s42010-020-00074-8

Helmke, A. (2014). *Kompetenzorientierter, an den Bildungsstandards orientierter Unterricht*. http://www.unterrichtsdiagnostik.de/media/files/Link%2029_Kompetenzorientierter%20-%20an%20Bildungsstandards%20orientierter%20Unterricht.pdf; 26.08.2017.

Helmke, A. (2015). *Unterrichtsqualität und Lehrerprofessionalität*. Kallmeyer.

Hudson, P. B. (2005). Identifying mentoring practices for developing effective primary science teaching. *International Journal of Science Education, 27*, 1723–1739.

Hudson, P. B. (2013). Mentoring as professional development: 'Growth for both' mentor and mentee. *Professional Development in Education*. https://doi.org/10.1080/19415257.2012.749415

Jasman, A. (2003). Initial teacher education: Changing curriculum, pedagogies and assessment, Change: Transformations. *Education, 6*(2), 1–22.
Joyce, J., Gitomer, D. H., & Iaconangelo, C. J. (2018). Classroom assignments as measures of teaching quality. *Learning and Instruction, 54*, 48–61.
Koballa, T. R., Jr., Kittleson, J., Bradbury, L. U., & Dias, M. J. (2010). Teacher thinking associated with science-specific mentor preparation. *Science Education, 94*, 1072–1091.
Korver, B., & Tillema, H. (2014). Feedback provision in mentoring conversation-differing mentor and student perceptions. *Journal of Education and Training Studies, 2*(2), 167–175.
Kreis, A. (2012). *Produktive Unterrichtsbesprechungen: Lernen im Dialog zwischen Mentoren und angehenden Lehrpersonen*. Haupt.
Langdon, F., & Ward, L. (2015). Educative mentoring: A way forward. *International Journal of Mentoring and Coaching in Education, 4*, 240–254.
Mayer, J. (2018). Erkenntnisse mit naturwissenschaftlichen Methoden gewinnen. In H. Gropengießer, U. Harms, & U. Kattmann (Eds.), *Fachdidaktik Biologie* (pp. 114–124). Aulis.
Malmberg, I., Nestler, E., & Retzlaff-Fürst, C. (2020). Qualitäten der Mentor*innenqualifizierung M-V. Eine Design Based Research Studie zu einem Lernbegleitungsprogramm an der Schnittstelle zwischen Schule und Hochschule. In F. Hesse & W. Lütgert (Eds.), *Auf die Lernbegleitung kommt es an!* Konzepte und Befunde zu Praxisphasen in der Lehrerbildung. Klinkhardt.
McComas, W. F. (2018). Grand challenges in biology education research: Some conclusions. In N. Gericke & M. Grace (Eds.), *Challenges in biology education research. A selection of papers presented at the XIth conference of European Researchers in Didactics of Biology (ERIDOB)* (pp. 388–392). Karlstad University Printing Office.
Meetze-Hall, M. (2018). *Educating educative mentors: Video as instructional tool*. Thesis, Concordia University, St. Paul.
Nestler, E., Malmberg, I., Heinrich, G., & Retzlaff-Fürst, C. (2020). Praxisphasen als Räume der Konstruktion von Fachlichkeit. In: T. Leonhard, P. Herzmann & J. Košinàr (Hrsg.), *„Grau, theurer Freund, ist alle Theorie"? Theorien und Erkenntniswege schul- und berufspraktischer Studien 2020, Schulpraktische Studien und Professionalisierung* (S. 163–178). Waxmann.
Nestler, E., & Retzlaff-Fürst, C. (2020). Die Mentor*innenqualifizierung im Fach Biologie zur Unterstützung der Reflexion von fachwissenschaftlichen und fachdidaktischen Konzepten Studierender. In Y. Völschow & K. Kunze (Hrsg.), *Reflexion und Beratung in der Lehrerinnen- und Lehrerausbildung. Beiträge zur Professionalisierung von Lehrkräften* (S. 365–380). Budrich.
Niggli, A. (2005). *Unterrichtsbesprechungen im Mentoring*. Pädagogik bei Sauerländer. Retrieved from. https://doc.rero.ch/record/234685/files/KopieBuchNI_Ment_2005.pdf
Paul, J., Lederman, N. G., & Groß, J. (2016). Learning experimentation through science fairs. *International Journal of Science Education, IJSE, 38*(15), 2367–2387.
Prediger, S., Leuders, T., & Rösken-Winter, B. (2017). Drei-Tetraeder-Modell der gegenstandsbezogenen Professionalisierungsforschung: Fachspezifische Verknüpfung von Design und Forschung. *Jahrbuch für Allgemeine Didaktik, 7*, 159–177.
Reynolds, R., Howley, P., Southgate, E., & Brown, J. (2015). Just add hours? An assessment of pre-service teachers' perception of the value of professional experience in attaining teacher competencies. *Asia-Pacific Journal of Teacher Education, 44*, 455–469. https://doi.org/10.1080/1359866X.2015.1086971
Roesler, M., Wellnitz, N., & Mayer, J. (2018). The role of interesting and motivating contexts in the assessment of content knowledge and decision-making. In N. Gericke & M. Grace (Eds.), *Challenges in biology education research. A selection of papers presented at the XIth conference of European Researchers in Didactics of Biology (ERIDOB)* (pp. 135–150). Karlstad University Printing Office.
Steffensky, M., & Neuhaus, B. J. (2018). Unterrichtsqualität im naturwissenschaftlichen Unterricht. In D. Krüger, I. Parchmann, & H. Schecker (Eds.), *Theorien in der naturwissenschaftsdidaktischen Forschung*. Springer. https://doi.org/10.1007/978-3-662-56320-5_18

Tal, T., & Yarden, A. (Eds.). (2016). *The future of biology education research: Proceedings of the 10th conference of European researchers in didactics of biology*. Technion.
Tillema, H. H. (2009). Assessment for learning to teach: Appraisal of practice teaching lessons by mentors, supervisors, and student teachers. *Journal of Teacher Education, 60*, 155–167.
Walan, S. (2020). Pre-service teachers' reflections when drama was integrated in a science teacher education program. *Journal of Biological Education, 54*, 1–14.

# Chapter 11
# The Assessment of an Educational Proposal to Address the Relationship Between Genetic Information and Protein Synthesis

**Patricia Esteve-Guirao, Isabel Banos-González, and Magdalena Valverde**

## 11.1 Introduction

Genetics may be able to explain situations that form part of students' sociocultural surroundings, such as the existence of food intolerances associated with the inability to synthesize certain proteins and enzymes, such as lactase (Ugidos-Rodríguez et al., 2018).

Lactose intolerance has received much media attention. There is widespread advertising of lactose-free products, which sometimes proclaims that they are light and easily digestible products for the general public. So, it is considered an issue familiar to secondary school students.

Despite the quotidian nature of these topics, students have significant learning difficulties with the contents of genetics in their courses and a lack of connection with their everyday life (Rodríguez-Gil et al., 2018; Vlckova & Kubiatko, 2018). New approaches for teaching these issues have therefore been proposed, which should be related to the students' ideas, to contexts relevant to the students, and to observable phenomena (Gerçek, 2018).

In this work, the problem of lactose intolerance, and the solutions proposed by the producers of dairy products, represent the starting point used to design and implement an educational proposal about the expression of genetic information and its link with protein synthesis. Moreover, we wanted to compare the results of a group of students who were taught based on this proposal with those of another group who received a different educational intervention, based on the transmission of theoretical knowledge.

---

P. Esteve-Guirao · I. Banos-González (✉) · M. Valverde
Departamento de Didáctica de las Ciencias Experimentales – Facultad de Educación, University of Murcia, Murcia, Spain
e-mail: p.esteve@um.es; ibbg1@um.es; mvp@um.es



K. Korfiatis, M. Grace (eds.), *Current Research in Biology Education*, Contributions from Biology Education Research,
https://doi.org/10.1007/978-3-030-89480-1_11

## 11.2 Theoretical Background

In recent years, issues related to genetics have had strong media and social repercussions, giving rise to some controversies, with ethical, economic, and social implications (Dimopoulos & Koulaidis, 2002; Eijck, 2010). Their increasing impact on our lives shows the importance of adequate citizen training, to allow critical and informed participation in these debates (Tsui & Treagust, 2010). However, for secondary school students, the genetics contents are one of the most difficult parts of their biology courses and, in general, they do not see the relationship with their lives (Alozie et al., 2010).

From educational research, numerous students' ideas about genetics have been identified as well as great learning difficulties, which may hinder its proper understanding (Ayuso & Banet, 2002). Hence, important efforts to determine the factors involved in these difficulties and to find adequate educational strategies for the teaching of this subject have been claimed.

Different authors have analyzed the learning difficulties related to this subject and have indicated that various factors contribute to them (Bahar et al., 1999; Gerçek, 2018; Rodríguez-Gil et al., 2018; Vlckova & Kubiatko, 2018). Among these factors, the abstract nature of the concepts and processes involved has been highlighted (Cimer, 2012), and this determines whether students can give a concrete meaning to their study. For Frieder et al. (1987), a level of abstract thinking is necessary to build knowledge that leads to the understanding of the relationship between "micro" and "macro" systems. This could explain the difficulties detected by Duncan and Reiser (2007) in 10th grade students with respect to reasoning through different levels of organization of the structures related to genetic information, and to understanding how mechanisms and interactions at the micro level (genes, proteins) produce effects at the macro level (organism). In addition, these authors found that students were often unaware of the different functions of proteins, their relationship to genes, and the role proteins have in mediating the effects of the genetic information. They concluded that these students knew little about the specific functions of proteins in our bodies. Without an understanding of proteins, the link between the genetic information and its physical effects could remain a black box. Regarding the processes involved in protein synthesis, Gerçek (2018, p. 20) points out that, even among future biology teachers, the learning about nucleic acids, proteins, and enzymes has been "incorrect and insufficient". Furthermore, the influence of epigenetic processes in protein synthesis, is usually oversimplified when these contents are taught (Tsui & Treagust, 2010).

It has also been found that a large part of the student population between 14 and 17 years of age does not know how energy is obtained to carry out cellular processes, such as protein synthesis. There seem to be serious difficulties in relation to respiration as a process that occurs at the cellular level and which is connected to nutritional processes (García, 1991; Hmelo-Silver & Azevedo, 2006). These difficulties may be linked to the reductionist and compartmentalized approach, frequent in the study of the processes that occur in the human body (Assaraf et al., 2013).

The way in which these contents are organized and presented in textbooks, curricular and in most classrooms (Aznar & Puig, 2014), might further complicate their learning. In general, these textbooks do not provide holistic approaches nor situations that are familiar to the students, and that mediate the construction of learning, and they tend to ignore the students' ideas (Íñiguez & Puigcerver, 2013). This could explain the lack of interest of secondary school students in these issues, since they do not recognize their importance in explaining everyday phenomena (Ayuso & Banet, 2002).

Therefore, a change in the teaching approach to these contents seems necessary, with the use of situations close to the students as a starting point from which to mobilize, share, and contrast their ideas. In this sense, during the educational sequence, it seems appropriate that observable phenomena appear first, followed by a descent to the cellular level to facilitate the interpretation of the mechanisms that explain these phenomena (Esteve et al., 2017).

In this approach, to address protein synthesis and its relationship with genetic information, the scenarios related to lactose intolerance could be interesting. Specifically, a relevant starting point could be the advertising messages that present lactose-free products as more digestible, to encourage a generalized consumption among the population. These controversial scenarios make it possible to relate the scientific sphere to socioeconomic, ethical, and environmental issues (Jiménez-Aleixandre, 2010). A critical analysis of this advertising could make sense of the inheritance mechanism of an intolerance and how genetic information is used to synthesize proteins, conceptualizing the transcription and translation processes, as well as the role of nutrition. So, it can be considered as a useful learning experience for students to scientifically interpret events close to their lives, and build clear relationships between them and the underlying cellular processes.

## 11.3 Key Objectives

In this work, we assess an educational intervention designed to address the relationship between the expression of genetic information and protein synthesis, employing an observable and familiar context related to lactose intolerance. This intervention is compared with one that addresses this relationship by means of a more traditional form of transmission of knowledge.

Specifically, we consider how our proposal may help students to overcome some of their ideas and difficulties about these particular contents of their course:

- The location of genetic information for protein synthesis.
- The inheritance of genetic information for protein synthesis.
- The use of genetic information in the protein synthesis process.
- The role of cellular respiration in the protein synthesis process.

## 11.4 Research Design and Methodology

This study follows a pretest/posttest-control design, recording values before and after the intervention, which is considered a useful framework for the evaluation of educational interventions (Zelenika et al., 2018).

### *11.4.1 Participants*

The participants were 43 10th grade students (15–16 years old), who were studying Biology and Geology as optional subjects in a Secondary school in southeast Spain. They were organized in 2 groups, taught by the same teacher: 18 students in the control group (CG) and 25 in the experimental group (EG). These students usually receive a theory-based teaching of the course contents, with little time dedicated to enquiry and problem solving. For all the students, this was the first opportunity to address the content on protein synthesis, although they had studied cells, their organelles, and the organization of genetic material prior the intervention. The assignment to each group was determined by the high school.

### *11.4.2 Data Collection*

Data were gathered by using pretest/posttest questionnaires comprising four questions – one for each of the contents analyzed- and 16 items (Table 11.1). These were related to the ideas about genetic information and protein synthesis extended among the secondary school students. The items were not mutually exclusive, so the students could select as many as they considered appropriate in each question.

### *11.4.3 Data Treatment and Analysis*

For the analysis of the answers in both tests, the SPSS program was used. Descriptive statistics of each item were generated, including the frequency with which it was selected by the participants. Non-parametric statistical tests were applied: the Mann-Whitney U test, to compare CG and EG, and the Wilcoxon signed-rank test, to compare the results within each group.

**Table 11.1** Description of the items included in the questionnaires, for each of the contents

| Content | Item | References |
|---|---|---|
| Location of genetic information to synthesize proteins | I1: Only in the gametes, which unite and form the new individual. | Banet and Ayuso (1995, 2000) |
| | I2: In the sex chromosomes, which replicate before meiosis for the inheritance of characters. | |
| | I3: The location depends on the inherited character. If it is a food intolerance, it will be in the cells of the digestive tract. | |
| | I4[a]: In all cells: they all transmit inherited traits when they form new cells. | |
| Inheritance of genetic information to synthesize proteins | I5[a]: Information about intolerance is found in parts of DNA, called genes. | Lewis and Wood-Robinson (2000) and Wood-Robinson et al. (2000) |
| | I6[a]: To be intolerant, both parents have to provide the information. | |
| | I7[a]: Inheritance of information is needed to synthesize proteins to break down lactose. | |
| | I8: A descendant of both intolerant parents, may be non-intolerant. | |
| Use of genetic information to synthesize proteins | I9: DNA carries information directly to ribosomes. | Duncan and Reiser (2007) and Gercek (2018) |
| | I10[a]: RNA carries information from DNA to ribosomes. | |
| | I11: Ribosomes use information from DNA as it undergoes a small mutation to form RNA. | |
| | I12: Ribosomes use the information contained in amino acids. | |
| Role of cellular respiration in the protein synthesis process. | I13: No energy is needed, just genetic information. | Flores et al. (2003) and García (1991) |
| | I14[a]: Protein synthesis involves the mitochondria, where the energy for the process is obtained. | |
| | I15: The energy obtained through nutrition enters the cell for protein synthesis. | |
| | I16[a]: To obtain energy, amino acids and simple sugars -from digestion- and oxygen -from respiration- are required. | |

[a]Correct items

## 11.4.4 Description of the Interventions

The educational intervention in the CG was organized according to a traditional teaching scheme, without special emphasis on familiar contexts or a systemic perspective. The students performed the tasks individually and, although their initial ideas were not considered, they did hold small discussions to solve the exercises within the entire group. The textbook was the curricular material, used along the three distinguished stages:

- In the first stage, the teacher explained the structure and organization of DNA, and the concepts of the gene and allele were defined, as well as their implication in the inheritance of genetic information. Next, using lactase synthesis as an example, the processes of transcription, translation, and replication of genetic material were described, paying attention to how RNA and ribosomes are involved.
- In the second stage, the students carried out classic inheritance exercises proposed in the book, after the theoretical content, into which other specific ones on lactose intolerance were incorporated. Among other aspects, they addressed dominance/recessivity relationships between alleles for the inheritance of this character.
- In the third stage, the Human Genome Project was presented to the students, as an international effort that shows the importance of sequencing DNA and mapping genes. In this context, the identification of the lactase gene and its importance for humans were described.

In the educational intervention aimed at the EG, the students, organized in five small groups, carried out descriptive and extrapolative tasks to solve a problem (Fig. 11.1). This was aimed at providing them with opportunities to contrast the sufficiency of their ideas, to discuss them, to generate new knowledge, and to be able to make a critical appraisal of messages about intolerances, especially with regard to food advertising. This proposal was also organized in three stages:

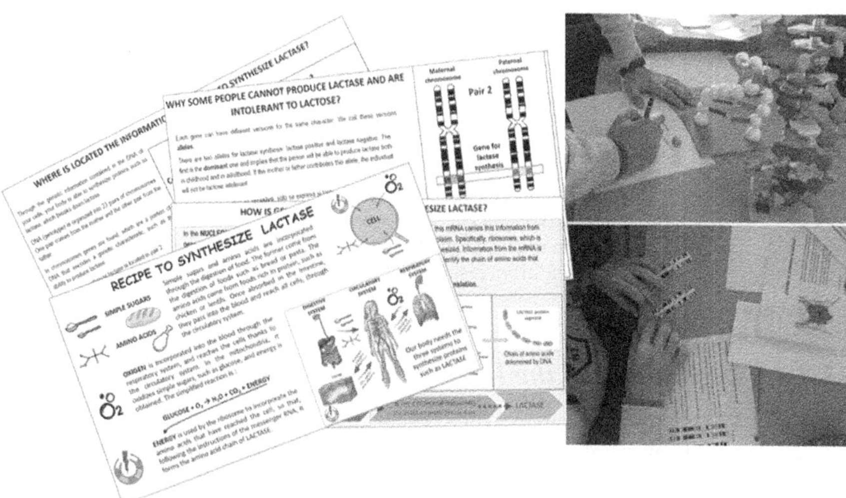


**Fig. 11.1** Tasks for the EG (left). Students from the EG working on the proposal (right)

- The first stage corresponded to the identification of the problem, which arose from the discussion among non-identical twins of why lactose intolerance can affect one twin and not the other. In addition, they considered whether the intake of lactose-free products by the whole family would improve the digestion for all the members. The objective of this stage was that the students share and organize their ideas about the causes of this intolerance and its possible inheritance, to reach an agreement on the key questions to investigate.
- The second stage, aimed at generating new knowledge, was organized to answer the four key questions established:

  - What is lactose and why is it important to my diet? This question made it possible to relate the intolerance with the inability to synthesize lactase. On this issue, the students wrote small reports about lactose, its nutritional interest, and the side effects of its ingestion by intolerant people.
  - Where is the information required for the synthesis of lactase located? This question was aimed at linking lactase synthesis with genetic information. This meant that the students had to address the organization of genetic information in the cell: DNA, chromosomes, and genes. Through diagrams and models, the students located the lactase gene on chromosome pair 2, giving meaning to the study of this organization. In addition, several diagrams were available that emphasized the presence of the same genetic information in all cells, although the expression of the lactase gene is associated with enterocytes.
  - Why are some people unable to produce lactase and are thus lactose-intolerant? From this question, the students had to address the concept of the allele. They were provided with the information about the alleles of the lactase gene of the twins, and had to determine which one of the pair of chromosomes 2 each of them had inherited. In the process, the students characterized the twins as homozygous or heterozygous individuals and may have recognized dominance/recessivity relationships between the alleles. In this way, the students could have associated observable phenomena, such as the side effects suffered by the intolerant sibling, with their explanation at the genetic (micro) level.
  - How does our body use the information for lactase synthesis and what is needed for this? With access to diagrams and models as well as brief interviews with specialists, the aim was to allow students to construct explanations about the entire process. Based on information regarding DNA in the cell nucleus, RNA, and ribosomes, three basic elements were considered: amino acids, $O_2$, and energy. Thus, the students addressed transcription, translation, and protein synthesis as a continuous process connected to nutrition, making explicit the relationships among the digestive, circulatory, and respiratory systems at the cellular level.

- The last stage was aimed at obtaining conclusions. Each group organized the knowledge acquired and completed a final report, in which they had to explain why only one sibling is intolerant and the processes – at the organic and cellular levels – that occur in each sibling after ingesting lactose. Finally, they structured their learning, comparing their initial and final ideas.

The ethics of the intervention and data collection included written formal consent and agreement to use encoded and anonymized data for further scientific purposes.

## 11.5 Results and Discussion

This section is organized according to the contents selected to address the relationship between protein synthesis and genetic information (Table 11.1), comparing the results obtained in the pretest and posttest and in the two groups (Fig. 11.2).

### *11.5.1 Locating the Genetic Information Necessary to Synthesize Proteins*

In the pretest, the ideas in the EG and CG were very similar to each other, without statistically significant differences, so they are described together. It is noteworthy that a total of 28 students of both groups located the genetic information in the gametes (I1) and a third in the sex chromosomes (I2), in line with the findings of Banet and Ayuso (2000). More specifically, they placed it in cells of the digestive tract, since it is a food intolerance (I3). Unlike in previous studies (Caballero, 2008), it seems that the idea that the genetic information present in each cell depends on its functionality did not dominate among our students. Less than one-fifth of the students in each group thought that the genetic information is located in all cells (I4).

In the posttest, an improvement was observed in both groups, decreasing the number of students who selected items I1 and I2. In other words, the frequency of students who considered that genetic information is found in gametes (I1) or chromosomes (I2) was reduced significantly in both groups (I1: $p = .006$ and $p = .005$; I2: $p = .014$ and $p = .037$, for EG and CG, respectively). The proportion of students who linked this information with cell function (I3) also fell, without statistically significant differences for any of the groups.

The number of students who accepted that genetic information is present in all cells (I4) increased significantly in both groups in the posttest, in comparison to the pretest ($p = .027$ and $p = .041$, for the EG and CG, respectively). However, this progression was not equivalent. In the case of the EG, 64% of the students accepted this idea, compared to 36% in the CG, representing a significant difference between the two groups ($p = .042$).

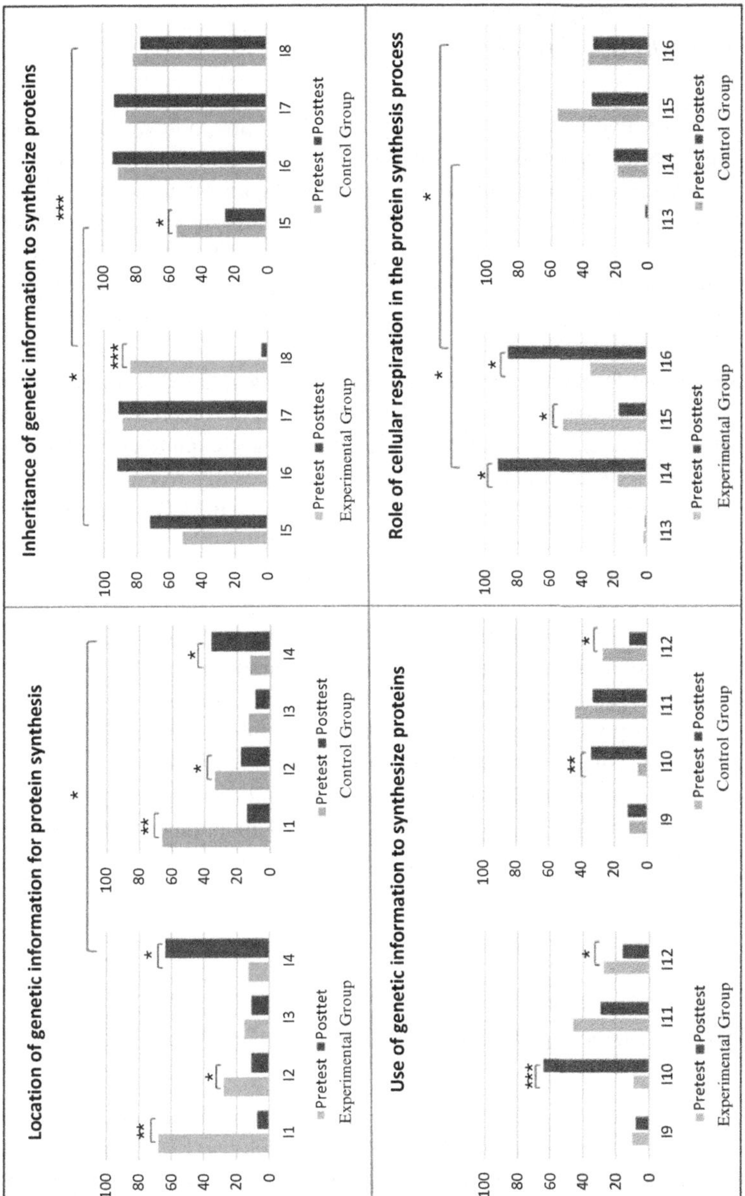


**Fig. 11.2** Percentage of the answers and statistical significance to each content

### 11.5.2 Inheritance of Genetic Information to Synthesize Proteins

In the pretest, more than half of the students of both groups indicated that genes provide the information to synthesize proteins (I5). Furthermore, more than four-fifths of the participants in both groups recognized that the information determining lactose (in)tolerance must be inherited (I6, I7). Nonetheless, it seems that for the students it was not clear how the genotypes of the parents determine the ability of their children to synthesize lactase (I8), since only 8 of them responded adequately to this item in both groups. Thus, the vast majority considered that children without intolerance can have intolerant parents.

In the posttest, the frequency of EG students who indicated that it is the genes which carry the information needed to synthesize proteins (I5) increased, although without significant differences from the pretest. In the CG, the number of students who selected I5 fell by more than one-half, giving significantly worse results than in the pretest ($p = .036$). Furthermore, the difference between the EG and the CG for I5 was also significant ($p = .039$).

Regarding the need to inherit genetic information to be able to synthesize lactase or not (I6, I7), the EG and CG exhibited similar percentages in the pretest and post-test, so no significant differences were found between the groups or between the tests. Concerning how the genotypes of the parents condition the ability of their children to synthesize proteins such as lactase (I8), in the EG almost all the students indicated that it is not possible for a child of intolerant parents to inherit the ability to synthesize lactase. This was a statistically significant progression for this group ($p <.001$), which suggests that allele conceptualization and understanding of allelic dominance/recessivity relationships improved. The CG maintained a percentage similar to that of the initial test for I8, so there were still many students who considered that intolerant parents can have tolerant offspring. This led to a significant difference between the EG and the CG for I8 ($p <.001$).

### 11.5.3 Use of Genetic Information to Synthesize Proteins

In the pretest, in both groups, there were few students who pointed out that DNA carries information to the ribosome for the synthesis of proteins (I9), and even fewer indicated that it is RNA (I10). In addition, 17 students of EG and CG thought that the ribosome uses information from DNA that is transformed into RNA due to a mutation (I11). This reflects a misunderstanding about the participation of DNA and RNA in protein synthesis. Furthermore, more than a quarter of the students, in both groups, considered that amino acids provide the information used by the ribosome (I12), which indicates that they established a certain link between these molecules and proteins.

In the posttest, the proportion of students who indicated that it is the DNA that carries the information to the ribosome (I9) remained low in EG and CG, so there were no differences (neither pretest-posttest nor between the two groups). Regarding the participation of RNA (I10), a significant improvement occurred in both groups in relation to the identification of the role of this molecule in protein synthesis ($p = .004$ and $p = .008$, for EG and CG, respectively), although the frequency was much higher in EG. However, around a third of the EG and CG students stuck to the idea of mutation as a process to transform DNA into RNA (I11). This does not necessarily imply that they ignored the transcription process, but it seems that they considered it the consequence of a mutation. This result seems to correspond to the fact that students tend to associate the idea of mutation with that of change, without recognizing that it must involve modifications in genetic information (Albaladejo & Lucas, 1988). Finally, regarding amino acids (I12), there was an equivalent decrease in both groups in the frequency of students who considered that these molecules carry information to the ribosome; so, there was a significant change for both groups ($p = .021$ and $p = .033$, for EG and CG, respectively), but no significant difference between them.

### *11.5.4 Role of Cellular Respiration in Protein Synthesis*

In the pretest, more than half of the EG and CG students recognized the need for energy in the protein synthesis process (I15), and just over a third linked it to nutrition (I16). However, there was a marked tendency to ignore the intervention of the mitochondria (I14), since only around 10% in both groups indicated their role. This result suggests that the idea that nutrition directly provides energy dominates, rather than the idea that the cell has to generate it through cellular respiration.

In the posttest, greater recognition of the role of nutrition in the protein synthesis process was observed in the EG. On the one hand, the frequency of students who considered that energy enters the cell directly (I15) was reduced to less than a half; whereas, a vast majority managed to identify the substances involved in cellular respiration (I16) and the role of mitochondria (I14). So, in the EG there were significant pretest-posttest differences ($p = .039$; $p = .026$; $p = .043$) for I14, I15, and I16. On the other hand, the proportion in all items remained similar to the pretest in the CG, without significant differences between the two tests. The differences between the EG and the CG were statistically significant for I14 and I16 ($p = .017$; $p = .023$).

## 11.6 Conclusions

We are aware that the limitations in the number of participants and the time of the interventions do not allow us to generalize the results. Besides, the use of tests that include multiple-choice answers may have represented a difficulty for students to solve them; thereby, future research should avoid this type of questioning.

As a tentative conclusion, the results seem to show the relevance of teaching protein synthesis in relation to problems present in the daily life of students. This allows the cellular processes that occur at the molecular level to be linked with their consequences at the macroscopic level, so that certain learning difficulties associated with these contents can be overcome (Gerçek, 2018). In both groups, the students' ideas before receiving the interventions were very similar. In the posttest, certain improvements were observed in both groups, although they were not equivalent.

Both educational proposals seem to be adequate to improve students' understanding of descriptive aspects, such as the location of genetic information. Whereas, for those aspects which require greater cognitive skills, such as achieving a systemic vision and relating nutrition and protein synthesis, the designed intervention seems more effective. In addition, the use of a daily context such as lactose intolerance and its inheritance would have helped students to make sense of the organization and function of genetic information from observable and everyday phenomena, as seen in the better identification of the role of alleles in the determination of the ability to synthesize proteins of the EG. Whereas among the CG students, even after receiving specific training, there remained some confusion about the role of genes, in line with the results of Lewis and Wood-Robinson (2000). It seems that the traditional approach does not help students to clarify the functionality of genes as carriers of genetic information. Furthermore, future interventions should emphasize the influence of environmental factors in these processes.

Finally, difficulties regarding the transcription of DNA into RNA still remain, which may correspond to its level of abstraction (Cimer, 2012). Therefore, greater attention to this process is necessary, including more specific tasks during the intervention.

## References


Albaladejo, C., & Lucas, A. M. (1988). Pupils' meanings for «mutation». *Journal of Biological Education, 22*(3), 215–219.

Alozie, N., Eklund, J., Rogat, A., & Krajcik, J. (2010). Genetics in the 21st century: The benefits & challenges of incorporating a project-based genetics unit in biology classrooms. *The American Biology Teacher, 72*(4), 225–230.

Assaraf, O. B. Z., Dodick, J., & Tripto, J. (2013). High school students' understanding of the human body system. *Research in Science Education, 43*(1), 33–56.

Ayuso, G. E., & Banet, E. (2002). Alternativas a la enseñanza de la Genética en Educación Secundaria. *Enseñanza de las Ciencias, 20*(1), 133–157.

Aznar, V., & Puig, B. (2014). ¿Cómo se presentan las enfermedades infecciosas en los libros de texto? *Revista Eureka sobre Enseñanza y Divulgación de las Ciencias, 11*(2), 135–144.

Bahar, M., Johnstone, A. H., & Hansell, M. H. (1999). Revisiting learning difficulties in biology. *Journal of Biological Education, 33*(2), 84–86.

Banet, E., & Ayuso, G. E. (1995). Introducción a la genética en la enseñanza secundaria y bachillerato: I. Contenidos de enseñanza y conocimientos de los alumnos. *Enseñanza de las Ciencias: Revista de investigación y experiencias didácticas, 13*, 137–153.

Banet, E., & Ayuso, G. E. (2000). Teaching genetics at secondary school: A strategy for teaching about the location of Inheritance information. *Science Education, 84*(3), 313–351.

Caballero, A. M. (2008). Algunas ideas del alumnado de secundaria sobre conceptos básicos de Genética. *Enseñanza de las ciencias, 26*(2), 227–244.

Cimer, A. (2012). What makes biology learning difficult and effective: Students' views. *Educational Research and Reviews, 7*(3), 61.

Dimopoulos, K., & Koulaidis, V. (2002). The socio-epistemic constitutions of science and technology in the Greek press: An analysis of its presentations. *Public Understanding of Science, 15*, 5–29.

Duncan, R. G., & Reiser, B. J. (2007). Reasoning across ontologically distinct levels: Students' understandings of molecular genetics. *Journal of Research in Science Teaching, 44*(7), 938–959.

Eijck, M. (2010). Addressing the dynamics of science in curricular reform for scientific literacy: The case of genomics. *International Journal of Science Education, 32*, 2429–2449.

Esteve, P., Banos-González, I., & Moreno, P. P. (2017). ¿Qué pasa a nivel microscópico? *Alambique: Didáctica de las ciencias experimentales, 88*, 27–33.

Flores, F., Tovar, M. E., & Gallegos, L. J. (2003). Representation of the cell and its processes in high school students: An integrated view. *International Journal of Science Education, 25*, 269–286.

Frieder, Y., Amir, R., & Tamir, P. (1987). High school students' difficulties in understanding osmosis. *International Journal of Science Education, 9*, 541–551.

García, A. M. (1991). Estudio llevado a cabo sobre representaciones de la respiración celular en los alumnos de Bachillerato y COU. *Enseñanza de las Ciencias, 9*(2), 129–134.

Gerçek, C. (2018). Prospective teachers' cognitive structures concerning protein synthesis and their degree of understanding. *Journal of Baltic Science Education, 18*, 19–30.

Hmelo-Silver, C. E., & Azevedo, R. (2006). Understanding complex systems: Some core challenges. *The Journal of the Learning Sciences, 15*(1), 53–61.

Íñiguez, F. J., & Puigcerver, M. (2013). Una propuesta didáctica para la enseñanza de la genética en la Educación Secundaria. *Revista Eureka sobre Enseñanza y Divulgación de las Ciencias, 10*(3), 307–327.

Jiménez Aleixandre, M. P. (2010). *10 Ideas Clave. Competencias en argumentación y uso de pruebas*. Editorial Graó.

Lewis, J., & Wood-Robinson, C. (2000). Genes, chromosomes, cell division and inheritance-do students see any relationship? *International Journal of Science Education, 22*(2), 177–195.

Rodríguez Gil, S. G., Fradkin, M., & Castañeda-Sortibrán, A. N. (2018). Conceptions of meiosis: Misunderstandings among university students and errors. *Journal of Biological Education, 53*(2), 191–204.

Tsui, C. Y., & Treagust, D. (2010). Evaluating secondary students' scientific reasoning in genetics using a two-tier diagnostic instrument. *International Journal of Science Education, 32*, 1073–1089.

Ugidos-Rodríguez, S., Matallana-González, M. C., & Sánchez-Mata, M. C. (2018). Lactose malabsorption and intolerance: A review. *Food & Function, 9*(8), 4056–4068.

Vlckova, J., & Kubiatko, M. (2018). Perception of genetics by using of semantic differential at high school students: Preliminary results. *Eurasia Journal of Mathematics, Science and Technology Education, 14*(1), 311–322.

Wood-Robinson, C., Lewis, J., & Leach, J. (2000). Young people's understanding of the nature of genetic information in the cells of an organism. *Journal of Biological Education, 35*(1), 29–36.

Zelenika, I., Moreau, T., Lane, O., & Zhao, J. (2018). Sustainability education in a botanical garden promotes environmental knowledge, attitudes and willingness to act. *Environmental Education Research, 24*(11), 1581–1596.

# Chapter 12
# Evaluating the Effectiveness of a Teaching Intervention in a Marine Biology Course: The Case of Greek Vocational Students

**Athanasios Mogias, Eleftheria Peskelidou, and Theodora Boubonari**

## 12.1 Introduction

The ocean covers 71% of our planet and holds 97% of the Earth's water. It supports life on Earth and provides us with tremendous economic, social, and environmental benefits (Costanza, 1999). Moreover, the ocean has intrinsic value for its own sake and its inhabitants. However, human activities continue to threaten the health and integrity of the marine environment (e.g. Derraik, 2002; Lotze et al., 2006; Worm et al., 2006). Since this fact can be attributed to the lifestyles, decision-making, and choices of individuals, as well as governments and industry, among others, the involvement of each and every person in understanding the importance of the ocean and the need to protect it is essential. Therefore, to ensure sustainable use of ocean resources, there is a need for individual responsible behaviour by ocean-literate citizens. Ocean-literate citizens have some level of knowledge on ocean sciences topics, understand how attitudes and values impinge upon a topic and are empowered to take action around the topic (Strang et al., 2007).

The need for ocean literate people is fully aligned with the recently introduced Education 2030 Agenda by UNESCO, a framework providing guidance for the implementation of seventeen Sustainable Development Goals (SDGs); SDG 14, "Life below Water", is among these goals and concerns the sustainable development of oceans, seas, and marine resources, highlights cognitive, socio-emotional, and behavioural learning objectives concerning the oceans, and notes the increase in student demand for a sustainability-centered education as a significant driver for changes in curriculum and teaching practice (UNESCO, 2017). To serve this need,

A. Mogias (✉) · E. Peskelidou · T. Boubonari
Department of Primary Education, Democritus University of Thrace, Alexandroupolis, Greece
e-mail: amogias@eled.duth.gr; elevpesk@eled.duth.gr; tmpoumpo@eled.duth.gr



K. Korfiatis, M. Grace (eds.), *Current Research in Biology Education*, Contributions from Biology Education Research,
https://doi.org/10.1007/978-3-030-89480-1_12

ocean literacy should be integrated into educational practice, research, curricula, textbooks, and assessments (Tran et al., 2010).

Promotion of ocean literacy in elementary and secondary education is vital since children represent the future citizens and consumers, who will make decisions that will inevitably affect the environment (Visbeck, 2018). Youth, apart from performing responsible environmental behaviours themselves directly, can also influence the environmental knowledge, attitudes, and behaviours of peers, family, and of the wider community (Hartley et al., 2015).

However, literature reviews reveal that knowledge on ocean sciences issues is inadequate (e.g. Boubonari et al., 2013; Chen & Tsai, 2015; Gelcich et al., 2014; Guest et al., 2015; Mogias et al., 2015; The Ocean Project, 2009). Specifically, published research on elementary and secondary school students revealed low levels of knowledge of ocean issues (e.g. Ballantyne, 2004; Brody, 1996; Brody & Koch, 1986; Guest et al., 2015; Lambert, 2006; Plankis & Marrero, 2010; Rodriguez-Martinez & Ortiz, 1999). This students' lack of familiarity with the ocean could be attributed to the fact that ocean concepts are rarely represented in the formal science education curriculum (e.g. Gough, 2017; Lal, 2017; McPherson et al., 2018; Mogias et al., 2019).

Therefore, research and assessment of teaching interventions concerning ocean literacy could help fill this gap. The literature also reveals that there are some learning activities, school programmes, and teaching interventions concerning ocean literacy, which underline students' knowledge improvement, especially after first-hand experiences on ocean-related topics (e.g. Cummins & Snivelly, 2000; Fortner & Teates, 1980; Hartley et al., 2015, 2018; Lambert, 2005, 2006; Stepath, 2007; Plankis & Marrero, 2010; Winks et al., 2020). Some of these studies discuss how specifically ICT tools and social media could probably enhance ocean literacy (Brennan et al., 2019; Fauville et al., 2019).

## 12.2 Aims of the Study

In this context, the aim of the present study is to (a) estimate vocational students' knowledge on a specific principle of ocean literacy before and after a teaching intervention, and (b) examine whether and how the teaching intervention actually helped the students. The results of this study will help researchers improve the teaching intervention and continue the study with a larger sample. This effort cumulatively will contribute to the design and implementation of programmes concerning principles of ocean literacy, and these could be integrated into the educational practice.

## 12.3 The Ocean Literacy Framework

Ocean literacy was launched at the beginning of this century. Although the interest in marine issues in education was much older, marine education became marginalized and often presented in a very local context (Strang, 2008). When the US

National Science Education Standards were published in 1996, various ocean scientists and marine educators realized that there was little mention of ocean topics. Furthermore, state standards did not include much about the ocean, coasts, or watersheds (Schoedinger et al., 2010). All the above have led to a new attempt for a rebirth of this educational area.

The ocean literacy movement was born in about 2004 in the U.S. To address the need for communication and a way to build community consensus on ocean literacy, a diverse group of representatives, with expertise in the ocean sciences, ocean education, and/or education policy, joined in a series of workshops and conferences to draft a common framework. This framework would define the meaning of ocean literacy and develop principles about a desirable state of knowledge for the ocean. The result of this extensive process was two documents comprising the Ocean Literacy Framework: (a) the Essential Principles and Fundamental Concepts of Ocean Sciences (National Oceanic and Atmospheric Administration, 2013), and (b) The Ocean Literacy Scope and Sequence (National Marine Educators Association, 2010).

The guide (NOAA, 2013) defines ocean literacy as "the understanding of the ocean's influence on you and your influence on the ocean", and identifies the seven essential ocean literacy principles, which all students should understand by the end of high school as follows:

1. The Earth has one big ocean with many features.
2. The ocean and life in the ocean shape the features of Earth.
3. The ocean is a major influence on weather and climate.
4. The ocean made Earth habitable.
5. The ocean supports a great diversity of life and ecosystems.
6. The ocean and humans are inextricably interconnected.
7. The ocean is largely unexplored.

Each of the principles is underpinned by a series of 45 fundamental concepts (NOAA, 2013), which support and explain them. According to the guide, an ocean literate person understands the essential principles about the ocean, can communicate about the ocean in a meaningful way and is able to make informed and responsible decisions regarding the ocean and its resources (NOAA, 2013).

The Ocean Literacy Scope and Sequence (NMEA, 2010) provides information and guidance as to what students need to comprehend in Grades K-2, 3–5, 6–8, and 9–12, in order to achieve a full understanding of these principles. These progressions show how students' thinking about the ocean may develop in ever more complex ways across many years of thoughtful, coherent science instruction. The Scope and Sequence, represented in a series of conceptual flow diagrams that include cross-references, also shows how concepts about the ocean are interconnected (NMEA, 2010). These guidelines, developed to help implement an ocean-dedicated curriculum in the USA, are now largely accepted and have been an inspiration for several initiatives worldwide (Fauville et al., 2019).

The present study focuses on Ocean Literacy Principle 5 (OLP5). This principle refers to marine biology issues, such as marine organisms, all marine ecosystems found in the ocean, the relationships among them, and which and how physicochemical factors influence them (Table 12.1).

**Table 12.1** The fifth principle of the Ocean Literacy Framework and its fundamental concepts. Concepts are given concisely

| Principle | Concepts |
|---|---|
| **5. The ocean supports a great diversity of life and ecosystems** | **5a.** ocean life ranges in size |
| | **5b.** microbes the most important primary producers |
| | **5c.** most major groups of organisms in ocean |
| | **5d.** important relationships among organisms |
| | **5e.** most of the living space in the ocean, unique ecosystems |
| | **5f.** ocean life not evenly distributed due to abiotic factors |
| | **5g.** deep ecosystems independent of sunlight |
| | **5h.** vertical zonation pattern along the coast and in the open ocean |
| | **5i.** estuaries |

## 12.4 Methodology

### *12.4.1 The Participants*

The current study was conducted in December 2019 with a convenience sample of 24 first-grade (15–16 year-old) male students of a public vocational senior high school in a provincial town in Northern Greece. The three-year course in Vocational High Schools offers general education knowledge, provides vocational training as a strong basis for future new professional occupations, and keeps pace with the advancement of technology and the constant changes in know-how. The first grade includes general education courses, as well as courses of related professional fields, in order to provide complete knowledge of the subjects of each field, while the other two grades are structured according to professional fields and specialties. The students of this study chose the field of mechanical engineering, which was the reason for the male gender dominance in the present study. The intervention took place within the course entitled "Research Work in Technology". The object of this course is to teach students the methodology of research and technologies.

Prior to the beginning of the intervention, the participants were informed about the purpose of the study, the voluntary basis of participation and assurance about anonymity, and the official approval of the school's principal was ensured. Therefore, the access authorization, the organizing of the field visits, and the engagement with the participants, played a key role in the selection of the sample. Their parents had not undertaken formal education beyond secondary school level. Most of the students (70.8%) had not attended a lesson or participated in a project concerning environmental education; only one student was a member of an environmental organization.

## 12.4.2 The Questionnaire

A structured questionnaire was administered to the students, in order to investigate their level of knowledge related to marine biology issues prior to and after the teaching intervention. The questionnaire contained a set of demographics, which concerned their parents' education and the source of information they mostly use for environmental issues, as well as 11 multiple choice questions (see Table 12.2) with five optional answers (including 'I don't know') per question concerning all fundamental concepts of OLP 5. The items of the research tool were borrowed from psychometrically valid and reliable questionnaires concerning ocean literacy (Fauville et al. 2019; Markos et al., 2017).

**Table 12.2** Relative frequencies of correct answers, and *p*-values per question prior and after the intervention

| | Percentage of correct answers | | |
|---|---|---|---|
| Questions/Statements | Pre-test (%) | Post-test (%) | *p*-value |
| 1. Ocean life ranges in size from the smallest living things, microbes, to the largest animal on Earth, **blue whales** | 33.3 | 100.0 | 0.000 |
| 2. Most of the organisms and biomass in the ocean are **microbes** | 37.5 | 66.7 | 0.050 |
| 3. Most of the major groups that exist on Earth are found exclusively **in the ocean** | 45.8 | 100.0 | 0.000 |
| 4. The ocean hosts **a great diversity of life** | 62.5 | 87.5 | ns |
| 5. The ocean provides a vast living space **from the surface through the water column, and down to, and below, the seafloor** | 58.3 | 70.8 | ns |
| 6. Ocean ecosystems are defined by environmental factors such as **oxygen, salinity, temperature, pH, light, nutrients, pressure, substrate, and circulation** | 8.3 | 20.8 | ns |
| 7. There are deep ocean ecosystems, independent of energy from sunlight, **supporting limited life** | 20.8 | 54.2 | 0.017 |
| 8. There are deep ocean ecosystems, independent of energy from sunlight, such as **hydrothermal vents, submarine hot springs, and methane cold seeps** | 12.5 | 25.0 | ns |
| 9. There are deep ocean ecosystems, independent of energy from sunlight, **relying only on chemical energy and chemosynthetic organisms** | 16.7 | 66.7 | 0.000 |
| 10. Factors causing vertical zonation patterns along the coast are **tides, waves, predation, substrate** among others | 12.5 | 66.7 | 0.000 |
| 11. Marine habitats providing important and productive nursery areas for many marine and aquatic species are called **estuaries** | 25.0 | 83.3 | 0.000 |
| **Mean** | **30.3** | **67.4** | **0.000** |

*ns* non-significant, *p*-level >0.05

### 12.4.3 The Intervention

A 6-h teaching intervention concerning concepts of OLP 5 was developed for the age group of the present study based on the Ocean Literacy Guide and the Scope and Sequence (NMEA, 2010; NOAA, 2013). The answers of the participants in the pre-test questionnaire about their most preferred source of information on environmental issues, as well as the teaching suggestions provided in the "One Ocean" (Mohan, 2013), a guide for teaching about the Ocean, were taken into account while structuring the intervention. One of the authors, who worked as a teacher at the senior high school of the experimental group and already knew the students, implemented the intervention.

The students worked in teams on case studies, e.g. the case study of an ecosystem, such as a coral reef, an estuary, or a deep ocean ecosystem. Case studies offer an in-depth look at how ocean concepts play out in particular locations or situations. Moreover, they provide real-life examples of how biotic and abiotic factors, as well as relationships among organisms affect natural communities. Videos and websites concerning the concepts of OLP 5 (e.g. marine organisms and ecosystems) were valuable resources to work with. These resources were sought and used by the groups with the support of the teacher. Additionally, each group of students used satellite technology to track and map migrations of ocean animals, e.g. beluga whale, and other internet resources so as to infer information about their life cycle, habitats, their diet, and reproduction. All these resources were used by each group to obtain data and answer specific questions concerning a marine ecosystem or an organism. Concept maps constructed with Cmap tools were used to summarize and synthesize all the information each group had managed to collate for its case study. Each group shared these maps and what they had learned about their topic with the rest of the groups during short feedback sessions.

### 12.4.4 Data Analysis

Descriptive and non-parametric inferential statistics (Wilcoxon T) were applied with the use of the Statistical Package for Social Sciences (SPSS v. 21); in terms of the latter case, results should be interpreted conservatively due to the small sample. The significance level was predetermined at a probability value of 0.05 or less.

## 12.5 Results

The respondents were found to possess a low level of marine biology knowledge before the teaching intervention (mean score 3.33 ± 1.523), while this was significantly increased immediately after the intervention (mean score 7.42 ± 1.717)

(Fig. 12.1). Each correct answer received a value of 1 and incorrect a value of 0; therefore, the score could vary between 0 and 11, as this was the total number of questions. Table 12.2 portrays the percentages of correct answers. Most answers showed a statistically significant difference between the two tests. As regards the source of information that students mostly use for environmental issues, the internet scored highest (mean ± SD: 3.63 ± 1.345), followed by formal education (2.33 ± 1.007), while books had the lowest score (1.58 ± 0.584). Their parents' generally did not study formally beyond secondary school and the effect of parents' education on students' knowledge was not significant at the 5% level.

## 12.6 Discussion

In the present study, according to the findings, the teaching intervention implemented with vocational high school students with regard to marine biology issues significantly increased their knowledge level on these issues, although they were found to be very low achievers before the intervention. Our findings seem to be in accordance with studies from other countries concerning secondary students' knowledge on marine biology issues (e.g. Brody, 1996; Brody & Koch, 1986; Greely, 2008; Guest et al., 2015; Hartley et al., 2018; Plankis & Marrero, 2010).

Participants' initial low level of knowledge in this study could be attributed to the fact that ocean sciences do not constitute a basic part of the Greek educational system (Mogias et al., 2019). Particularly, references on marine biology issues exist in the school textbooks in Greek secondary education, but these issues are only patchily and superficially touched on. In addition, there are no suitable educational

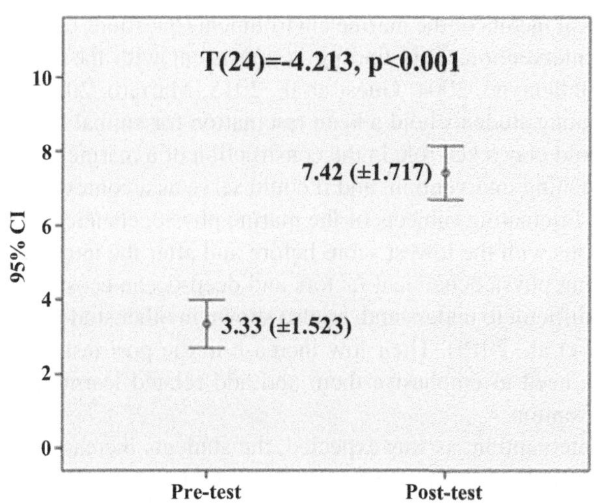


**Fig. 12.1** Mean scores at the pre- and post-tests

materials to work further on these subjects. Moreover, teachers and students participate voluntarily in environmental education programmes, which could serve as a source of knowledge. However, not all schools implement environmental education programmes, and not all students take part in them. Most importantly, only a few of these programmes are related to marine biology issues. Beyond these, teachers can hardly enrich their teaching with additional knowledge, because of the strict context of the formal curriculum (Liarakou et al., 2009).

The inadequate implementation of ocean science topics in secondary school curricula is supported by the finding that the respondents appear to derive little information from formal education about these issues. On the other hand, they declare that they acquire information on these issues mostly from the internet. Our respondents' preference for the internet as a source of information for marine biology issues, suggests that this media could be an effective way to educate them about the marine environment. Therefore, it should be utilized in a teaching intervention. Along the same lines, the internet is considered as an immediate, available anytime, medium for the access of environmental information, which delivers the exact piece of current information required faster and easier than other forms. It also has the potential to provide more details about topics that other media cannot (Haklay, 2002). However, it is uncertain whether all of the ocean-related literature available on the internet is trustworthy. It has been argued that misconceptions may arise from misinterpretation of information since children's ideas acquired mainly through the media are not properly tested (Boyes & Stanistreet, 1997; Çakır et al., 2010). Therefore, students need to be taught how to deal with this source of information in a constructive manner. Moreover, appropriate and reliable links concerning marine biology issues should also be provided.

Although the average score of the pre-test was low, there was a considerable variation among questions. Species-related or biological questions (questions 1, 3, 4, 11) were generally scored higher than abiotic-related questions concerning the physicochemical factors of the marine environment (questions 6, 8) before and after the teaching intervention. This finding is consistent with the results of previous studies (e.g. Ballantyne, 2004; Guest et al., 2015; Marrero, 2010), and underlines the fact that young students hold a keen fascination for animal life. This interest in marine life could play a key role in the construction of a marine environmental programme or teaching intervention, and it could serve as a context to teach more difficult and less fascinating subjects of the marine physicochemical sciences.

The questions with the lowest score before and after the intervention (questions 6, 8) concerning physicochemical factors and deep-ocean ecosystems are subjects which can be difficult to understand, as also shown in other studies (e.g. Guest et al., 2015; Mogias et al., 2015). Their low increase in the post-test achievement score underlines the need to emphasise them and add related learning activities in the teaching intervention.

After the intervention, as was expected, the students increased their knowledge of marine biology issues. The effect sizes were high, which might be the result of their low prior knowledge. This also indicates that the teaching intervention was effective enough to make a significant difference between the pre- and post-assessments and to develop participants' knowledge.

## 12.7 Conclusions

In light of the findings of the current study, we conclude that:

1. Students' pre-intervention knowledge was poor, but the much improved post-intervention scores show that they have the potential to make significant learning gains in this area.
2. Students struggle with concepts concerning physicochemical factors and deep-ocean ecosystems and these topics need particular support.
3. The fascination students show with marine life, as well as their preference for the internet as a basic source of information on marine biology issues, should be thoroughly considered as an appropriate context and media respectively to enhance their knowledge on these issues and generally improve their ocean literacy.

## References


Ballantyne, R. (2004). Young students' conceptions of the marine environment and their role in the development of aquaria exhibits. *GeoJournal, 60*, 159–163.

Boubonari, T., Markos, A., & Kevrekidis, T. (2013). Greek pre-service teachers' knowledge, attitudes, and environmental behavior toward marine pollution. *Journal of Environmental Education, 44*, 232–251.

Boyes, E., & Stanistreet, M. (1997). The environmental impact of cars: Children's ideas and reasoning. *Environmental Education Research, 3*, 269–282.

Brennan, C., Ashley, M., & Molloy, O. (2019). A system dynamics approach to increasing ocean literacy. *Frontiers in Marine Science, 6*, 1–20.

Brody, M. (1996). An assessment of 4th-, 8th-, and 11th-grade students' environmental science knowledge related to Oregon's marine resources. *Journal of Environmental Education, 27*, 21–27.

Brody, M., & Koch, H. (1986). *An assessment of 4th-, 8th-, and 11th-grade students' knowledge related to marine sciences and natural resource issues*. University of Maine, College of Education, ED273502.

Çakır, M., İrez, S., & Kıvılcan, D. Ö. (2010). Understandings of current environmental issues: Turkish case study in six teacher education colleges. *Educational Studies, 36*, 21–33.

Chen, C. L., & Tsai, C. H. (2015). Marine environmental awareness among university students in Taiwan: A potential signal for sustainability of the oceans. *Environmental Education Research, 22*, 958–977.

Costanza, R. (1999). The ecological, economic, and social importance of the oceans. *Ecological Economics, 31*, 199–213.

Cummins, S., & Snivelly, G. (2000). The effect of instruction on children's knowledge of marine ecology, attitudes toward the ocean, and stances toward marine resource issues. *Canadian Journal of Environmental Education, 5*, 305–324.

Derraik, J. G. (2002). The pollution of the marine environment by plastic debris: A review. *Marine Pollution Bulletin, 44*, 842–852.

Fauville, G., Strang, C., Cannady, M., & Chen, Y. F. (2019). Development of the international ocean literacy survey: Measuring knowledge across the world. *Environmental Education Research, 25*, 238–263.

Fortner, R. W., & Teates, T. G. (1980). Baseline studies for marine education: Experiences related to marine knowledge and attitudes. *Journal of Environmental Education, 11*, 11–19.

Gelcich, S., Buckley, P., Pinnegar, J. K., Chilvers, J., Lorenzoni, I., Terry, G., Guerrero, M., Castilla, J. C., Valdebenito, A., & Duarte, C. M. (2014). Public awareness, concerns, and priorities about anthropogenic impacts on marine environments. *Proceedings of the National Academy of Sciences of the United States of America, 111*, 15042–15047.

Gough, A. (2017). Educating for the marine environment: Challenges for schools and scientists. *Marine Pollution Bulletin, 124*, 633–638.

Greely, T. (2008). *Ocean literacy and reasoning about ocean issues: The influence of content, experience and morality*. PhD dissertation, University of South Florida, USA.

Guest, H., Lotze, H. K., & Wallace, D. (2015). Youth and the sea: Ocean literacy in Nova Scotia, Canada. *Marine Policy, 58*, 98–107.

Haklay, M. (2002). Public environmental information – Understanding requirements and patterns of likely public use. *Area, 34*, 17–28.

Hartley, B., Thompson, R. C., & Pahl, S. (2015). Marine litter education boosts children's understanding and self-reported actions. *Marine Pollution Bulletin, 90*, 209–217.

Hartley, B., Pahl, S., Holland, M., Alampei, I., Veiga, J. M., & Thompson, R. C. (2018). Turning the tide on trash: Empowering European educators and school students to tackle marine litter. *Marine Policy, 96*, 227–234.

Lal, N. (2017). Oceans and rivers literacy in Fiji's Social Science Curriculum: An analysis of Primary school textbooks. *Pacific Journal of Education, 1*, 47–56.

Lambert, J. (2005). Students' conceptual understandings of science after participating in a high school marine science course. *Journal of Geoscience Education, 55*, 531–539.

Lambert, J. (2006). High school marine science and scientific literacy: The promise of an integrated science course. *International Journal of Science Education, 28*, 633–654.

Liarakou, G., Gavrilakis, C., & Flouri, E. (2009). Secondary school teachers, knowledge and attitudes towards renewable energy sources. *Journal of Science Education and Technology, 18*, 120–129.

Lotze, H. K., Lenihan, H. S., Bourque, B. J., Bradbury, R. H., Cooke, R. G., Kay, M. C., Kidwell, S. M., Kirby, M. X., Peterson, C. H., & Jackson, J. B. C. (2006). Depletion, degradation, and recovery potential of estuaries and coastal seas. *Science, 312*, 1806–1809.

Markos, A., Boubonari, T., Mogias, A., & Kevrekidis, T. (2017). Measuring ocean literacy in preservice teachers: Psychometric properties of the Greek version of the survey of ocean literacy and experience (SOLE). *Environmental Education Research, 23*, 231–251.

Marrero, M. (2010). Uncovering student interests in the ocean. *Current: The Journal of Marine Education, 26*, 2–5.

McPherson, K., Wright, T., & Tyedmers, P. (2018). Examining the Nova Scotia science curriculum for international ocean literacy principle inclusion. *International Journal of Learning, Teaching and Educational Research, 17*, 1–16.

Mogias, A., Boubonari, T., Markos, A., & Kevrekidis, T. (2015). Greek pre-service teachers' knowledge of ocean sciences issues and attitudes toward ocean stewardship. *Journal of Environmental Education, 46*, 251–270.

Mogias, A., Boubonari, T., Realdon, G., Previati, M., Mokos, M., Koulouri, P., & Cheimonopoulou, M. T. (2019). Evaluating ocean literacy of elementary school students: Preliminary results of a cross-cultural study in the Mediterranean Region. *Frontiers in Marine Science, 6*, 396.

Mohan, L. (Ed.). (2013). *One Ocean. A guide for teaching the ocean in grades 3–8* (Environmental Literacy Teacher Guide Series). National Geographic.

National Marine Educators Association. (2010). *Ocean literacy scope and sequence for grades K-12*. Published in the National Marine Educators Association, U.S.A. Special Report #3 on The Ocean Literacy Campaign Featuring the Ocean Literacy Scope and Sequence for Grades K-12.

National Oceanic and Atmospheric Administration. (2013). *Ocean literacy: The essential principles and fundamental concepts of ocean sciences for learners of all ages, Version 2, a brochure resulting from the 2-week On-Line Workshop on Ocean Literacy through Science Standards.*

Published by National Oceanic and Atmospheric Administration, U.S.A.; Published June 2005, revised March 2013.
Plankis, B. J., & Marrero, M. E. (2010). Recent ocean literacy research in United States public schools: Results and implications. *International Electronic Journal of Environmental Education, 1*, 21–51.
Rodriguez-Martinez, R., & Ortiz, L. M. (1999). Coral reef education in schools of Quintana Roo, Mexico. *Ocean and Coastlal Managment, 42*, 1061–1068.
Schoedinger, S., Tran, L. U., & Whitley, L. (2010). From the principles to the scope and sequence: A brief history of the ocean literacy campaign. *NMEA Special Report, 3*, 3–7.
Stepath, C. M. (2007). Marine education: Learning evaluations. *Journal of Marine Education, 23*, 45–51.
Strang, C. (2008). Education for ocean literacy and sustainability: Learning from elders, listening to youth. *Current: The Journal of Environmental Education, 24*, 6–10.
Strang, C., DeCharon, A., & Schoedinger, S. (2007). Can you be science literate without being ocean literate? *Current: The Journal of Marine Education, 23*(1), 7–9.
The Ocean Project. (2009). *America, the ocean, and climate change: New research insights for conservation, awareness, and action.*
Tran, L. U., Payne, D. L., & Whitley, L. (2010). Research on learning and teaching ocean and aquatic sciences. *National Marine Educators Association Special Report, 3*, 22–26.
UNESCO. (2017). *Education for sustainable development goals. Learning objectives*. Paris.
Visbeck, M. (2018). Ocean science research is key for a sustainable future. *Nature Communications*. https://doi.org/10.1038/s41467-018-03158-3
Winks, L., Ward, M., Zilch, J., & Woodley, E. (2020). Residential marine field-course impacts on ocean literacy. *Environmental Education Research, 26*, 969–988.
Worm, B., Barbier, E. B., Beaumont, N., Duffy, J. E., Folke, C., Halpern, B. S., Jackson, J. B. C., Lotze, H. K., Micheli, F., Palumbi, S. R., Sala, E., Selkoe, K. A., Stachowicz, J. J., & Watson, R. (2006). Impacts of biodiversity loss on ocean ecosystem services. *Science, 314*, 787–790.

# Chapter 13
# Promoting Students' Understanding of Gene-Environment Interaction in Genetics Education

**Johannes Zang and Marcus Hammann**

## 13.1 Introduction

Trait formation is complex. The causal models underlying current explanations increasingly consider this complexity and recognize that both genes and environment as well as multiple interrelationships between these factors influence the formation of most traits. (Halldorsdottir & Binder, 2017; Hunter, 2005; Tabery, 2014).

Students' understanding about what causes variation in traits, however, has been described as rather simplistic and often gene-oriented. Research in biology education particularly addressed belief in genetic determinism (BGD), an explanatory model emphasizing genetic influence while neglecting other sources of variation (Carver et al., 2017; Kampourakis, 2017). Recent findings further indicate BGD as a rather trait-specific phenomenon. For example, students intuitively tend to neglect the influence of environmental factors on the development of body-related characteristics, and they tend to neglect the influence of genetic factors on the development of mind-related characteristics (Zang et al., 2021). Thus, educators are challenged to particularly point out different sources of variation when addressing mind-related or body-related traits in the classroom. The fact that students' genetic cognitions are often related to agency and fate concepts (Hammann et al., 2021) adds to the complexity of this topic.

To facilitate students' understanding of genetics, different instructional strategies emphasize the relevance of thinking across multiple levels of biological organization (Knippels & Waarlo, 2018), to increase students' ability to reason mechanistically about the molecular processes linking genes to traits (Duncan & Tseng, 2011), or to build upon gene-environment interaction as a central theme of the curriculum

J. Zang (✉) · M. Hammann
Centre for Biology Education, University of Münster, Münster, Germany
e-mail: johannes.zang@uni-muenster.de; hammann.m@uni-muenster.de



K. Korfiatis, M. Grace (eds.), *Current Research in Biology Education*, Contributions from Biology Education Research,
https://doi.org/10.1007/978-3-030-89480-1_13

(Jamieson & Radick, 2017). Here we focus particularly on students' causal cognition – that is how students think about causal relationships underlying perceived phenomena. From a cognitive psychology perspective, two aspects render the comprehension of complex causal relationships especially difficult: (i) Students often have limited knowledge about possible causal relationships and particularly struggle to understand interactive, time-delayed, or relational causality; and (ii) cognitive processes support the development of linear, reductionist causal assumptions. Such reductions are helpful in everyday life (Kahneman, 2012). However, scientific phenomena are often complex, students are often unaware of their reductionist assumptions (Grotzer & Mittlefehldt, 2012), and furthermore, insufficiently familiar with strategies that help to critically reflect upon causal models underlying explanations (Perkins & Grotzer, 2000). Different instructional strategies have been described to address these issues.

Epistemic strategies help to analyse the causal structures of an explanation. For example, learners are encouraged to look closely into given evidence to find a causal explanation for the observed phenomenon, to uncover flawed evidence and build upon counterevidence for the construction of causal explanatory models. Metacognitive strategies aim at evoking a cognitive engagement with the causal structures of an explanation or idea. Prompts encourage, for example, the examination of one's own thoughts (Should I believe this assumption?), or given explanations (Should I believe the concept of ...?) (Grotzer & Mittlefehldt, 2012; Perkins & Grotzer, 2000). With reference to Mendelian genetics, the school curriculum often emphasizes a monocausal model that solely offers genetic variation as a source of variance. However, the emergence of trait variation often results from complex relationships of genetic and environmental factors, which may be additive or interactive. Especially interactive relations between genetic and environmental factors (GxE) are in clear contradiction to reductionist explanations only considering genetic or environmental variation as causes of trait differences. If GxE adequately describes the emergence of variation in a trait, the effect of environmental influence depends on the genotype and carriers of different genotypes react differently to different environments. Thus, GxE is an example of relational causality, and norm-reaction graphs are particular suited to visualize the interrelation of genetic and environmental factors. This kind of graph depicts both the effects of genetic and environmental variation on a trait, and is a central element of the teaching materials piloted here. We believe that addressing GxE in the classroom complements Mendelian-based teaching practices and represents a promising approach to foster students' understanding of causal relationships.

The river dwarf model (RDM) (Zang, 2021) was developed for this purpose. The RDM aims at fostering students' understanding of causal relationships. It deploys instructional strategies that support the detection and investigation of causal structures and provides knowledge about causal models that portray the interrelation of genetic and environmental variation in the development of characteristics. The RDM first introduces the concept of GxE (Phase 1) using a fictitious model organism (river dwarfs). GxE is further explored by focussing on specific traits (Phase 2). Recurring instructional elements (e.g. tasks and prompts) and the general material design ensure continuity between the two phases. The RDM's main principle builds upon strategies developed to foster causal cognition (see above) and confronts students with opposing monocausal explanations (only genes, only environment) for trait differences. Epistemic and metacognitive prompts then encourage a critical examination of the causal structures of these explanations. Within this process, norm-reaction graphs are used to develop different causality models. Comparing these models, students gain the insight that neither genetic nor environmental explanations alone best explain variation within the respective traits, but a model of relational causality (GxE), and all materials address different levels of biological organisation. The self-study materials evaluated here are part of the RDM Phase 2 and consist of a sequence of worksheets with tasks, prompts, and templates for diagrams. They illustrate GxE by addressing the development of three characteristics with high social relevance and complex aetiology – being overweight (A), depression (B) and phenylketonuria (PKU) (C).

## 13.2 Research Question

The broader aim of this study was to evaluate whether an instructional strategy addressing causal cognition promotes students' understanding of relational causality in genetics. More specifically, we asked to what extent the use of trait-specific materials (the RDM Phase 2 teaching materials) support students' understanding of GxE interactions.

Since GxE strongly contradicts monocausal explanations, we expected students to:

(a) place more emphasis on the interrelation of genetic and environmental factors when asked to indicate sources for variation in a trait; and
(b) refrain from statements providing monocausal explanations for variation in a trait after working with the materials.

A further aim of the study was to explore to what extent established scale types, designed to operationalize BGD, display changes in students' responses after the intervention.

## 13.3 Methods

### 13.3.1 Sample, Design & Instrument

A total of 54 German high school students (25 female; age: $M = 16.54$, $SD = 0.61$) from four schools (qualification phase 1) were randomly assigned to three groups. Following an in-part switching replication design (Fig. 13.1a), Group A (n = 17; 9

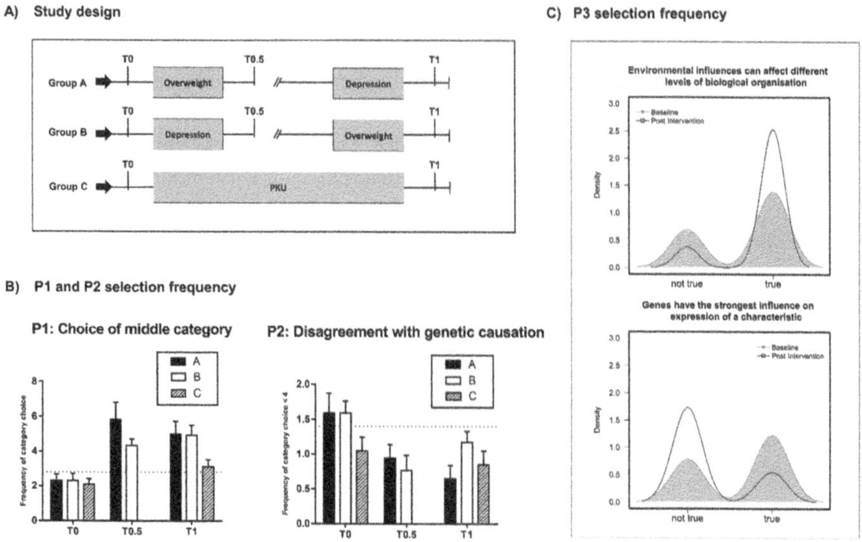


**Fig. 13.1** Study design and frequency response parameter
The study design is visualized in (**a**). Bar plots depict group-specific mean choice of the middle category in P1 traits and disagreement with genetic statements in P2 items (**b**) across time. Error bars represent SEM. Students' responses to the two most time sensitive P3 statements are depicted by density plots (**c**). Grey areas show students' response distribution at T0, red curves show students' response distribution at T1

female) first worked with material A (being overweight) and with material B (depression) the next day. Group B (n = 17; 9 female) worked first on material B and then with material A. Group C (n = 20; 7 female) worked on material C (PKU). All students worked with the paper-based self-study materials within 90-min school lessons (Fig. 13.1a). Taking the teachers' role during the study, one investigator handed out the materials, guided the students through all working and survey phases and was available to answer questions. To evaluate the materials' impact on students' cognition about GxE, we deployed a four-parted questionnaire containing different scale types (P1–4) to assess: (i) students' attribution of trait variation to environmental or genetic causes (P1); (ii) agreement with gene-deterministic or gene-related statements (P2); and (iii) selection frequencies in multiple-choice tasks (P3 & P4). All participants completed a paper-and-pencil version of this questionnaire before (T0) and after (T1) intervention. To disentangle intervention-specific effects, groups A & B additionally filled in a short version of the questionnaire (P1 & P2, see below) after having worked with the first materials (T0.5). The first part (P1) of the questionnaire represents an adapted version of the PUGGS BGD Scala (Carver et al., 2017), comprising 12 traits (six body-related). We used a seven-pointed Likert scale to assess the students' attribution of trait differences from 1 = only environmental influences to 7 = only genetic factors. The second part (P2) was constructed in reference to the BGD scale (Keller, 2005) and contains 11 items. Students were asked to indicate their agreement with the given statements on a seven-pointed Likert scale ranging from 1 = not at all true to 7 = completely true. Five of the presented items express a strong link between genes and traits (e.g. *In my opinion, being overweight is mainly caused by genetic factors*), three items relate causes to perceived agency (e.g. *I can affect characteristics that are caused by environmental influences*), two items relate genetic or environmental causation to fate concepts (e.g. *The fate of every person is defined by environmental influences*), and one item emphasized gene-environmental interplay (*Genetic and environmental influences always have a joint effect on trait differences*). The third part (P3) is inspired by the PUGGS knowledge scale and has been designed as a multiple-choice format. Students were asked to indicate their agreement with a total of ten statements by ticking them. Statements focussed on the perceived relationship between causal factors and traits (*A gene always codes for a specific trait; An environmental factor always has the same influence on a specific trait),* preference for either genetic causation (*Genes have the strongest influence on the expression of a characteristic - environmental influence is of minor importance*) or environmental causation (*Environmental factors have the strongest influence on the expression of a characteristic – genetic influence is of minor importance*) and consideration of the levels of biological organisation (*Environmental influences can affect different level biological of organisation*). The fourth part (P4) comprises four multiple-choice tasks. The tasks asked for causes of variation in the development of obesity (A), body weight in mice (B), spatial learning in either mice (C) or plant growth (D), and the choices presented reflected genetic or environmental causation as well as an understanding of gene-environment interplay. At T0, students were asked to indicate their sex and their last mark in biology, and they were asked to rate their interest in biology (IB) and genetics (IG) on a seven-pointed Likert scale ranging from 1 = not at all interested to 7 = very interested.

### *13.3.2 Data Preparation and Analysis*

Focussing our analysis mainly on P1 and P2, we investigated students' responses to the intervention at the level of selection frequencies, mean scores and single item ratings.

First, we looked into selection frequencies of P1 and P2 scale levels. Choice of the P1 middle category (4) and P2 disagreement (<4) orientation served as a model response parameter, and students' responses to P3–P4 of the questionnaire were analysed based on selection frequency as well. We used Poisson generalised linear models (GLM) with log-link function or binomial models with logit-link function to investigate intervention effects on these frequency-based response parameters, and we modelled the factor time as a random effect in all analyses.

Next, we subsumed body-related (n = 6) and mind-related (n = 6) traits (P1) and items (n = 5) reflecting genetic causation (P2) by calculating mean scores. Cronbach's alpha is reported as an indicator of internal consistency. We used analysis of variance (ANOVA) and Tukey's HSD test to compare groups and repeated measure ANOVA to analyse time effects on mean scores. Greenhouse-Geisser corrections were applied where necessary, and $\eta^2$ is given as a measure of effect size.

Subsequently, we built itemwise linear models to: (i) estimate the influence of intervention (T0 to T1); and (ii) to compare material-specific effects (T0 to T0.5 or T1) on students' attribution (P1) and agreement (P2) ratings. For the latter, we calculated a response-variable delta by subtracting baseline values (T0) from post-intervention values (T0.5 or T1) and estimated the effects of group (A–C) as regressors. Bonferroni-Holm corrections were applied to account for multiple testing. Prior to model generation, we partialed out the influence of school as a source of variability using a generalized linear model (GLM) to residualize the data. For a more in-depth analysis of intervention effects (T0 to T1), we constructed mixed linear models to test the influence of time, group, sex, mark, interest and a time x group interaction effect on variation in students' trait-specific attribution (P1) or agreement to given statements (P2), by simultaneously controlling for the influence of all other variables. All analyses were performed within the R statistical environment v.3.4.2 (R Core Team, 2017) at a significance level of .05.

## 13.4 Results

Groups were comparable in most characteristics, but differences in students' interest in genetics became apparent $F(2, 51) = 3.83$, $p < .05$, $\eta^2 = .13$ (Table 13.1). Analysis of response frequencies showed the following: in P1 ratings, students rarely chose one of the extreme categories as the cause of trait differences (only genes = 7.3%, only environment = 9.6%), and rather tended to environmental (42.89% < 4) instead of genetic (33.08% > 4) attribution across all traits at T0 and T1. As expected, the choice of the middle category (4) increased significantly after studying the materials (T0 to T1) $\chi^2(1) = 36.1$, $p < .001$, and we found no substantial evidence for group differences $\chi^2(2) = 1.95$, $p = 0.37$ or an interaction effect (group × time) $\chi^2(2) = 2.70$, $p = 0.25$ (Fig. 13.1b). In P2 ratings, students tended to agree with statements deterministically linking genes to traits across all points of measurement (61.29% >4 across all items at T0 and T1). We found no effect of intervention on agreement tendencies $\chi^2(1) = 1.33$, $p = 0.24$ and no group x time interaction effect $\chi^2(2) = 0.05$, $p = 0.97$. Contrary to our expectations, students' tendency to disagree with given statements (22.77% < 4) slightly, but significantly decreased in all groups $\chi^2(1) = 5.97$, $p < .05$ after the intervention (Fig. 13.1b). Among the items presented in P3, two showed a sensitivity to change in response to the intervention. In particular, the students' agreement with the statement "*Environmental influences can affect different levels of biological organisation*" increased from T0 to T1 $\chi^2(2) = 6.71$, $p < .01$. In contrast, agreement with the statement "*Genes have the strongest influence on the formation of a trait - environmental influence is of minor importance*" decreased $\chi^2(2) = 18.37$, $p < .001$ after working with the GxE materials (Fig. 13.1c).

**Table 13.1** Sample characteristics and mean response parameter

| | | Group A (being overweight/depression) | | | | | Group B (depression/being overweight) | | | | | Group C (PKU) | | | | | T0 Group diff. | |
|---|---|---|---|---|---|---|---|---|---|---|---|---|---|---|---|---|---|---|
| | | | | | | | | | | | | | | | | | *p* | Post hoc |
| n[a] | No. (♀) | 17 (9) | | | | | 17 (9) | | | | | 20 (9) | | | | | .44 | – |
| Age[b] | *M (SD)* | 16.41 (0.51) | | | | | 15.65 (0.70) | | | | | 16.55 (0.60) | | | | | .53 | – |
| Mark[c, d] | *M (SD)* | 1.82 (0.64) | | | | | 2.12 (0.72) | | | | | 2.1 (0.72) | | | | | .36 | – |
| IB[b] | *M (SD)* | 6.24 (1.03) | | | | | 6.41 (0.74) | | | | | 6.25 (0.64) | | | | | .77 | – |
| IG[b, e] | *M (SD)* | 5.06 (1.39) | | | | | 5.12 (1.45) | | | | | 6.05 (0.83) | | | | | .03 | C < A, B |
| | | *M (SD)* | | α | | δ *M (SD)* | *M (SD)* | | α | | δ *M (SD)* | *M (SD)* | | α | | δ *M (SD)* | | |
| | | T0 | T1 | T0 | T1 | | T0 | T1 | T0 | T1 | | T0 | T1 | T0 | T1 | | | |
| P1: Mind[b] | | 2.91 (0.64) | 3.39 (0.20) | 0.45 | 0.61 | 1.02 (0.48) | 3.24 (0.88) | 3.40 (0.56) | 0.55 | 0.57 | 1.16 (0.79) | 3.52 (0.91) | 3.33 (0.86) | 0.65 | 0.81 | 1.02 (0.84) | .09 | – |
| P1: Body[b] | | 4.38 (0.49) | 4.47 (0.48) | 0.41 | 0.50 | 0.89 (0.29) | 4.19 (0.51) | 4.20 (0.57) | 0.31 | 0.56 | 1.15 (0.82) | 4.41 (0.66) | 4.26 (0.58) | 0.39 | 0.42 | 0.95 (0.45) | .48 | – |
| P2: GC[b] | | 4.66 (0.87) | 5.07 (0.73) | 0.48 | 0.49 | 1.14 (0.62) | 4.48 (0.51) | 4.49(0.47) | 0.53 | 0.53 | 1.07 (0.41) | 4.93 (0.68) | 5.19 (0.72) | 0.49 | 0.46 | 1.12 (0.41) | .15 | – |

[a]Chi-square tests were used to test for sex distribution across groups
[b]Analysis of variance (ANOVA) was used to test for group differences
[c]Kruskal-Wallis Test was used to test for group differences in biology mark
[d]A total of 53 participants reported their last mark in biology with lower numbers indicating a better performance
[e]Post-hoc comparisons were performed using Tukey's HSD Test
*α* Cronbach's alpha
*δ* absolute differences (T1–T0) in student ratings
*GC* genetic causation

Next, we focussed our analysis at the level of scale means. Before the intervention, students attributed mind-related P1 traits ($M$ = 2.91, $SD$ = 0.64) stronger to environmental influences compared to body-related characteristics ($M$ = 4.38, $SD$ = 0.47) $F(1, 53) = 75.54$, $p < .001$, $\eta^2 = .36$, and these orientations were stable across groups and time (Table 13.1). After the intervention, mind mean scores increased within group A & B, decreased within group C (Table 13.1), and the group x time interaction was significant $F(2, 51) = 3.91$, $p < .05$, $\eta^2 = .03$.

Intervention had no overall influence on students' body-related traits mean scores $F(2, 51) = 0.61$, $p = .55$. Analysis of P2 genetic causation scores showed that students tended to agree with gene-deterministic statements (Table 13.1). This tendency slightly but significantly increased after the intervention $F(1, 51) = 4.49$, $p < .05$, $\eta^2 = .03$, and group C students exhibited higher genetic causation scores compared to students in group A and B $F(2, 53) = 5.09$, $p < .05$, $\eta^2 = .12$. Overall, small Cronbach's α value showed mean scores to insufficiently reflect variation in students' ratings. α value increased after intervention and particularly in P1 mind ratings of group A and B, indicating rising degrees of homogeneity in students' attribution. (Table 13.1).

Subsequently, we investigated the trait (P1) and item (P2) level to further characterize how students responded to the GxE materials. All students attributed variation in depression ($\delta_{T1-T0} = 0.75$) significantly stronger to genetic causes, and they attributed variation in PKU ($\delta_{T1-T0} = -0.48$) significantly stronger to environmental causes after the intervention (T0 to T1) (Fig. 13.1a). After controlling for the influence of all other variables, the intervention still explained variance of students' attribution of depression (9.93%) and PKU (2.08%). Accounting for a group x time interaction explained an additional percentage of variance in students' attribution of depression (4.89%), empathy (4.22%) and overweight (12.44%).

Students who worked with the depression-focused and overweight-focused materials (group A and B) but not group C students (PKU) attributed variation in both these traits less to environmental causes and shifted towards the P1 middle category after the intervention. To unravel material-specific effects we compared the trait-specific baseline (T1) to post (T0.5 or T1) delta values between groups (Fig. 13.2e). After working with the depression material (B), the overall increase of gene orientation in depression attribution was significantly stronger compared to students who worked with the PKU (C) material and slightly stronger compared to students who worked with the overweight material (A). Working with the overweight material had no particular influence on how students attributed metabolism-related traits, and working with the PKU material had no specific influence on how students attributed variation in PKU. Interestingly, particularly group A students showed a group-specific increase in gene orientation for a set of mind-related traits (gambling addiction, empathy, interest in fashion) and increasing environmental orientation when stating causes for variation in height (Fig. 13.2e).

We observed an overall effect of the intervention on six P2 items. One of these items was part of the genetic causation subset (item 7). After the intervention, students' agreement with the statements "*I can affect characteristics that are caused by environmental influences*" (Item 2, $\delta_{T1-T0} = 0.77$), "*The fate of any person is in their genes*" (Item 5, $\delta_{T1-T0} = 0.6$) and "*Genetic and environmental influences always*

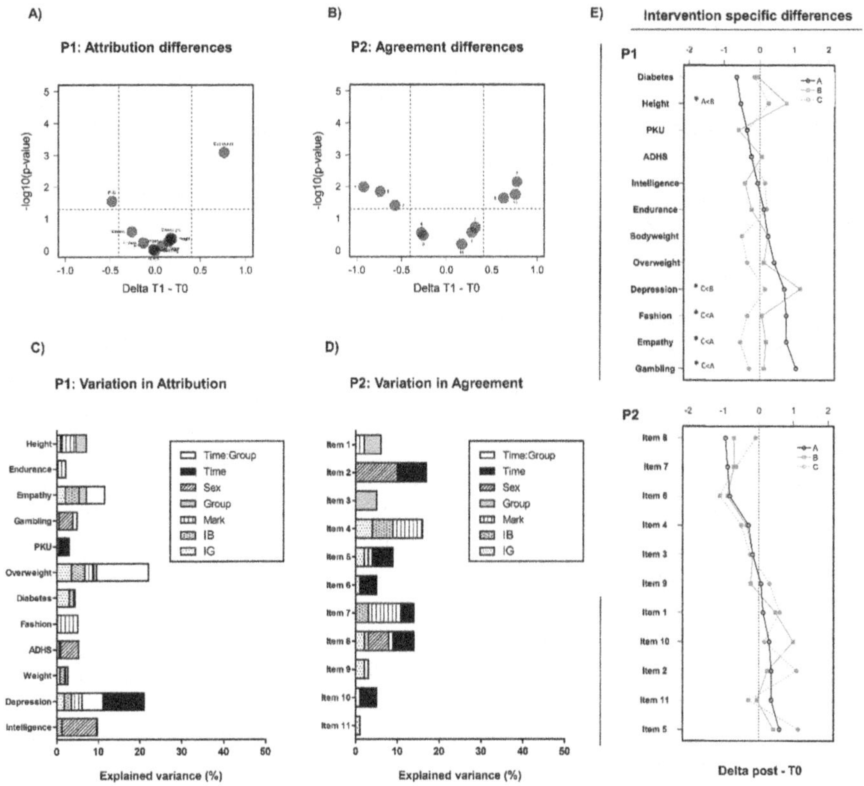


**Fig. 13.2** Attribution and agreement differences after intervention
Volcano plots depict general T1–T0 differences in students' cross group responses to P1 traits (**a**) or P2 items (**b**) against – log10 *p*-values. Dashed horizontal lines indicate the $p = .05$ threshold, vertical lines indicate a delta of 0.4. Stacked bar plots visualize the proportion of variance of P1 traits (**c**) or P2 items (**d**) as explained by listed variables. Intervention-specific mean differences (delta T1/T0.5–T0) in students' trait (P1) or item (P2) response are shown as line plots. * indicate significant baseline to post response differences ($p < .05$) between groups according to Tuckey HSD post hoc analysis

*have a joint effect on trait differences"* (Item 10, $\delta_{T1-T0} = 0.75$) increased significantly. Students' agreement with the statements *"I can affect characteristics that are caused by genetic factors"* (Item 6, $\delta_{T1-T0} = -0.92$), *"I believe that the genetic predisposition has no influence on the development of intellectual abilities"* (Item 7, reverse coded, $\delta_{T1-T0} = -0.57$) and *"The fate of any person is defined by environmental influences"* (Item 8, $\delta_{T1-T0} = -0.74$) on the other hand significantly decreased after the intervention (Fig. 13.2b). Further analyses supported these findings. Time explained 3–7% of variance in students' responses to the reported statements even after controlling for further variables (Fig. 13.2d). Time × group interaction, in contrast, did not account for significant proportions of variance in students' responses to P1 statements, and we did not observe material-specific effects on students' response differences (Fig. 13.2e).

## 13.5 Discussion

Learning to think about causal relations is important for understanding complex phenomena. Focusing on *depression*, *being overweight* and *PKU*, the materials piloted in this study addressed body-related as well as mind-related traits and further encouraged students to reflect upon causal relationships. By comparing different causal models, students gained the insight that GxE interaction as an example of relational causality – rather than monocausal explanations – best explains trait variation within the presented traits. We note that our study is based on a rather small sample and we could not control for intervention-independent variation over time. We therefore consider our findings as preliminary and in need of replication, but also informative, and several parameters indicated that the instructional strategy of addressing causal cognition promoted students' understanding of relational causality in genetics.

At the level of selection frequencies, we observed clear but contradictory intervention effects. As expected, students chose the middle category significantly more often when rating P1 traits, and agreed more often with the P3 statement emphasizing biological organization, and less often with the P3 statement emphasizing genetic influence over environmental influence after the intervention. However, contrary to our expectations, students' disagreement with gene-deterministic statements (P2) declined after working with the materials, and these rather opposing effects are as well reflected at the level of mean scale values.

In line with previous findings on students' trait-specific attributions (Hammann et al., 2021; Zang et al., 2021), P1 mean scores depicted students' preference for environmental causation of mind-related traits. Interestingly, intervention specifically impacted on students' mind mean scores, which increased within group A, and B and decreased within group C. Body mean scores on the other hand remained rather stable and this gave a first indication of trait-specific and group-specific responses to the provided materials.

The fact that P2 mean genetic causation scores slightly increased after the intervention suggests that providing genetic information – even if presented within a clear GxE context – potentially fosters agreement with gene-deterministic statements. Yet, rather small $\alpha$ value demonstrate that mean scores only insufficiently reflect variation in students' ratings, and that is why we focused subsequent analyses on the trait and item level.

Analyses at the trait and item level revealed further insights. All students attributed variation in PKU stronger to environmental causes and variation in depression stronger to genetic causes after the intervention. Attribution of depression "leveled in" at the scales' middle category, and increase in gene orientation was strongest in students who worked with depression-focused materials. We observed no general but group-specific effects on how students attributed variation in overweight, and students who studied the overweight-focused material took genetic influences stronger into account and shifted towards the scale's middle category. Thus, we found the expected attributional shifts in all traits addressed by the materials and some indications for material-specific effects. Yet group A students in particular, exhibited increasing gene orientations in a set of mind-related traits and to a lesser extent increasing environmental orientations in some body-related traits after working with the overweight-focused materials.

Thus, the intervention potentially affects the students' attribution orientations beyond the addressed trait, and students seemed rather prone to acknowledge genetic factors as an additional source of variation in mind-related traits. Although genetic causation mean values increased after the intervention, little evidence supported this finding on the item level. More controlled analyses showed that students tended to agree with P2 genetic causation items, but ratings for most of these items did not change after the intervention, which likely renders findings at the mean level as artefacts.

Moreover, responses to P2 statements indicated divergent material effects. On the one hand, the intervention seemed to further reinforce existing cognitive concepts about fate and agency. As such, students' agreement with statements linking agency to the idea of environmental causation and students' disagreement with statements linking agency to the idea of genetic causation increased. Further, students were more inclined to agree with the link between genes and fate but less inclined to agree with the link between environmental factors and fate after the intervention. On the other hand, students also tended to agree more strongly with joint causation by genetic and environmental factors after the intervention.

Taken together, we observed indications for general and material-specific effects on different levels of the provided questionnaire. Overall, our results suggest that addressing causal cognitions is a promising approach to promote students' understanding of GxE. Yet, they also shed some light on the diverse and complex landscape of students' genetic cognition. Future research should address the question to what extent trait-specific attribution orientations and agreement-based ratings reflect students' understanding of genetic complexity, and particularly investigate the relation between genetic causation, fate and agency concepts. We recommend that educators address students' causal cognitions when teaching about genetics, and complement the teaching of monocausal models by introduction of GxE as a form of relational causality. Further, although we observed some evidence for cross-trait effects, we suggest addressing both mind-related and body-related traits when teaching genetics. Educators should pay attention to environmentally-centred beliefs and particularly point out environmental variation as a source of variation in the development of body-related traits. Furthermore, instructional materials should reflect the complexity of gene-related cognition, and focusing on fate and agency concepts might be a strategy worth pursuing to foster students' understanding of genetics. Since we previously observed the students' tendency for mind-related and body-related attribution orientations, the material evaluated here focused on both, body and mind traits. Adding traits to the collection is a task for development efforts, and future studies will need to evaluate to what extent the successive processing of phase 1 & 2 of the RDM affects students' cognition about genetics. The RDM will be available at thinkingaboutgenes.com.

**Acknowledgments** We thank Lea Weber and Nadine Pander for supporting this study.

# Figure 13.2 Graphic Components

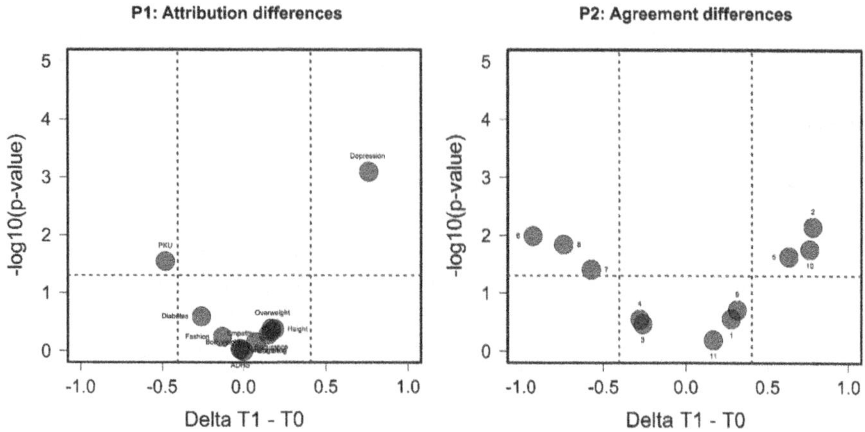


Volcano plots depict general T1–T0 differences in students' cross group responses to P1 traits or P2 items against – log10 *p*-values. Dashed horizontal lines indicate the $p = .05$ threshold, vertical lines indicate a delta of 0.4.

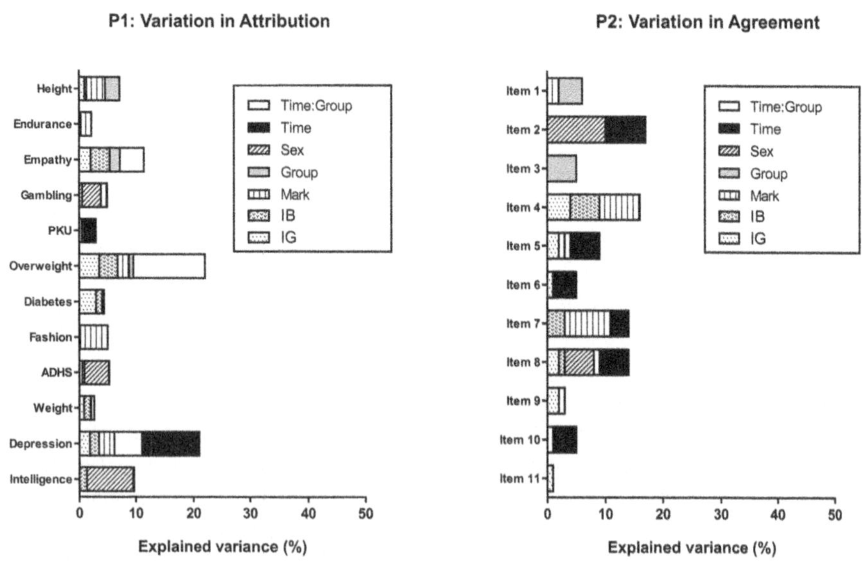


Stacked bar plots visualize the proportion of variance of P1 traits or P2 items as explained by listed variables.

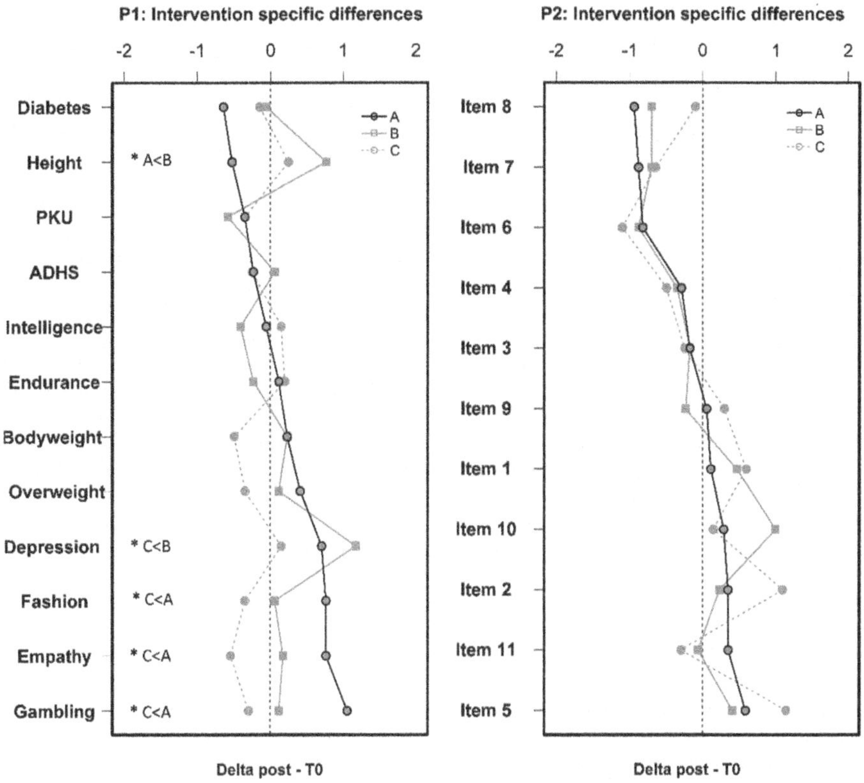


Intervention-specific mean differences (delta T1/T0.5–T0) in students' trait (P1) or item (P2) response are shown as line plots. * indicate significant baseline to post response differences ($p < .05$) between groups according to Tuckey HSD post hoc analysis.

## References


Carver, R. B., Castera, J., Gericke, N., Evangelista, N. A. M., & El-Hani, C. N. (2017). Young adults' belief in genetic determinism, and knowledge and attitudes towards modern genetics and genomics: The PUGGS questionnaire. *PloS One, 12*(1), e0169808. https://doi.org/10.1371/journal.pone.0169808

Duncan, R. G., & Tseng, K. A. (2011). Designing project-based instruction to foster generative and mechanistic understandings in genetics. *Science Education, 95*(1), 21–56. https://doi.org/10.1002/sce.20407

Grotzer, T. A., & Mittlefehldt, S. (2012). The role of metacognition in students' understanding and transfer of explanatory structures in science. In A. Zohar & Y. J. Dori (Eds.), *Contemporary trends and issues in science education* (Metacognition in science education: Trends in current research) (Vol. 40, pp. 79–99). Springer. https://doi.org/10.1007/978-94-007-2132-6_5

Halldorsdottir, T., & Binder, E. B. (2017). Gene × Environment Interactions: From molecular mechanisms to behavior. *Annual Review of Psychology, 68*, 215–241. https://doi.org/10.1146/annurev-psych-010416-044053

Hammann, M., Heemann, T., & Zang, J. C. S. (2021). Why does multiple and interactive causation render comprehension of genetic phenomena difficult and what should genetics educators do about it? In M. Haskel-Ittah & A. Yarden (Eds.), *Genetics education for the 21st century*. Springer.

Hunter, D. J. (2005). Gene-environment interactions in human diseases. *Nature Reviews Genetics, 6*(4), 287–298. https://doi.org/10.1038/nrg1578

Jamieson, A., & Radick, G. (2017). Genetic determinism in the genetics curriculum. *Science & Education, 76*(9), 477. https://doi.org/10.1007/s11191-017-9900-8

Kahneman, D. (2012). *Thinking, fast and slow*. Penguin Books.

Kampourakis, K. (2017). *Making sense of genes*. Cambridge University Press.

Keller, J. (2005). In genes we trust: The biological component of psychological essentialism and its relationship to mechanisms of motivated social cognition. *Journal of Personality and Social Psychology, 88*(4), 686–702. https://doi.org/10.1037/0022-3514.88.4.686

Knippels, M.-C., & Waarlo, A. (2018). Development, uptake, and wider applicability of the yo-yo strategy in biology education research: A reappraisal. *Education Sciences, 8*(3), 129. https://doi.org/10.3390/educsci8030129

Perkins, D. N., & Grotzer, T. A. (2000, April). *Models and moves. Focusing on dimensions of causal complexity to achieve deeper scientific understanding*. Presented at the American Educational Research Association annual conference, April 24–28. New Orleans.

R Core Team. (2017). *R: A language and environment for statistical computing*. R Foundation for Statistical Computing. https://www.R-project.org/

Tabery, J. (2014). *Beyond versus: The struggle to understand the interaction of nature and nurture. Life and mind*. The MIT Press.

Zang, J. C. S. (2021). Facilitating students understanding of causal complexity in genetics: The river dwarf model. In P. Schmiemann, S. Kaperlari & A. Möller (Eds.), *Lehr- und Lernforschung in der Biologiedidaktik* (Vol. 9, pp.145–161). StudienVerlag.

Zang, J. C. S., Heemann, T., & Hammann, M. (2021). High school student's causal attribution of traits to genes and environmental factors: A quantitative approach towards the mind-body split. Manuscript in preparation.

# Chapter 14
# Students' Conceptions as a Neglected Perspective in Trainee Teachers' Biology Lesson Plans

**Leroy Großmann and Dirk Krüger**

## 14.1 Introduction

Following a constructivist view on learning processes, good teaching practice requires a thoroughly defined adjustment of instructional strategies to match the unique requirements of the group of students in question (Duit & Treagust, 2012). In that sense, a constructivist learning environment must fit specifically to the current individual needs of particular students. To satisfy those needs in instructional practice, teachers have to be aware of them by analyzing the group prior to teaching with regard to cognitive, motivational and social aspects (Hardy et al., 2019). This corresponds to the definition of pedagogical content knowledge (PCK) as the core-part of science teachers' expertise, i.e.

> …the knowledge of, reasoning behind, and planning for teaching a particular topic in a particular way for a particular purpose to particular students for enhanced student outcomes. (Gess-Newsome, 2015, p. 36).

Consequently, teachers are expected to explicitly consider the general learning context and characteristics of their students as well as topic-specific learning difficulties and misconceptions when planning a lesson.

Notwithstanding the vast amount of research in the field of teachers' expertise and the diverse approaches to investigate science teachers' PCK (Chan & Hume, 2019), there is a lack of research on what aspects of their students' understanding in science they take into consideration when planning a lesson in an authentic context (Weingarten, 2019). So far, research focusing science teachers' considerations and implementations of students' conceptions has assessed teachers' views basically in

L. Großmann (✉) · D. Krüger
Department of Biology, Chemistry, Pharmacy, Freie Universität Berlin, Berlin, Germany
e-mail: leroy.grossmann@fu-berlin.de; dirk.krueger@fu-berlin.de



K. Korfiatis, M. Grace (eds.), *Current Research in Biology Education*, Contributions from Biology Education Research,
https://doi.org/10.1007/978-3-030-89480-1_14

interview-based research designs with rather small sample sizes (Lin, 2016; Lucero et al., 2017; Moodley & Gaigher, 2019). To the best of our knowledge, there is no exploration of authentic biology teachers' lesson plans and their rationales provided focusing on the challenges that occur when using students' understanding in science for instructional decisions. To close this gap, our study provides further representative insights into the PCK dimension *Students' Understanding in Science* and its relatedness to *Instructional Strategies* (Park & Oliver, 2008).

## 14.2 Theoretical Background

### *14.2.1 Biology Teachers' Enacted Pedagogical Content Knowledge in Lesson Planning (ePCK$_p$)*

Teachers have to access knowledge that helps them to plan and conduct lessons that foster students' competencies as effectively as possible. For that particular knowledge, Shulman (1986) introduced the notion of *pedagogical content knowledge* (PCK) as the knowledge base that distinguishes teachers from other professions. In spite of differences between various conceptualizations of PCK in detail (Neumann, Kind and Harms, 2019), the Refined Consensus Model reflects the current representation of PCK that is held in the science educational community (Carlson & Daehler, 2019). For the purpose of this paper, we explicitly focus on the enacted pedagogical content knowledge during planning (ePCK$_p$), i.e.

> ...the knowledge in action generated during, and made visible in, science teachers' planning. (Alonzo et al., 2019, p. 274).

There is consensus that knowledge about students' needs and their prior understanding is a constitutive part of ePCK$_p$ (e.g. Gess-Newsome, 2015). For instance, in their Pentagon model, Park and Oliver (2008) stress the importance of interplay between the five PCK dimensions *Orientations to Teaching Science, Science Curriculum, Assessment of Science Learning, Students' Understanding in Science* and *Instructional Strategies*. It was shown that particularly knowledge of students' understanding (including motivation and interest, learning difficulties, need and students' conceptions), and knowledge of instructional strategies, were central in the teaching practice of (biology) teachers (Park & Chen, 2012; Reynolds & Park, 2020).

These findings support Shulman's (1986) theoretical assertion and are in line with later empirical findings (e.g. Aydin & Boz, 2013) that those components are most crucial for science teachers' PCK development. However, conflicting findings which showed hardly any influence of teachers' assertions about their students' understanding on the choice of instructional strategies (Walan et al., 2017), raise the question about how far science teachers actually consider students' understanding for instructional decisions. Research on teachers' planning skills has shown that particularly the consideration of students' cognitive abilities and the anticipation of learning difficulties is challenging for teachers (e.g. Borko & Livingston, 1989;

Westerman, 1991). Therefore, it is important to further investigate this very interplay in detail and based on authentic non-experimental enacted planning of in-service teachers (Weingarten, 2019).

### *14.2.2 Leveraging Students' Conceptions for Instruction*

A variety of aspects (tasks, activities etc.) needs to be considered when planning a lesson that is intended to be tailored to students' understanding in science. In current PCK studies, aspects like motivation and interest are barely focused upon, whereas particularly students' conceptions are regarded as crucial for engaging students in science classes and designing appropriate learning processes (Duit et al. 2012; Larkin, 2012).

Yet, as Krell (2020) points out, the notion of "conception" itself is ambiguously defined. In a constructivist sense, it is defined as the internal representation of an entity shaped by external representations shaped by other people. This view leads directly to the conceptual change theory according to which students are confronted with a phenomenon, become dissatisfied with their prior conceptions, and are provided an alternative, plausible and fruitful conception that can permanently or temporarily replace or complement their conception of the world (Duit & Treagust, 2012).

Diagnosing students' conceptions would require teachers' willingness to deliberately include the learning context and base instruction on those conceptions. Hence, Lin (2016) states that teachers can plan their instruction more effectively if they gain insights into their students' minds. Therefore, teachers are expected not only to be aware of the nature of their students' conceptions (Lin, 2016), but also of their sources (Taber & Tan, 2011). Yet, this is often not the case (Lucero et al., 2017). In their analysis of novice science teachers' teaching practice, Windschitl, Thompson and Braaten (2011) distinguish three levels of working with students' ideas: (1) *monitoring, checking, reteaching ideas*, i.e. not considering or working with students' ideas; (2) *eliciting students' understanding*, i.e. considering students' ideas but not using them for instructional design; (3) *referencing students' ideas and adapting instruction*, i.e. shaping instruction based on particular students' ideas.

Beyond that, even if students' conceptions are considered by teachers when planning a lesson, the practical implementation of instructional strategies is often difficult for teachers. This might be because they do not apply useful assessment tools to elicit students' understanding (Morrison & Lederman, 2003), or because teachers tend to unilaterally focus on transmitting subject matter knowledge rather than working with students' conceptions (Moodley & Gaigher, 2019).

Additionally, experience-based concepts and academic concepts are regarded as distinct, and thus neither one of them is considered instead of using their synergies (Otero & Nathan, 2008). Moreover, teachers prefer to restate their own understanding of a scientific topic instead of representing the content for students' learning, even if they believe to have insights into their students' thinking (Halim & Meerah, 2002).

## 14.3 Research Questions

The present study contributes to research on in-service trainee teachers' PCK by gaining further insights into the planning processes of biology lessons.

Two research questions are addressed:

RQ 1: Which facets of knowledge about *Students' Understanding in Science* as part of their $ePCK_p$ are most taken into account by German trainee teachers in written biology lesson plans?

RQ 2: To what extent do German trainee teachers align their planned instructional strategies in biology lessons to their analysis of students' conceptions?

## 14.4 Research Design and Methods

### *14.4.1 Data Material and Collection*

The sample consists of 107 biology lesson plans (for students in grades 5–13) that were obtained from the Ministry for Education, Youth and Family in a federal state of Germany in an anonymized form. They were written by German trainee teachers during their final exam in the second phase of teacher education. German teacher education is organized in two phases: the first pre-service phase consists of 3 years of Bachelor's study and 2 years of Master's study at university; the second phase consists of 1.5 years of in-service teacher training (Neumann et al., 2017). It is a complete sample since it includes all lesson plans of trainee teachers who finished their training in the years 2018/2019. Therefore, those lesson plans can be regarded as representative indicators of the state-of-the-art of biology lesson planning in this federal state. Since those lesson plans have been written in the course of a formal examination, they had to follow certain structural regulations. They contain a description of the learning objectives, an analysis of the subject matter, an analysis of the group of students, a description of the planned lesson including rationales for planning decisions (activities, tasks, methods, etc.) and materials (e.g. worksheets including solutions). There were no more concrete instructions for writing the lesson plan. Due to data protection we were not allowed to collect further data such as grade in the exam, gender, teaching experience, interviews and so forth.

### *14.4.2 Data Analysis*

RQ 1: Biology teachers' $ePCK_p$

The lesson plans were analyzed within the methodological framework of qualitative content analysis (Elo et al., 2014; Schreier, 2012). We applied a mainly

**Table 14.1** Categories of trainee teachers' analysis of *Students' Understanding in Science*

| Category | Subcategories | Examples |
|---|---|---|
| 1. Cognition | A. prior knowledge | *My students know the terms natural and sexual selection as well as the related driving factors. [...] They do not know about the interplay of those factors yet.* (BT43, p. 4) |
| | B. students' conceptions | *My students understand evolution basically as a teleological process that does not continue once the intended goal is achieved.* (BT43, p. 8) |
| | C. learning difficulties | *As it turned out repeatedly, my students struggle to distinguish the essential from the non-essential information in texts. The more abstract the topic, the more problems they have.* (BT87, p. 4) |
| 2. Methodological knowledge | | *Some of my students have already participated in panel discussions concerning bioethical topics like the use of glyphosate.* (BT35, p. 3) |
| 3. Motivation/interest | | *My students told me that they like experimenting in groups.* (BT97, p. 5) |
| 4. Social aspects | | *Some of my students lack the ability to collaborate, which can be the source of severe disturbances in cooperative learning.* (BT21, p. 4) |

*BT* biology teacher

deductive approach derived from the above-mentioned Pentagon model of PCK (Park & Oliver, 2008), i.e. the categories 1B, 1C and 3 (Table 14.1) have been directly adopted from this model. Category 4 was included since it is an important aspect in adaptive teaching (Hardy et al., 2019). Two categories have been added inductively to fully cover the trainee teachers' considerations of their *Students' Understanding in Science*, i.e. to ensure empirical validity (Schreier, 2012): Prior Knowledge (1A) is often referred to in order to explain the selection of content, and Methodological Knowledge (2) describes students' knowledge about and experiences with certain activities and materials.

For coding, the software MAXQDA 20 was used. The first author coded 20% of the sample twice to ensure reliability (intrarater-agreement), all lesson plans were coded by the first author and a student assistant to ensure objectivity (inter-rater agreement). Cohen's Kappa indicated very high intra-rater agreement ($M_{\kappa} = 0.85$) and moderate to very high inter-rater agreement ($M_{\kappa} = 0.76$; Wirtz & Caspar, 2002).

RQ 2: Leveraging students' conceptions for instruction.

To gain further insight into how biology teachers align their instructional planning to their students' conceptions, two cases that include students' conceptions and address the same conception have been purposefully selected (Patton, 1990) to compare their ePCK$_p$ and deduce similarities and differences. Therefore, rationales for the use of instructional strategies referring to students' conceptions are analyzed.

## 14.5 Results

### 14.5.1 *Science Teachers' ePCK$_p$ [RQ 1]*

Table 14.2 illustrates the distribution of codings in the lesson plans that include information from the six dimensions described above.

Nearly all trainee teachers analyze their students' prior knowledge and approximately two thirds refer to methodological, motivational and social aspects. Strikingly, students' conceptions are barely considered.

On average, prior knowledge was addressed eight times per lesson plan whereas all the other categories were coded approximately twice per lesson plan (if they were considered at all). Further statistical analyses revealed no significant differences between school types, grades or topics.

### 14.5.2 *Leveraging Students' Conceptions for Instruction [RQ 2]*

Three of those six trainee teachers considering students' conceptions only speculated on their students' ideas, two used empirical findings from literature without concrete application to their own students and only one explicitly used a strategy to elicit students' thinking. However, neither types nor sources of students' conceptions are analyzed in those lesson plans. Two lesson plans in the sample focus on the same topic and refer to students' conceptions, specifically to the idea that ecosystems are static instead of dynamic. The following sections compare those two cases in more detail.

*Case 1: The lake* (Biology Teacher: BT87)

In this lesson, students are intended to learn how it is possible that there are constantly sufficient minerals in the epilimnetic zone of a lake even though they tend to sink down into the hypolimnetic zone. Right at the beginning, BT87 claims that

**Table 14.2** Quantity of codings per lesson plan addressing facets of *Students' Understanding in Science*

| Category | | Lesson plans referring to the facets at least once (N = 107) | Relative frequency |
|---|---|---|---|
| 1.A. | prior knowledge | 106 | 99.1% |
| 1.B. | students' conceptions | 6 | 6.5% |
| 1.C. | learning difficulties | 23 | 21.5% |
| 2. | methodological knowledge | 75 | 70.1% |
| 3. | motivation/interest | 71 | 66.4% |
| 4. | social aspects | 73 | 68.2% |

solving ecological problems requires systems thinking skills, yet there is no analysis at all in how far students have already experienced systems thinking. Instead, the focus lies on what activities particular students are familiar with and on how motivating this is for them:

> Leonhard is a friendly student who has sophisticated knowledge in biology [...]. However, he does not participate in lessons if writing is necessary. [...] He tends to interrupt constantly when he has to work individually. Therefore, I prefer group work since due to the interdependencies, he will feel more responsible. [...] I decided to let them produce a graphic representation which requires little writing effort. (BT87, p. 3)

It is claimed that instructional decisions like group work and avoidance of writing are made in order to specifically support certain students in their individual learning abilities. However, analyses of individual students' requirements are usually used to legitimise instructional decisions for the whole group of students rather than for individual students. Hence, the instructional design often only appears to be differentiated.

As far as the relevance of systems thinking and the consideration of students' conceptions are concerned, BT87 refers to the importance of analyzing cyclic cause-and-effect relationships as well as possibly occurring side effects. In this context, empirical research is consulted to point out that 16/17-year-old students perceive ecosystems as static systems:

> At the beginning of the teaching unit, I analyzed my students' prior knowledge (Hammann & Asshoff, 2014, S.194) [...]. I diagnosed that my students have similar conceptions. Since ecosystems change even without a harmful human influence, it is important to address this topic in biology classes. (BT87, p. 3)

Two aspects are noteworthy here. First, an instructional decision is made based on the consideration and use of empirical findings. However, this is only used as a rationale for the choice of topic, not for the instruction itself. Second, neither concrete misconceptions are named nor any kind of differences within the group of students are mentioned. Hence, all instructional decisions to follow refer to the whole group instead of addressing the needs of particular subgroups of students.

The reasons for the structure of the lesson are only partly related to the extensive analysis of *Students' Understanding in Science*:

> At the beginning, I present the question and ask students to hypothesize about it. The purpose is to reduce linguistic inhibition and activate all students in the class. (BT87, p. 5)

Interestingly, there is neither any reference to the importance of hypotheses as far as scientific reasoning competencies are concerned, nor are those hypotheses used in a diagnostic way to tailor the further learning process based on the variety of students' ideas.

To summarize, BT87 has analyzed many aspects of his/her students' understanding in science and provides a clear insight into the classroom situation. However, most of those aspects are not used in the planning of instruction. Particularly, the claimed consideration of students' conceptions is not reflected in the tasks, the activities or in the learning outcome.

*Case 2: Succession of plants in the Berlin Wall death strip* (Biology Teacher: BT60)

In this lesson, students are intended to learn about ecological succession of plants. The context chosen is the Berlin Wall death strip that has been colonized by many plant species since the German reunification in 1990. To justify the choice of topic, BT60 claims two relevant things right at the beginning:

> In the last unit, students experienced how ecosystems work by the example of forests. Now, we will deal with changes in ecosystems over time. Even though students generally notice changes in their environment, natural ecosystems are regarded as relatively static. (BT60, pp. 3–4)

First, the decision for the topic is made based on the prior knowledge about ecosystems in general. Second, BT60 refers to students' conceptions but without any reference to literature or without having diagnosed particular students' conceptions about ecosystems. Instead, this rationale rather appears more like an intuitive speculation than a valid diagnosis. In the following, since a variety of students' products is intended, the prerequisite for that is explained:

> The classroom climate is positive and supportive. Problems and questions […] can be presented and discussed in a constructive manner. The students work in groups in a friendly, respectful […] way. (BT60, p. 3)

Based on that analysis, BT60 plans an instructional setting in which groups of students have to arrange descriptions and pictures of ten plants in a certain order based on how students imagine the chronological order in which those plants colonized the Berlin Wall death strip. Up to three groups have to present their ideas, the students then have to compare the different results and discuss their findings.

Afterwards, the students read informative texts about ecological succession of plants and compare the plants in early and late stages of this process. Based on this newly acquired knowledge, the students then get the opportunity to re-check their order systems from the beginning, change them if necessary and reflect on the criteria they used when ordering. Two reasons are given for that procedure. First, BT60 can diagnose whether the learning objective has been achieved, and second, the students themselves get some feedback on their ideas as well.

To summarize, BT60 has analyzed some aspects of his/her students' understanding in science and designed an instructional setting that is evidently based on students' initial conceptions. However, it should be noted that the setting (ten plants only) is rather strict so that the variety of students' conceptions cannot be visualized fully.

## 14.6 Discussion

This study aimed to identify which aspects of *Students' Understanding in Science* as one component of ePCK$_p$ are considered by German trainee teachers in written biology lesson plans, and how far these analyses have been used to design the learning process, particularly focusing the consideration of students' conceptions.

### 14.6.1 Biology Teachers' ePCK$_p$ [RQ 1]

The results show that trainee teachers rarely describe learning difficulties and students' conceptions. This could be explained by the fact that more long-term effort and in-depth analysis is needed to diagnose students' thinking, to analyze a topic thoroughly and to find certain difficulties that would need to be considered in the instructional setting (Hammann & Asshoff, 2014). It is far more convenient to describe the knowledge gained in the previous lessons as well as general methodological and social aspects that could be referenced for each lesson in that class anyway. Hence, those aspects could be regarded as personal PCK (pPCK) elements, i.e.

> ...the "reservoir of knowledge and skills that the teacher can draw upon during the practice of teaching" (Carlson & Daehler, 2019, p. 85).

But the more specific assertions are part of the ePCKp (Alonzo et al., 2019). However, it should be considered that trainee teachers might know about students' conceptions in general (pPCK), but don't recognize and acknowledge them in the enacted planning practice (ePCK$_p$).

Moreover, about one third of the sample neglected to consider motivational aspects. This is surprising since sufficient motivation is considered crucial for students' learning (Hidi & Harackiewicz, 2000). This raises the question how trainee teachers perceive the task of writing a biology lesson plan. In most cases, they seem to be regarded as a bureaucratic proposition that has to be filled out tidily, rather than focusing on the aspects really needed to satisfy the unique needs of the particular group of students concerning the particular topic. Although it was not observed how the lesson plans were translated into teaching practice, we argue that this analysis can contribute valid insights into biology teachers' ePCK$_p$ since those lesson plans give a representative insight into the state-of-the-art lesson planning in this federal state. Since the lesson plans were not written under scientific guidance but in the course of formal examination, we could not control whether other persons (colleagues, friends etc.) have proofread and made contributions to those lesson plans. Therefore, we might have gained only partial insights into the ePCK$_p$ of individual trainee teachers rather than into collective PCK (cPCK) within a community of experts (Carlson & Daehler, 2019).

### 14.6.2 Use of Students' Conceptions in Instruction Planning [RQ 2]

Considering that those biology lesson plans have been produced in the course of the highest formal exam at the end of German teacher education (Neumann et al., 2017), it is remarkable that only 6 out of 107 trainee teachers referred to students' conceptions at all. This supports other empirical findings concerning the insensitivity of science teachers to students' conceptions (Lin, 2016; Lucero et al., 2017) and raises the question for further research regarding why biology teachers apparently do not see the necessity to include students' thinking into their lesson planning processes. This is particularly striking since there is plenty of literature on students' conceptions and how to elicit and use them within the framework of conceptual change in instruction available in Germany (e.g. Hammann & Asshoff, 2014; Kattmann, 2016). Moreover, in the few lesson plans that include students' conceptions, neither types nor sources of students' conceptions are analyzed in those lesson plans confirming the claimed insensitivity of science teachers to students' conceptions (Lin, 2016; van Driel et al., 2002).

The comparison of two lesson plans about ecosystem dynamics illustrates the difference between a closed scenario in which the emphasis is put on learning certain subject matter (case 1) on the one hand, and providing the opportunity for students to put forward their initial ideas and reflect on them after an instructional input (case 2) on the other hand. The first case is in line with Halim and Meerah (2002), since BT87 actually diagnosed his/her students' ideas in advance but planned the lesson with more focus on the subject matter than on the students' thinking (Moodley & Gaigher, 2019). This corresponds to the second level in terms of Windschitl, Thompson and Braaten (2011). The second case can be understood as a partially responsive setting that uses formative assessment and a reflection phase on the learning process to some degree (Otero & Nathan, 2008). Unlike Walan et al.'s findings (2017), BT60 in case 2 explicitly referred to aspects of students' understanding and used it for an instructional decision. This might correspond with the highest level of working with students' ideas (Windschitl et al., 2011). However, we regard this example as a rather superficial analysis lacking an in-depth view of conceptions (Morrison & Lederman, 2003) that would allow an instructional setting to be associated with conceptual change (Duit & Treagust, 2012). This does not occur in the whole sample. Yet, as Weingarten (2019) points out, the analysis of lesson plans written in the context of a formal exam must consider the possibility that the trainee teachers' decisions are at least somehow influenced by what the examiners desire in the sense of symbolic interactionism (Mead, 1934). Consequently, trainee teachers might tend to avoid any risks in their planning and rather prefer whole-class discussions or teacher-centered consolidation phases to convey the biological content.

Since there are only six lesson plans considering students' conceptions in the complete sample, the scope of this study is limited. The selection of the two cases aimed to exemplify contrasting approaches in educational practices when working with students' conceptions. Broader research is needed to further investigate biology teachers' challenges and needs when leveraging students' conceptions for instruction on a larger sample of lesson plans that include students' conceptions. Therefore, we plan to explicitly ask pre-service teachers and trainee teachers to plan a lesson and take students' conceptions into account. By doing so, a much larger sample can be analyzed in a comparative case study.

## 14.7 Conclusion

In summary, this study reveals two major findings:

1. Trainee biology teachers analyze student groups basically on a general level, i.e. by merely naming declarative prior knowledge, methodological knowledge or motivational and social aspects. In-depth learning difficulties and students' conceptions concerning the topics to be taught are mostly neglected.
2. If in-depth aspects like students' conceptions are considered, they are hardly explicitly used for planning instruction, but rather named as considerations of minor relevance.

Both are astonishing since the lesson plans were produced in the course of the highest final exam in German teacher education after 3 years of Bachelor's study, 2 years of Master's study (both pre-service) and 1.5 years of in-service teacher training (Neumann et al., 2017). Hence, we might expect trainee teachers to show sophisticated and highly interconnected ePCK$_p$. However, unlike constructivist learning theory and conceptual change theory suggest (Duit et al., 2012; Larkin, 2012), trainee teachers at the end of their teacher training either do not even consider students' conceptions or struggle with designing instruction based on students' conceptions. These findings are consistent with results from earlier studies (Lucero, 2017; Morrison & Lederman, 2003; Moodley & Gaigher, 2019; Otero & Nathan, 2008).

It should be noted that in most lesson plans, trainee teachers used differentiated instruction and obviously tried to include their analysis of their students' needs into the instructional design, e.g. by providing bonus tasks for high performing students or further help for low performing students (Hardy et al., 2019). Hence, trainee teachers do take their students' abilities and needs into account, but they fail to fundamentally design instructional processes based on their students' conceptions.

The present study demonstrates that prospective biology teachers need more support in eliciting students' conceptions in an appropriate manner and using them explicitly in the process of designing instruction. Teacher training programmes need to actively and repeatedly emphasize the importance of students' conceptions in biology classes. Pre-service teachers and trainee teachers need help to bridge the gap between theoretical assertions (e.g. conceptual change theory; Duit et al., 2012) as part of their PCK and practical strategies to elicit and use students' conceptions in planning and teaching biology.

**Acknowledgments** The project *K2Teach* is part of the "*Qualitätsoffensive Lehrerbildung*", a joint initiative of the German Federal Government and the Länder, which aims to improve the quality of teacher training. The programme is funded by the Federal Ministry of Education and Research (grant number 01JA1802). The authors are responsible for the content of this publication.

## References


Alonzo, A. C., Berry, A., & Nilsson, P. (2019). Unpacking the complexity of science teachers' PCK in action. In A. Hume, R. Cooper, & A. Borowski (Eds.), *Repositioning pedagogical content knowledge in teachers' knowledge for teaching science* (pp. 271–286). Springer.

Aydin, S., & Boz, Y. (2013). The nature of integration among PCK components: A case study of two experienced chemistry teachers. *Chemistry Education Research and Practice, 14*(4), 615–624.

Borko, H., & Livingston, C. (1989). Cognition and improvisation: Differences in mathematics instruction by expert and novice teachers. *American Educational Research Journal, 26*(4), 473–498.

Carlson, J., & Daehler, K. R. (2019). The Refined Consensus Model of pedagogical content knowledge in science education. In A. Hume, R. Cooper, & A. Borowski (Eds.), *Repositioning pedagogical content knowledge in teachers' knowledge for teaching science* (pp. 77–92). Springer.

Chan, K. K. H., & Hume, A. (2019). Towards a consensus model: literature review of how science teachers' Pedagogical Content Knowledge is investigated in empirical studies. In A. Hume, R. Cooper, & A. Borowski (Eds.), *Repositioning Pedagogical Content Knowledge in teachers' knowledge for teaching science* (pp. 3–76). Springer. https://doi.org/10.1007/978-981-13-5898-2_1

Duit, R., & Treagust, D. F. (2012). How can conceptual change contribute to theory and practice in science. In B. J. Fraser, K. Tobin, & C. McRobbie (Eds.), *Second international handbook of science education* (pp. 107–118). Springer.

Duit, R., Gropengießer, H., Kattmann, U., Komorek, M., & Parchmann, I. (2012). The model of educational reconstruction – A framework for improving teaching and learning science. In D. Jorde & J. Dillon (Eds.), *Science education research and practice in Europe: Retrospective and prospective* (pp. 13–37). Sense Publishers.

Elo, S., Kaariainen, M., Kanste, O., Polkki, T., Utriainen, K., & Kyngas, H. (2014). Qualitative content analysis. *SAGE Open, 4*, 1–10.

Gess-Newsome, J. (2015). A model of teacher professional knowledge and skill including PCK: Results of the thinking from the PCK Summit. In A. Berry, P. J. Friedrichsen, & J. Loughran (Eds.), *Re-examining pedagogical content knowledge in science education* (pp. 28–42). Routledge.

Halim, L., & Meerah, S. M. M. (2002). Science trainee teachers' pedagogical content knowledge and its influence on physics teaching. *Research in Science & Technological Education, 20*(2), 215–225.

Hammann, M., & Asshoff, R. (2014). *Schülervorstellungen im Biologieunterricht. Ursachen für Lernschwierigkeiten* [Students' conceptions in biology classes. Causes of learning difficulties]. Stuttgart.

Hardy, I., Decristan, J., & Klieme, E. (2019). Adaptive teaching in research on learning and instruction. *Journal for Educational Research Online, 11*(2), 169–191.

Hidi, S., & Harackiewicz, J. M. (2000). Motivating the academically unmotivated: A critical issue for the 21st century. *Review of Educational Research, 70*(2), 151–179.

Kattmann, U. (2016). *Schüler besser verstehen: Alltagsvorstellungen im Biologieunterricht; zusätzliche Stichwörter zum Download* [Better understanding students: Students' conceptions in biology classes]. Aulis Verlag.

Krell, M. (2020). Vorstellung und Kompetenz [Conception and competence]. In B. Reinisch, K. Helbig, & D. Krüger (Eds.), *Biologiedidaktische Vorstellungsforschung: Zukunftsweisende Praxis* (pp. 69–82). Springer Spektrum.

Larkin, D. (2012). Misconceptions about misconceptions: Preservice secondary science teachers' views on the value and role of student ideas. *Science Education, 96*(5), 927–959.

Lin, J. W. (2016). Do skilled elementary teachers hold scientific conceptions and can they accurately predict the type and source of students' preconceptions of electric circuits? *International Journal of Science and Mathematics Education, 14*(2), 287–307.

Lucero, M. M., Petrosino, A. J., & Delgado, C. (2017). Exploring the relationship between secondary science teachers' subject matter knowledge and knowledge of student conceptions while teaching evolution by natural selection. *Journal of Research in Science Teaching, 54*(2), 219–246.

MAXQDA. (n.d.). (Version 20) [Computer software]. VERBI Software GmbH.

Mead, G. H. (1934). *Mind, self and society* (Vol. 111). University of Chicago Press.

Moodley, K., & Gaigher, E. (2019). Teaching electric circuits: Teachers' perceptions and learners' misconceptions. *Research in Science Education, 49*(1), 73–89.

Morrison, J. A., & Lederman, N. G. (2003). Science teachers' diagnosis and understanding of students' preconceptions. *Science Education, 87*, 849–867.

Neumann, K., Härtig, H., Harms, U., & Parchmann, I. (2017). Science teacher preparation in Germany. In J. Pedersen, T. Isozaki, & T. Hirano (Eds.), *Model science teacher preparation programs* (pp. 29–52) IAP.

Neumann, K., Kind, V., & Harms, U. (2019). Probing the amalgam: the relationship between science teachers' content, pedagogical and pedagogical content knowledge. *International Journal of Science Education, 41*(7), 847–861.

Otero, V. K., & Nathan, M. J. (2008). Preservice elementary teachers' views of their students' prior knowledge of science. *Journal of Research in Science Teaching, 45*, 497–523.

Park, S., & Chen, Y. C. (2012). Mapping out the integration of the components of pedagogical content knowledge (PCK): Examples from high school biology classrooms. *Journal of Research in Science Teaching, 49*(7), 922–941.

Park, S., & Oliver, J. S. (2008). National board certification (NBC) as a catalyst for teachers' learning about teaching: The effects of the NBC process on candidate teachers' PCK development. *Journal of Research in Science Teaching, 45*(7), 812–834.

Patton, M. Q. (1990). *Qualitative evaluation and research methods*. Sage.

Reynolds, W. M., & Park, S. (2020). Examining the relationship between the Educative Teacher Performance Assessment and preservice teachers' pedagogical content knowledge. *Journal of Research in Science Teaching*. https://doi.org/10.1002/tea.21676

Schreier, M. (2012). *Qualitative content analysis in practice*. Sage.

Shulman, L. S. (1986). Those who understand: Knowledge growth in teaching. *Educational Researcher, 15*(2), 4–14.

Taber, K. S., & Tan, K. C. D. (2011). The insidious nature of 'hard-core' alternative conceptions: Implications for the constructivist research programme of patterns in high school students' and pre-service teachers' thinking about ionization energy. *International Journal of Science Education, 33*(2), 259–297.

Van Driel, J. H., Jong, O. D., & Verloop, N. (2002). The development of preservice chemistry teachers' pedagogical content knowledge. *Science Education, 86*(4), 572–590.

Walan, S., Nilsson, P., & Mc Ewen, B. (2017). Why inquiry? Primary teachers' objectives in choosing inquiry-and context-based instructional strategies to stimulate students' science learning. *Research in Science Education, 47*(5), 1055–1074.

Weingarten, J. (2019). *Wie planen angehende Lehrkräfte ihren Unterricht? Empirische Analysen zur kompetenzorientierten Gestaltung von Lernangeboten* [How do prospective teachers plan their lessons? Empirical analyses on competency-oriented instructional design]. Waxmann Verlag.

Westerman, D. A. (1991). Expert and novice teacher decision making. *Journal of Teacher Education, 42*(4), 292–305.

Windschitl, M., Thompson, J., & Braaten, M. (2011). Ambitious pedagogy by novice teachers: Who benefits from tool-supported collaborative inquiry into practice and why? *Teachers College Record, 113*(7), 1311–1360.

Wirtz, M., & Caspar, F. (2002). *Beurteilerübereinstimmung und Beurteilerreliabilität* [Rater agreement and rater reliability]. Hogrefe.

# Part III
# Perceptions of Biology and Biology Education

# Chapter 15
# Students' Opinions About Interdisciplinary Lessons

**Annkathrin Wenzel and Norbert Grotjohann**

## 15.1 Introduction

Scientifically educated citizens are able to make informed decisions about their health and wellbeing. Moreover, in order to find solutions to the complex problems of our time, it is necessary that all people, not only scientists, deal with science-related dilemmas (OECD, 2018). To support the development of scientific competences is the aim and purpose of science education (Oliver et al., 2019). In addition, it is becoming increasingly important to promote a positive attitude towards science, as this knowledge has an economic benefit (Osborne et al., 2003).

For this reason, interest in the natural sciences should be supported (Oliver et al., 2019). Nevertheless, students' interest in the natural sciences steadily declines during their time at school. This reduction in interest basically affects all school subjects, even biology is affected (Berck & Graf, 2018; Killermann et al., 2018; Ruppert, 2015). In biology, the decline in interest is particularly significant among young men, while young women tend to become less interested in the other natural sciences (Ruppert, 2015). About 40 percent of all fifteen-year-old students generally experience only little or no enthusiasm for scientific themes (Schiepe-Tiska et al., 2016). At the moment, fewer and fewer students are choosing to study natural sciences or other STEM subjects (Osborne et al., 2003; Statistisches Bundesamt, 2020).

Interdisciplinary course offerings could be a good way to deal with relevant problems (Huber, 1998; Labudde, 2014; Moegling, 2010) and prevent a decline in interest. Thus, there is increasing demand for interdisciplinary course offerings and lessons within school systems (Duncker & Popp, 1998; Krause, 2015). Since Huber (1998) has been explicitly calling for interdisciplinary instruction at the secondary

A. Wenzel (✉) · N. Grotjohann
Biology Didactics, Faculty of Biology, Bielefeld University, Bielefeld, Germany
e-mail: annkathrin.wenzel@uni-bielefeld.de; norbert.grotjohann@uni-bielefeld.de



K. Korfiatis, M. Grace (eds.), *Current Research in Biology Education*, Contributions from Biology Education Research,
https://doi.org/10.1007/978-3-030-89480-1_15

school level, but it is still not prevalent (Caviola et al., 2011). Interdisciplinary programmes are of course not new and not only being developed in Germany or German-speaking countries. For example, Walker (2017) outlines some of the successes and challenges relating to integrated STEM learning experiences in K-12 education in the United States over the past twenty years.

## 15.2 Theoretical Background

### *15.2.1 Interdisciplinary Instruction*

In research and school practice, the term "interdisciplinary instruction" is defined and used very differently as a general term (Labudde et al., 2005). Interdisciplinary instruction is a general term for different types of teaching that goes beyond the barriers of a single subject to connect two or more different subjects.

There are various forms of interdisciplinary instruction. According to Huber (1994), these should be considered from two different perspectives: from the subject level as well as from the lesson-schedule level. This was reaffirmed by Labudde (2003, 2014). In terms of the content, a difference can be made between three forms of interdisciplinary instruction: intradisciplinary teaching, multidisciplinary teaching, and interdisciplinary teaching.

In intradisciplinary teaching, a connection is established between a subject and one or more other subjects, for example, through overlapping topics or practice-oriented examples. This is the simplest form of interdisciplinary instruction. The term "intradisciplinary" is used because it comes from a single subject that makes small connections to other subjects.

Multidisciplinary teaching deals with one topic in different subjects at the same time. The teachers should have some knowledge of the content of the other subjects. Therefore, they should coordinate and adapt the contents if necessary.

For interdisciplinary teaching, the focus should be on a question or a problem, and various disciplines should be involved in answering it. The involved teachers should coordinate intensively with each other (Huber, 1995; Labudde, 2003, 2014).

This can be supplemented at the lesson-schedule level by modifying the selection of subjects. The first possibility is to create another course offering in addition to the individual disciplines. This includes forming clubs or conducting project weeks.

In contrast to this, integrated teaching is another possibility. The individual subjects are combined into a single subject. One example of this would be the school subject *Sachkunde* in elementary schools in Germany, which includes biology, physics, chemistry, geography, and history. Another example is the school subject science in German comprehensive schools, which comprises biology, physics, and chemistry (Huber, 1995; Labudde, 2014). Often it is not possible to clearly separate the various forms of interdisciplinary instruction from one another as there are flowing boundaries.

### *15.2.2 Opportunities and Advantages of Introducing Interdisciplinarity*

Interdisciplinary programmes offer many advantages and opportunities. They particularly promote holistic, problem-oriented, and reflective learning (Huber, 1995). In addition, higher thinking skills can be developed (Jones, 2010). Students can:

> ...see relationships among content areas and understand principles that cross curricular lines [...] and to look at situations from various viewpoints" (Duerr, 2008, p. 177).

This gives hope for a future with many discoveries and innovations (Jones, 2010).

Klafki (1998) points out the importance of addressing the key problems facing mankind and therefore supports interdisciplinary instruction. Only in this way can the world's major and complex problems be addressed intelligently through holistic approaches (Woods, 2007). Furthermore, interdisciplinary instruction solidly prepares students to participate in social life (Duncker & Popp, 1998). This is supported by the fact that interdisciplinary course offerings, especially in academic secondary schools, promote academic competences, such as clarifying the ways of thinking and working as well as the limits of the subject (Huber, 1998).

Today, modern work structures increasingly require multi-professional teamwork in which work is done on an interdisciplinary basis (Woods, 2007). Interdisciplinary course offerings allow students to develop interdisciplinary competences (Labudde, 2003) and to improve their communication skills (Jones, 2010). Moreover, such course offerings can be used to learn how to obtain information in the ICT (Information and Communication Technologies) Age.

Furthermore, the PISA and TIMSS results show that German students do not perform well enough in applying and transferring mathematical and scientific knowledge to new situations. This can be trained by interdisciplinary teaching (Labudde, 2003). Through contextual references and real-life problems, teaching can also be designed in such a way that it appeals to more students and is more gender balanced (Labudde, 2014). Because students construct new knowledge through context and the connection to different prior knowledge, constructivist learning theory is also applicable. Although the project method is often suitable for interdisciplinary teaching and can thus be learned (Labudde, 2003), interdisciplinary learning must not always be reduced to the project method.

Another argument for interdisciplinary education is that it raises learners' interest in science education and in science in general. There are various reasons for this. For example, interdisciplinary teaching content is often more interesting for many students than subject-specific content alone. Or the students can be motivated by questions, phenomena, and contexts even though they are not interested in the actual subject (Labudde, 2014).

> For some disciplines, the interdisciplinary approach is said to not only be preferred but needed. (Jones, 2010, p. 78)

Of course, interdisciplinary course offerings also have disadvantages. For example, developing the project or lessons is very time intensive and often requires intensive teamwork. At the same time, there can also be confusion due to the combination of subjects (Jones, 2010). This study describes an example of an interdisciplinary project week designed for students who are already interested in science.

## 15.3 The Study

### 15.3.1 Research Approach

In the past few years, there have been various empirical studies on interdisciplinary course offerings, for example, Buse (2017) and Rodenhauser (2016), who were researching at a pupil laboratory, or Klos (2008) and Åström (2008), who compared integrated teaching with normal lessons. However, the studies mainly deal with the effects of interdisciplinary interventions and consider, for example, the increase in knowledge or interest. But to our knowledge, no study has examined the ideas and opinions of students about interdisciplinary course offerings and school lessons. For this reason, we developed a study on the following research questions within the context of an interdisciplinary project week:

1. How do students define interdisciplinary teaching? Can the different forms of interdisciplinary teaching, as described by Labudde (2014), be found in the students' answers?
2. How do they perceive interdisciplinary teaching compared to "normal" teaching?
3. Why did the students register for an interdisciplinary project week? What advantages and disadvantages do they see?

### 15.3.2 The Systems Biology Project Week

The study was conducted during a project week, which is described below.

The Systems Biology project week is oriented towards secondary school students who are interested in biology and mathematics. Over the course of the week, they work in an interdisciplinary manner, experiencing the connection between the different natural sciences (Joachim Herz Stiftung, 2016). This is particularly important because systems biology is a key technology that is representative of the STEM subjects (science, technology, engineering, and mathematics). Young scientists can improve their interdisciplinary thinking within the context of systems biology (Wanka, 2015). Although systems biology already has many fields of application, it is a young research discipline in which Germany now has an internationally leading position. Various programmes now exist to inspire and support more young people in this area (BMBF, 2018). The combination of biology and mathematics seems especially useful because mathematics is a useful tool in the life sciences (Keller,

2011). But mathematics is not only a tool: a basic mathematical education is essential in the education of responsible citizens and qualified specialists, serving as a foundation of learning ability across all subjects (Loos & Ziegler, 2015). In Germany, there are some mathematics components in all degree programmes.

The Systems Biology project week deals with the *lac* operon. Various biological experiments are carried out using the bacterium *Escherichia coli*. These include a growth experiment with the use of different sugars as a carbon source. In addition to that, a quantitative and a qualitative enzyme test as well as an investigation of the promoter activity with green fluorescent protein are performed. Finally, different parameters are calculated based on the experimentally determined data and compared with the literature. The students carry out calculations independently with the help of instructions. Furthermore, students perform simulations with the modelling and simulation software *CellDesigner*.

The project week comprises additional programme items, such as specialist talks and guided tours. Through this varied programme, participants should be able to repeat, expand, deepen, and apply their existing knowledge from school. Students are also provided the opportunity to get in contact with scientists and academics and to get information about work and research. A detailed programme and presentation of the contents can be found in Wenzel and Grotjohann (2019).

## 15.4 Materials & Methods

This study evaluates the transcribed interviews, which is qualitative data. The semi-structured interviews were carried out during the project week described above. They dealt with interdisciplinary course offerings and the opinions of the participants. A total of 19 students (female: 47.4%, male: 52.6%; $M_{Age}$ = 16.5 years, $SD_{Age}$ = 1.1) were interviewed. They were interested students who voluntarily took part in a project week during their holidays. All were upper secondary school students and lived in the area near the student laboratory. The interviews were conducted by two researchers in biology didactics, who used an interview guide as a basis for their work. The guide contained questions about the project week, the subject combination, biology, and mathematics, all of which were related to school lessons. The interview started by requesting the students to "tell me why you signed up for the project week". Afterwards, the interdisciplinary aspect of the project week was discussed in detail. The students were asked to define the term "interdisciplinary" ("What do you think of when you hear the term interdisciplinary teaching?"). After that, they were asked for examples from school and whether they saw advantages and disadvantages in interdisciplinary teaching. In addition, they were asked about their normal maths and biology lessons for comparison. At the end of the interview, they were asked to name any other aspects they could think of ("Is there anything else you would like to add?").

The method of the guided interview was chosen because although there is a rough structure for the interview, there is still free space in the interview process. Thus, the questions can be adapted or modified according to the situation. However, standardisation is achieved by structuring the content of the topics (Döring & Bortz, 2016; Kruse, 2014).

The interviews were evaluated by content analysis. According to Mayring (2015), communication that is fixed in content analyses is examined systematically, rule guided, and theory driven. The goal is to draw conclusions regarding selected aspects or questions. According to Kuckartz (2016), qualitative content analysis is an interpretative form of evaluation that is linked to a human ability to understand and interpret. For the present material, the content-structuring qualitative content analysis according to Kuckartz (2016) was used.

For this purpose, all process steps were run through and supplemented by iteration and feedback steps. The formation of categories was largely inductive, that is, statements that could be assigned to the questions were grouped according to content. A codebook was created for the fixation, which, according to Kuckartz (2016, pp. 29–55), contains the description of the content and the application of the category as well as corresponding anchor examples and special comments.

To check the quality, the intercoder reliability between the first and the second coder was determined. The second coder coded approximately 35% of the text material based on the codebook. According to Döring and Bortz (2016, p. 558) at least 10–20% of the data material should be coded by two coders. Cohen's kappa was determined as the reliability coefficient. It was 0.76 and can be classified as "substantive" according to Landis and Koch (1977). The entire analysis was carried out using the MAXQDA 2020 software.

To illustrate the results, anchor examples were translated into English. This procedure does not represent a falsification of the material.

## 15.5 Results

All 19 interviews were included in the data analysis. A total of 208 statements were coded. Five categories were identified for the first research question with a total of 26 statements. The number of mentions in each category can be seen in Table 15.1.

There were also five categories found in the benefits (see Table 15.2), where there was a total of 52 statements. Some disadvantages were also mentioned (10 statements).

In addition to the advantages, the students gave a variety of reasons for attending the project week (97 statements). These can be seen in Table 15.3.

In addition, there are 23 statements that are important in this context but could not be assigned to any of the categories. These will be included in the discussion.

**Table 15.1** Five definitions of interdisciplinary teaching (given by students)

| Subcategory | Anchor example | Number of mentions |
|---|---|---|
| Subject level | | |
| Intradisciplinary teaching | "Somehow one subject area from the other areas just gets mixed in with another." | 7 |
| Multidisciplinary teaching | "Topics are covered in different lessons, and they can be somehow be linked together." | 9 |
| Interdisciplinary teaching | "This means projects that don't just have a single subject but rather focus on the project itself. Then it is just a mixture of different disciplines. There is something about everything." | 3 |
| Lesson-schedule level | | |
| Supplementary teaching | "project courses" | 5 |
| Integrated teaching | "science teaching" | 2 |

**Table 15.2** Students' views on the advantages and disadvantages of interdisciplinary teaching

| Subcategory | Anchor example | Number of mentions |
|---|---|---|
| Advantages | | |
| Connections | "One can see more connections" | 20 |
| More effective/better understanding | "It is much more effective"<br>"Facts and circumstances are easier to understand" | 13 |
| Improve knowledge | "To gain more expertise and broaden one's knowledge" | 8 |
| Interest/motivation | "I think teaching that way definitely increases a student's motivation even more" | 7 |
| Relevance to everyday life | "That one can establish the relevance to everyday life" | 4 |
| Disadvantages | | |
| Missing competences | "If you are not so good in one subject, then of course you will not do so well in the interdisciplinary subjects" | 6 |
| Ignoring fundamental knowledge | "Fundamental knowledge is perhaps not taught so well" | 4 |

**Table 15.3** Students' reasons for registering on the interdisciplinary project week

| Subcategory | Anchor example | Number of mentions |
|---|---|---|
| Interest | "Because I am definitely interested in maths" | 23 |
| Benefits in the school | "I believe that this will also help at school" | 18 |
| Career orientation | "I can (...) imagine something like this in my future" | 16 |
| Practical work | "Work in a laboratory" | 12 |
| fun | "I also think it's a lot of fun" | 8 |
| Training for A-levels | "It helps me prepare for maths lessons as well as for the final secondary school examinations." | 8 |
| Identify applications | "But one also learns how one can apply the knowledge" | 8 |
| Subject combination | "I think this combination is quite interesting" | 4 |

## 15.6 Discussion

In the interviews, the students described what they understood by interdisciplinary teaching and what advantages and disadvantages they saw in such a course offering. First of all, it was apparent from the interviews that all students have an idea of interdisciplinary instruction or other course offerings. However, they do not make a conscious distinction between the different forms. Moreover, different elements of the individual gradations can be found.

Statements such as "I now imagine that there is a connection between the subjects. For example, if you need a certain calculation or something in biology", or "Somehow one subject area from the other areas just gets mixed in with another" show that there is a small overlap between subjects. These students know or describe intradisciplinary teaching. More statements are made about multidisciplinary teaching. Here students describe dealing with the same topic in different subjects. "That we connect two subjects, for example, and then cover certain topics in both subjects", or "Topics are covered in different lessons, and they can be somehow be linked together".

Relatively little attention is focused on interdisciplinary teaching. For example, "This means projects that don't just have a single subject but rather focus on the project itself. Then it is just a mixture of different disciplines. There is something about everything". Here the common question or the corresponding context is also dealt with; e.g. "Keep it mostly related to a context".

The students can cite many examples from their own experience for the supplementary teaching of subjects. For example, "project courses", the "differentiation course in biochemistry", or the supplementary subject "natural sciences". In this supplementary subject "Math, biochemistry, chemistry, and physics are taught together". This means that these subjects make a contribution, but it is a supplement to the single subjects. In contrast, for integrated teaching, there was only the example of "science teaching" in secondary school. This includes biology, physics and chemistry and is a regular part of the schedule.

In summary, all three forms of interdisciplinary instruction according to Labudde (2014) were mentioned by the students. Based on the frequency of the corresponding statements, intra- and multidisciplinary teaching are most frequent at the subject level and supplementary teaching at the lesson-schedule level (see Table 15.1).

The students stated a variety of advantages (see Table 15.2) that they see in the combination of subjects. Regardless of how they defined it beforehand. Thus, they find the combination "much more interesting" than normal lessons, which "often seem quite dry". Also "I think teaching that way definitely increases a student's motivation even more". Labudde (2014) already points this out (see Sect. 15.2.2). In addition, almost all of them say that they can better recognise "what the connection is" and "you can really implement this on something".

Moreover, they perceive it as "much more effective" since "Facts and circumstances are easier to understand". "For example, I have not learned a formula by heart now, but I now have really understood what it says, what this formula actually means".

Another "advantage is that if you can do both well, you can expand your knowledge and go much deeper into these areas of information". "That one can establish the relevance to everyday life" is also often positively highlighted. These statements underline the advantages of networking content but also the advantages of interdisciplinary competences and science propaedeutics, as formulated by Labudde (2014) and Huber (1998).

One student voiced the view of several by saying "I would like to see more interdisciplinary lessons in schools or, alternatively, school clubs in the afternoon or other course offerings from schools". This means that they would like to see more lessons combining subjects but also free course offerings through project weeks or something similar.

However, they raise some limited concerns that some of the subject matter would be dropped compared to their normal teaching. "Fundamental knowledge is perhaps not taught so well". Uncertainties also become clear in relation to their own competences, "If you are not so good in one subject, then of course you will not do so well in the interdisciplinary subjects".

For this reason, a project week combining different subjects could be a good opportunity to complement the school. In the project week, the connection between subjects is emphasised very positively, as "that simply belongs together". The explicit combination of biology and mathematics was also positively highlighted here, "I think this combination is quite interesting". However, many of the reasons for registering for the project week have nothing to do with the subject combination (see Table 15.3). Students reported that they were very interested in one subject: "because I am simply very interested in biology" or "because I am definitely interested in maths". Most of the statements, however, were related to their interest in biology, which was enough reason to register. In addition to their own interests, career guidance also had an influence on their registration. "I can (...) imagine something like this in my future. I wanted to have a closer look". Many of the participants can imagine working in the STEM fields, and they want to use the project week "to gain experience".

Importance is attached to practical and laboratory work. The interviewed students wanted to know "what it is like to work in a laboratory" and to be able to "gain practical experience", as this is often not possible in schools. According to Frey (2010), practical work is often part of project work or project teaching, which in turn often contains interdisciplinary aspects.

The students also hope that they will be able to apply the contents of the course at school: "I believe that this will also help at school". In particular, the preparation and the repetition for the central school examination is discussed: "I had to repeat something like that anyway" or "It helps me prepare for maths lessons as well as for the final secondary school examinations". This wish necessitates links to the corresponding curricula.

Overall, when evaluating the results, we must consider that these are highly interested and possibly talented students. Their motives for participating and their general attitudes towards project weeks may differ from those of other students. The same applies to the awareness of advantages and disadvantages of interdisciplinary course offerings. It should be noted that the students became more and more talkative over the course of the interview. This could explain the continuum in the number of statements during the interview. Moreover, this is a study with a very specific question. Next time, the research field should be further examined. Although Becker and Park (2011) were able to show that interdisciplinary connection between the STEM subjects has a positive effect on learning, other subjects should not be excluded in this context. Through the current study design, this could be the case.

## 15.7 Conclusion and Outlook

Our results indicate that all students have an idea of interdisciplinary instruction. All statements made can be classified as a sub-form according to Labudde (2003, 2014). However, the students are not aware of the different forms. In their view, there are also many different advantages to be gained from interdisciplinary instruction or course offerings. These are the same as those found in literature. Therefore, it is not surprising that they would like to see more such course offerings, even though, at the moment, they are rarely implemented in German schools. These two statements could be further confirmed by a quantitative questionnaire survey. This raises the question of why interdisciplinary lessons or courses at school do not take up more space in the lesson schedule. For a deeper insight into the topic, the next step could be to interview teachers about interdisciplinary course offerings. In this way, structures and help could be provided to support more interdisciplinary teaching in schools. As a further step, the actual effectiveness of interdisciplinary teaching should be evaluated.

## References


Åström, M. (2008). *Defining integrated science education and putting it to test. Studies in science and technology education (Print): Vol. 26*. Swedish National Graduate School in Science and Technology Education, FontD; Department of Social and Welfare Studies, Linköping University.

Becker, K., & Park, K. (2011). Effects of integrative approaches among science, technology, engineering, and mathematics (STEM) subjects on students' learning: A preliminary meta-analysis. *Journal of STEM Education, 12*(5), 23–37.

Berck, K.-H., & Graf, D. (2018). *Biologiedidaktik: Grundlagen und Methoden* (5th ed.). Quelle & Meyer Verlag.

Bundesministerium für Bildung und Forschung (BMBF). (2018). *Systembiologie – Moderne Forschung zur Entschlüsselung des Lebens*. https://www.bmbf.de/de/systembiologie-moderne-forschung-zur-entschluesselung-des-lebens-411.html

Buse, M. (2017). *Bilinguale englische experimentelle Lehr-Lernarrangements im Fach Biologie: Konzeption, Durchführung und Evaluation der kognitiven und affektiven Wirksamkeit* [Dissertation]. University of Wuppertal, Wuppertal.

Caviola, H., Kyburz-Graber, R., & Locher, S. (2011). *Wege zum guten fächerübergreifenden Unterricht: Ein Handbuch für Lehrpersonen* (1. Aufl.). hep verla ag.

Döring, N., & Bortz, J. (2016). *Forschungsmethoden und Evaluation in den Sozial- und Humanwissenschaften* (5th ed.). Springer-Lehrbuch. Springer.

Duerr, L. (2008). Interdisciplinary Instruction. *Educational Horizons, 86*(3), 173–180.

Duncker, L., & Popp, W. (1998). Formen fächerübergreifenden Unterrichts auf der Sekundarstufe – eine Einleitung. In L. Duncker & W. Popp (Eds.), *Fächerübergreifender Unterricht in der Sekundarstufe I und II: Prinzipien, Perspektiven, Beispiele* (pp. 7–17). Klinkhardt.

Frey, K. (2010). *Die Projektmethode* (11th ed.). Beltz.

Huber, L. (1994). Wissenschaftspropädeutik – Eine unerledigte Hausaufgabe der allgemeinen Didaktik. In M. A. Meyer & W. Plöger (Eds.), *Pädagogik: Bd. 10. Allgemeine Didaktik, Fachdidaktik und Fachunterricht* (pp. 243–253). Beltz.

Huber, L. (1995). Individualität zulassen und Kommunikation stiften – Vorschläge und Fragen zur Reform der gymnasialen Oberstufe. *Die Deutsche Schule, 87*(2), 161–182.

Huber, L. (1998). Fächerübergreifender Unterricht - auch auf der Sekundarstufe II? In L. Duncker & W. Popp (Eds.), *Fächerübergreifender Unterricht in der Sekundarstufe I und II: Prinzipien, Perspektiven, Beispiele* (pp. 18–33). Klinkhardt.

Joachim Herz Stiftung. (2016). Projektwoche Systembiologie. *LeLa Magazin* (15), 18–19.

Jones, C. (2010). Interdisciplinary approach – Advantages, disadvantages, and the future benefits of interdisciplinary studies. *ESSAI, Article 26*(7), 76–81.

Keller, G. (2011). *Mathematik in den Life Sciences: Grundlagen der Modellbildung und Statistik mit einer Einführung in die Statistik-Software R. UTB Biologie, Agrarwissenschaften, Ernährungswissenschaften: Vol. 3493*. Ulmer.

Killermann, W., Hiering, P., & Starosta, B. (2018). *Biologieunterricht heute: Eine moderne Fachdidaktik* (17th ed.). Didaktik. Auer.

Klafki, W. (1998). Fächerübergreifender Unterricht - Begründungsargumente und Verwirklichungsstufen. In S. Popp (Ed.), *Grundrisse einer humanen Schule: Festschrift für Rupert Vierlinger* (pp. 41–57). Studien-Verlag.

Klos, S. (2008). *Kompetenzförderung im naturwissenschaftlichen Anfangsunterricht – der Einfluss eines integrierten Unterrichtskonzepts. Studien zum Physik- und Chemielernen: Vol. 89*. Logos-Verl.

Krause, E. (2015). Fächerverbindende Didaktik am Beispiel von subjektiven Lernvoraussetzungen im Mathematik- und Physikunterricht. In Gesellschaft für Didaktik der Mathematik (Ed.), *Beiträge zum Mathematikunterricht 2015, 49. Jahrestagung der Gesellschaft für Didaktik der Mathematik vom 09.02. bis 13.02.2015 in Basel* (pp. 492–495).

Kruse, J. (2014). *Qualitative Interviewforschung: Ein integrativer Ansatz* (2nd ed.). Grundlagentexte Methoden. Beltz Juventa.

Kuckartz, U. (2016). *Qualitative Inhaltsanalyse: Methoden Praxis Computerunterstützung* (3rd ed.). Grundlagentexte Methoden. Beltz Juventa.

Labudde, P. (2003). Fächer übergreifender Unterricht in und mit Physik: eine zu wenig genutzte Chance. *Physik Und Didaktik in Schule Und Hochschule, 2*(1), 48–66.

Labudde, P. (2014). Fächerübergreifender naturwissenschaftlicher Unterricht – Mythen, Definitionen, Fakten. *Zeitschrift Für Didaktik Der Naturwissenschaften, 20*(1), 11–19.

Labudde, P., Heitzmann, A., Heiniger, P., & Widmer, I. (2005). Dimensionen und Facetten des fächerübergreifenden naturwissenschaftlichen Unterrichts: ein Modell. *Zeitschrift Für Didaktik Der Naturwissenschaften, 11*, 103–115.

Landis, J. R., & Koch, G. G. (1977). The measurement of observer agreement for categorical data. *Biometrics, 33*(1), 159–174. https://doi.org/10.2307/2529310

Loos, A., & Ziegler, G. M. (2015). Gesellschaftliche Bedeutung der Mathematik. In R. Bruder (Ed.), *Handbuch der Mathematikdidaktik* (pp. 3–17). Springer Spektrum.

Mayring, P. (2015). *Qualitative Inhaltsanalyse: Grundlagen und Techniken* (12th ed.). Beltz Pädagogik.

Moegling, K. (2010). *Kompetenzaufbau im fächerübergreifenden Unterricht: Förderung vernetzten Denkens und komplexen Handelns; didaktische Grundlagen Modelle und Unterrichtsbeispiele für die Sekundarstufen I und II. Theorie und Praxis der Schulpädagogik: Vol. 2*. Prolog-Verl.

OECD. (2018). *PISA 2015: PISA results in focus*. https://www.oecd.org/pisa/pisa-2015-results-in-focus.pdf

Oliver, M., McConney, A., & Woods-McConney, A. (2019). The efficacy of inquiry-based instruction in science: A comparative analysis of six countries using PISA 2015. *Research in Science Education*. Advance online publication. https://doi.org/10.1007/s11165-019-09901-0

Osborne, J., Simon, S., & Collins, S. (2003). Attitudes towards science: A review of the literature and its implications. *International Journal of Science Education, 25*(9), 1049–1079. https://doi.org/10.1080/0950069032000032199

Rodenhauser, A. (2016). *Bilinguale biologische Schülerlaborkurse Elektronische Ressource: Konzeption und Durchführung sowie Evaluation der kognitiven und affektiven Wirksamkeit* [Dissertation]. University of Wuppertal, Wuppertal.

Ruppert, W. (2015). Welches Interesse haben Schüler an biologischen Themen? In U. Spörhase-Eichmann (Ed.), *Biologie-Didaktik: Praxishandbuch für die Sekundarstufe I und II* (7th ed., pp. 94–111). Cornelsen Scriptor.

Schiepe-Tiska, A., Simm, I., & Schmidtner, S. (2016). Motivationale Orientierungen, Selbstbilder und Berufserwartungen in den Naturwissenschaften in PISA 2015. In K. Reiss, C. Sälzer, A. Schiepe-Tiska, E. Klieme, & O. Köller (Eds.), *PISA 2015: Eine Studie zwischen Kontinuität und Innovation* (pp. 99–132). Waxmann.

Statistisches Bundesamt. (2020). *Studienanfänger/-innen im 1. Fachsemester nach Fächergruppen*. https://www.destatis.de/DE/Themen/Gesellschaft-Umwelt/Bildung-Forschung-Kultur/Hochschulen/Tabellen/studierende-erstes-fs-faechergruppen.html

Walker, W. S. (2017). Integrated STEm or Integrated STEM? *School Science and Mathematics, 117*(6), 225–227. https://doi.org/10.1111/ssm.12234

Wanka, J. (2015). Grußwort in Schülerausgabe (Systembiologie.de.Scholae). *Systembiologie. De*(Schülerausgabe).

Wenzel, A., & Grotjohann, N. (2019). Projektwoche Systembiologie im *teutolab*-biotechnologie. *Journal Für Didaktik Der Naturwissenschaften Und Der Mathematik, 3*, 99–113.

Woods, C. (2007). Researching and developing interdisciplinary teaching: Towards a conceptual framework for classroom communication. *Higher Education, 54*(6), 853–866. https://doi.org/10.1007/s10734-006-9027-3

# Chapter 16
# Correlation Between the Popularity and Difficulty of Secondary School Biology and Perceived Importance of Knowledge Acquired for Personal Wellbeing

**Vida Lang and Andrej Šorgo**

## 16.1 Introduction

Today, one cannot read a newspaper without being aware of the central importance of biology in people's lives. Whether it is understanding new developments in medicine or contributing to local environmental decisions, everyone needs a basic understanding of key biological concepts, and we should all value science as a special way of understanding the world, which should be reflected in biology education. Students are motivated to learn biology for a variety of reasons. Knowledge of biology increases their level of general education; they use the knowledge they acquire in their daily lives and in a variety of careers.

Globally, we observe a declining interest in STEM-oriented studies and lifelong careers in STEM-oriented fields (Glynn et al., 2009; Xie & Achen, 2009; Simon et al., 2015; Wiebe et al., 2018), including careers in biotechnology and agriculture (Šorgo & Špernjak, 2020). The importance of the advancement of basic scientific knowledge in the field of biology has not only led to an upgrading of explanations of objects and processes in the living world and its environment, but has also significantly impacted our personal and social life and careers (Bell et al., 2009; Šorgo & Špernjak, 2020; Šorgo et al., 2018; Tai et al., 2006). Therefore, in the future, we can expect even greater importance of general biological education for making informed personal, professional, and social decisions. It is reasonable to hypothesize that what is taught in school biology will have an impact on personal wellbeing and careers later in life. As proof of concept, we have recently seen the importance of

V. Lang (✉) · A. Šorgo
Faculty of Natural Sciences and Mathematics, University of Maribor, Maribor, Slovenia
e-mail: vida.lang1@um.si; andrej.sorgo@um.si



K. Korfiatis, M. Grace (eds.), *Current Research in Biology Education*, Contributions from Biology Education Research,
https://doi.org/10.1007/978-3-030-89480-1_16

knowledge of microbiology on protective behaviour during the outbreak of COVID-19. Therefore, the role of biology education does not end with making the school subject of biology self-sufficient and empowering, but can also influence everyday and career decisions.

Attitudes (Jidesjö, 2008; Usak et al., 2009) and motivation are considered to be important factors influencing not only outcomes and achievement in biology education (Koballa Jr. & Glynn, 2013), but also decisions about future study and career choices (Šorgo & Špernjak, 2020; Šorgo et al., 2018). Among many intrinsic, extrinsic, and motivational constructs (Ryan & Deci, 2000), the most important is content-based goal motivation (Nenniger, 1987). Research shows (Prokop et al. 2007a; Prokop et al. 2007b) that adolescents are not equally interested in all biological topics. Motivation is moderated by an almost unlimited number of factors, including the importance, difficulty, and popularity of the subject Biology considered in this study.

Relations between the perceived importance, difficulty, and popularity of Biology and its various biological topics are poorly studied. Prokop et al. (2007b) found that almost half of students considered Biology as important and agreed that they need biology knowledge. Biology is regularly considered the most popular subject among STEM subjects, where girls prefer Biology and boys are more attracted to Physics (Baram-Tsabari & Yarden, 2008; Prokop et al. 2007a). Temelli and Kurt (2013) investigated various factors that might influence students' attitudes towards STEM subjects. Uitto (2014) analysed students' interests and beliefs about their abilities and how these are related to their attitudes towards school science subjects. Incantalupo, Treagust, and Koul (2014) analysed the impact of technology on students' attitudes towards science subjects. In Slovenia, several bachelor's and master's theses showed that Biology is the most popular science subject, compared to Physics and Chemistry. In her MA work, Kotnik (2016) found that students consider biology knowledge to be important because there are living things everywhere around us. Students showed the lowest level of agreement for the statement that Biology knowledge is important for understanding other school subjects. Research (Jidesjö, 2008; Usak et al., 2009) shows that adolescents are interested in science and biology, and they are more interested in topics about animals, health problems, nutrition, and medicines. Boys are more interested in topics about evolution, cloning of animals, space, earth science and sports, while girls have more preferences for human biology and general biology (Dawson, 2000). Zoology has been found to be of most interest to both genders (Prokop et al., 2007b). According to Oscarsson et al. (2009), extracurricular experiences, experimental work, field trips, and reading science journals are significantly influenced by maintaining interest in biological content. However, the relationship between popularity and different aspects of importance has not yet been investigated. Therefore, we decided to investigate the potential correlation between the importance and popularity of Biology, and the potential correlation between the importance and difficulty of Biology.

## 16.2 Research Design and Method

### *16.2.1 Aims and Scope of the Research*

The main aim of the study was to provide empirical evidence that would enable us to initiate a debate on the position and necessary changes in Slovenian secondary school Biology teaching in relation to the importance, popularity, and difficulty of the subject. In addition, we wanted to determine the relative position of Biology among other school subjects.

The research questions were as follows:

1. How important are different dimensions of biology as perceived by upper secondary school students?
2. How popular is the subject of Biology in relation to other school subjects?
3. How difficult is the subject of Biology in relation to other school subjects?
4. Is there a correlation between perceived importance, popularity, and difficulty of the subject of Biology?
5. Are there any gender differences in the recognition of the importance, popularity and difficulty of Biology as a subject?

### *16.2.2 Research Methods*

#### 16.2.2.1 Data Collection and Instruments

An online survey in the form of a questionnaire was designed on the Slovenian open survey system 1KA (Enklik, 2017). The link, accompanied by a short flyer with a brief description of the research objectives and the survey instrument, was sent to the email addresses of 92 high school Biology teachers. It was also posted on teacher forums and online social networks. Participation in the survey was voluntary and complete anonymity was guaranteed. Teachers decided for themselves whether their students would participate in the survey. Our goal was to collect at least 500 complete surveys.

The limitations of such a research design are well known and well documented (Lefever et al., 2007). The main limitations are that respondents are self-selecting and, in our case, may include a higher number of them who are interested in biology, which may lead to biased results.

The online questionnaire questions contained subscales. The first part consisted of 10 items that assessed different aspects of the importance of biology (Table 16.1). In the second part of the questionnaire, students rated school subjects in terms of popularity and difficulty (Table 16.2), and demographic data were collected in the third part.

**Table 16.1** Measures of central tendencies for responses to the question "Why is knowledge of biology important?" and correlations of importance with popularity (*r1*) and difficulty (*r2*). The response format was a 7-point Likert scale (1 – completely unimportant; 7 – very important)

| # | Claim | M | SD | Me | Mo | PC | *r1* | *r2* |
|---|---|---|---|---|---|---|---|---|
| 1 | Because of my knowledge of biology, I am more generally broad-minded, | 5.54 | 1.37 | 6 | 6 | .80 | .45 | .07 |
| 2 | I know more about nature and living things, | 5.45 | 1.29 | 6 | 6 | .78 | .36 | .13 |
| 3 | I know more about the structure and functioning of living things, | 5.43 | 1.36 | 6 | 6 | .75 | .31 | .16 |
| 4 | I have a better understanding of sciences such as genetics, medicine and biotechnology, | 5.38 | 1.50 | 6 | 6 | .75 | .36 | .07 |
| 5 | I know more about human anatomy and physiology, | 5.35 | 1.51 | 6 | 6 | .73 | .34 | .09 |
| 6 | It helps me understand everyday processes in the environment, | 5.34 | 1.35 | 5 | 5 | .82 | .39 | .09 |
| 7 | I can use biology knowledge to protect nature, | 5.17 | 1.46 | 5 | 5 | .75 | .33 | .05 |
| 8 | Knowledge of biology is important for me to understand modern scientific discoveries. | 4.74 | 1.59 | 5 | 5 | .77 | .37 | .14 |
| 9 | It helps me to live in a healthy way, | 4.62 | 1.49 | 5 | 5 | .69 | .32 | .09 |
| 10 | Knowledge of biology helps me to solve everyday problems. | 4.03 | 1.69 | 4 | 4 | .69 | .33 | .12 |

Note: *M* arithmetic mean, *SD* standard deviation, *Me* median, *Mo* mode, *PC* principal component, *r1* correlation between importance and popularity (BIO), *r2* correlation between importance and difficulty (BIO)

**Table 16.2** Measures of tendencies in the popularity and difficulty of school subjects and correlations between them. The scales were from 1 (most popular/difficult) to 10 (least popular/difficult)

| | Popularity | | | | Difficulty | | | | | |
|---|---|---|---|---|---|---|---|---|---|---|
| School subject | M | SD | Mo | Me | M | SD | Mo | Me | *r* | *d* |
| Biology | 4.45 | 2.83 | 1 | 4 | 4.61 | 2.36 | 4 | 4 | −.16 | 0.06 |
| History | 4.85 | 2.90 | 1 | 5 | 6.31 | 2.83 | 10 | 7 | −.28 | 0.51 |
| Geography | 5.09 | 2.67 | 1 | 5 | 6.12 | 2.40 | 5 | 6 | −.23 | 0.41 |
| Chemistry | 5.42 | 2.73 | 7 | 5 | 4.95 | 2.46 | 3 | 5 | −.21 | −0.18 |
| Slovene | 5.44 | 2.68 | 6 | 6 | 5.06 | 2.66 | 4 | 5 | −.33 | −0.14 |
| Mathematics | 5.68 | 3.02 | 9 | 6 | 3.52 | 2.73 | 1 | 2 | −.33 | −0.75 |
| Physics | 6.71 | 2.94 | 10 | 7.5 | 3.45 | 2.69 | 1 | 2 | −.27 | −1.12 |

Note: *M* arithmetic mean, *SD* standard deviation, *Me* median, *Mo* mode, *r* correlation between popularity and difficulty; *d* Cohen's *d* calculated from means and SD's of popularity and difficulty

#### 16.2.2.2 Participants

The sample consisted of 598 students (72% girls and 28% boys) from several Slovenian upper secondary schools, who participated anonymously. 94% of the respondents were from general secondary schools and 6% of the respondents were from vocational secondary schools. Students from all four upper secondary grades participated in the survey (23% from 1st year, 38% from 2nd year, 27% from 3rd year, and 12% from 4th year). 31.2% of respondents answered that they will most probably choose to study in the field of natural sciences, 24.6% of students will study social sciences, and the remaining 44.2% have not yet decided or did not answer the question.

#### 16.2.2.3 Data Analysis

(a) The scale of importance of biology. We asked students why they considered biology knowledge important in different contexts (Table 16.1). The response format was a 7-point Likert scale (1 – completely unimportant; 7 – very important). Cronbach's alpha of the instrument was 0.92. Means, standard deviations, modes, and medians are reported. The adequacy of the data to perform Principal Component Analysis (PCA) was assessed using KMO (0.92) (Barlett's test = 3331.626, $df = 45$, $p < 0.001$). PCA revealed unidimensionality of the instrument, and a component explaining 57.5% of the variance was extracted.
(b) Popularity and difficulty of Biology scales. To assess the popularity of Biology as a subject, students were instructed to assign a number to each of the ten subjects listed, ranking them on a scale from most popular (1) to least popular (10), and similarly, they rated the difficulty of the subjects on the scale from most difficult (1) to the least difficult (10).

Spearman's rank correlation test was used to calculate the correlations between importance, popularity, and difficulty. Subjects such as Philosophy, Sociology, and Psychology were later excluded from the analyses because they did not appear in the curricula until later school years and some participants were not able to rate them due to their experiences. Gender differences were calculated using the Mann-Whitney test. All statistics were performed using SPSS 24.

## 16.3 Findings

The results of the importance of biology knowledge (Table 16.1), popularity, and difficulty level of Biology as a subject (Table 16.2) are presented in the tables with comments.

It can be seen from Table 16.1 that at the top of the importance list are statements that relate to knowledge, which shows that school Biology is not seen as something that gives one practical and usable knowledge. An overview of the students' statements shows that students are aware of the importance of biology knowledge, both for the comprehensive overview and for understanding the different processes of living things and the environment; but they do not find it useful. Nevertheless, the majority of the statements are on the positive side.

The results show a weak to moderate correlation between the importance of biology knowledge and the popularity of Biology as a school subject, and an almost non-existent correlation with the difficulty of Biology as a school subject. No statistically significant gender differences were detected (Table 16.1).

Among the seven subjects listed, students ranked Biology as the most popular subject (Table 16.2), with Physics at the bottom of the list and Chemistry in the middle. On a scale of 1 (most popular) to 10 (least popular), 18.2% of the students surveyed ranked Biology as the most popular subject in secondary school, 13.1% of the students ranked it as the second most popular subject, 10.7% ranked it as the third most popular subject, and only 5.1% of students ranked Biology last in terms of popularity.

On a scale of 1 (most difficult) to 10 (least difficult), 7.4% of the students surveyed ranked Biology as the most difficult subject in secondary school, and 4.2% of the students ranked it as the least difficult subject. The subjects with the highest level of difficulty include Physics, Mathematics, Biology, Chemistry, and Slovene. Students ranked Biology as both a popular and difficult subject at school.

When examining the correlation between the popularity of a subject and its perceived difficulty, it can be seen (Table 16.2) that all correlations are negative and lie in small or medium ranges. The differences presented as effect size are in ranges from insignificant (e.g. Biology) to large (e.g. Physics), which means that Biology, as the most popular subject, is balanced with its difficulty. History and Geography are more likely to be perceived as popular than difficult, while from Chemistry onwards difficulty is an indicator of unpopularity, with the extreme of Physics being perceived as the most difficult of all subjects and the least popular. No statistically significant differences were found by gender.

## 16.4 Conclusions

Looking at the importance of the different dimensions of biology as perceived by upper secondary school students in our study we can conclude that the students regard Biology as an important discipline. However, they rate the importance of biology more as an academic discipline rather than something that can help someone lead a healthy life or solve everyday problems. Similar conclusions were also reached by Prokop et al. (2007a), who found that Slovak students do not apply biology knowledge in their everyday life. According to Prokop et al. (2007a), a large proportion of students in Slovakia believe that biology knowledge is an important asset for their profession (career), but they least agree with the idea that biology knowledge helps them solve everyday problems. The probable solution to this is not to encourage them to feel they are generally more broadly positioned and know more about nature and living things, but to incorporate strategies that help them realise its value in solving everyday problems.

We can conclude that science should not be considered a monolithic discipline, so the question of popularity and difficulty – of various aspects of science – is justified only when someone examines Science as a single school subject.

Because the subject Biology is together with History and Geography very popular, these other two subjects are its main competitors. Biology is far ahead of Chemistry, Mathematics, and Physics, therefore we suggest that subjects include topics that allow interdisciplinarity and knowledge transfer between them (Šorgo, 2010). Popularity can arise from interest. For example, Krapp et al. (1992) found that interest in STEM professions is not based on achievement, but on curiosity and enjoyment in learning a particular subject or field. This curiosity and enjoyment are seen as a person focusing more on certain areas, followed by an inclination to seek out and participate in activities in the future (Hidi & Renninger, 2006). In order to increase interest, popularity, and awareness of the importance of biology knowledge in life and in general, it is important not to limit Biology to school, but to put more emphasis on students' experiences outside school, in nature, and involve them in non-formal education (Uitto et al., 2006). It seems that the advantages of non-formal science education can help overcome the disadvantages of formal science education and vice versa. Therefore, the collaboration between school time and learning environments outside the classroom can open up promising possibilities for the future of STEM education (Dabney et al., 2012).

Biology ranks third in place on the difficulty scale, behind Physics and Mathematics and ahead of Chemistry. The perceived difficulty of a school subject seems to have a strong influence on its popularity – the extreme example being the subject of Physics. However, our research shows that Biology is perceived by students as a balanced subject, i.e. equally popular and difficult. From this, we can conclude that it may not be fair to blame all science disciplines (in our case Biology) for the general unpopularity of the STEM professions. It can be regarded as a weak negative predictor of importance and moderate predictor of popularity, which raises the question of whether we would prefer to find a subject popular or important if we reduce its difficulty. Left unanswered is the question of whether reducing difficulty would keep Biology at the same level of importance.

We did not find any gender differences in perceptions of the importance, popularity, and difficulty of Biology as a subject, Which contrasts with findings that girls have a greater preference for Biology as a subject and boys for Physics (Baram-Tsabari & Yarden, 2008; Prokop et al. 2007a; Simon et al., 2015; Wiebe et al., 2018).

## References


Baram-Tsabari, A., & Yarden, A. (2008). Girls' biology, boys' physics: Evidence from free-choice science learning settings. *Research in Science & Technological Education, 26*(1), 75–92.

Bell, P., Lewenstein, B., Shouse, A. W., & Feder, M. A. (2009). *Learning science in informal environments: People, places, and pursuits* (Vol. 140). National Academies Press.

Dabney, K. P., Tai, R. H., Almarode, J. T., Miller-Friedmann, J. L., Sonnert, G., Sadler, P. M., & Hazari, Z. (2012). Out-of-school time science activities and their association with career interest in STEM. *International Journal of Science Education, Part B, 2*(1), 63–79.

Dawson, C. (2000). Upper primary boys' and girls' interests in science: Have they changed since 1980? *International Journal of Science Education, 22*(6), 557–570.

Enklik. (2017). 1KA [software] Ljubljana Faculty of Social Sciences. Available at https://www.1ka.si

Glynn, S. M., Taasoobshirazi, G., & Brickman, P. (2009). Science motivation questionnaire: Construct validation with nonscience majors. *Journal of Research in Science Teaching: The Official Journal of the National Association for Research in Science Teaching, 46*(2), 127–146.

Hidi, S., & Renninger, K. A. (2006). The four phase model of interest development. *Educational Psychologists, 41*(2), 111–127.

Incantalupo, L., Treagust, D. F., & Koul, R. (2014). Measuring student attitude and knowledge in technology-rich biology classrooms. *Journal of Science Education and Technology, 23*(1), 98–107.

Jidesjö, A. (2008). Different content orientations in science and technology among primary and secondary boys and girls in Sweden: Implications for the transition from primary to secondary school? *Nordic Studies in Science Education, 4*(2), 192–208.

Koballa, Jr., T. R., & Glynn, S. M. (2013). Attitudinal and motivational constructs in science learning. In *Handbook of research on science education* (pp. 89–116). Routledge.

Kotnik, A. (2016). *Odnos učencev do biologije in naravoslovnih predmetov v osnovni šoli* [Attitude towards Biology and other science subjects, of learners in primary school] (Master Thesis). Univerza v Mariboru, Fakulteta za naravoslovje in matematiko).

Krapp, A., Hidi, S., & Renninger, K. A. (1992). Interest, learning, and development. In K. A. Renninger, S. Hidi, & A. Krapp (Eds.), *The role of interest in learning and development* (pp. 3–25). Lawrence Erlbaum Associates, Inc..

Lefever, S., Dal, M., & Matthiasdottir, A. (2007). Online data collection in academic research: Advantages and limitations. *British Journal of Educational Technology, 38*(4), 574–582.

Nenniger, P. (1987). How stable is motivation by contents? In E. D. Corte, H. Lodjwiks, R. Parmentier, & P. Span (Eds.), *Learning and instruction, European research in an international context* (Vol. 1, pp. 159–168). Pergamon Press.

Oscarsson, M., Jidesjö, A., Strömdahl, H., & Karlsson, K. G. (2009). Science in society or science in school: Swedish secondary school science teachers' beliefs about science and science lessons in comparison with what their students want to learn. *Nordic Studies in Science Education., 5*(1), 18–34.

Prokop, P., Prokop, M., & Tunnicliffe, S. D. (2007a). Is biology boring? Student attitudes toward biology. *Journal of Biological Education, 42*(1), 36–39.

Prokop, P., Tuncer, G., & Chudá, J. (2007b). Slovakian students' attitudes toward biology. *Eurasia Journal of Mathematics, Science & Technology Education., 3*(4), 287–295.

Ryan, R. M., & Deci, E. L. (2000). Intrinsic and extrinsic motivations: Classic definitions and new directions. *Contemporary Educational Psychology., 25*(1), 54–67.

Simon, R. A., Aulls, M. W., Dedic, H., Hubbard, K., & Hall, N. C. (2015). Exploring student persistence in STEM programs: A motivational model. *Canadian Journal of Education, 38*(1), n1.

Šorgo, A. (2010). Connecting biology and mathematics: First prepare the teachers. *CBE – Life Sciences Education, 9*(3), 196–200.

Šorgo, A., & Špernjak, A. (2020). Biology content and classroom experience as predictors of career aspirations. *Journal of Baltic Science Education., 19*(2), 317–332.

Šorgo, A., Dojer, B., Golob, N., Repnik, R., Repolusk, S., Pesek, I., & Špur, N. (2018). Opinions about STEM content and classroom experiences as predictors of upper secondary school students' career aspirations to become researchers or teachers. *Journal of Research in Science Teaching, 55*(10), 1448–1468.

Tai, R. H., Liu, C. Q., Maltese, A. V., & Fan, X. (2006). Planning early for careers in science. *Science, 312*, 1143–1144. https://doi.org/10.1126/science.1128690

Temelli, A., & Kurt, M. (2013). Attitudes of primary education and science education student's towards science and science education. *International Journal of Academic Research, 5*(4).

Uitto, A. (2014). Interest attitudes and self-efficacy beliefs explaining upper-secondary school students' orientation towards biology related careers. *International Journal of Science and Mathematics Education, 12*(6), 1425–1444.

Uitto, A., Juuti, K., Lavonen, J., & Meisalo, V. (2006). Students' interest in biology and their out-of-school experiences. *Journal of Biological Education, 40*(3), 124–129.
Usak, M., Prokop, P., Ozden, M., Ozel, M., Bilen, K., & Erdogan, M. (2009). *Turkish university students' attitudes toward biology: The effects of gender and enrolment in biology classes.*
Wiebe, E., Unfried, A., & Faber, M. (2018). The relationship of STEM attitudes and career interest. *EURASIA Journal of Mathematics, Science and Technology Education, 14*(10).
Xie, Y., & Achen, A. (2009). Science on the decline? Educational outcomes of three cohorts of young Americans. *Population Studies Center Research Report, 9*(684), 1–30.

# Chapter 17
# Perception of Biological Disciplines by Upper Secondary School Students

**Vanda Janštová, Natálie Tichá, and Petr Novotný**

## 17.1 Introduction

Science is often perceived by students as neither interesting, relevant (Goodrum et al., 2001) nor creative (Salonen et al., 2017), which makes it a rather unpopular school subject (Riess, 2000). However, biology is usually a well-perceived subject among other science subjects (Osborne et al., 2003; Prokop et al., 2007a).

### *17.1.1 Biology Disciplines*

Biology consists of several disciplines (sometimes referred to as sub-disciplines or topics), such as human biology, zoology, botany, microbiology, cell biology, genetics, mycology, ecology and evolution, and there are quite big differences in students' perception of these individual biology disciplines (Cimer, 2012). Human biology and zoology are often seen as the favourite disciplines among students (Prokop et al., 2007a) and teachers (Bukáčková & Janštová, 2017), unlike botany (Prokop et al., 2007a; Strgar, 2007). Zoology is also perceived as an easy discipline, possibly due to the fact students often keep a pet, and observe animals with interest, which is a supporting factor that leads to a deeper knowledge and more positive feelings about them (Bjerke et al., 2001; Prokop & Tunnicliffe, 2010). Students often rate genetics, cell biology, endocrinology (human biology) among the most difficult disciplines (Cimer, 2012). One of the causes as described for genetics is

V. Janštová (✉) · N. Tichá · P. Novotný
Department of Biology Education, Faculty of Science, Charles University, Prague, Czech Republic
e-mail: vanda.janstova@natur.cuni.cz; tichan@natur.cuni.cz; petr.novotny@natur.cuni.cz



K. Korfiatis, M. Grace (eds.), *Current Research in Biology Education*, Contributions from Biology Education Research,
https://doi.org/10.1007/978-3-030-89480-1_17

probably the complexity and hierarchical arrangement of these disciplines (Knippels, 2002) leading to misconceptions (reviewed by (Kampourakis & Reiss, 2018)).

Interestingly, Janštová et al. (2015) reported that students with a deeper interest in biology who took part in the Biology Olympiad also rated zoology among the most interesting biology disciplines, but botany was the third most interesting discipline for them just after molecular and cell biology. Gender differences have also been reported; girls often prefer human health topics, while boys are interested in the biological processes (Uitto et al., 2006).

Educators have been examining how to increase motivation toward studying science, and several ways to deal with poor perception of biology disciplines have been proposed. One suggestion is to teach the non-popular ones using popular teaching methods and forms like field trips (e.g. Janštová et al., 2020), as enrichment experiences in science have the potential to engage students with science (Tytler & Osborne, 2012). But this does not solve the bigger problem which may lay in the curriculum or simply using the long-established practices which may not be ideal.

### *17.1.2 Biology Curriculum*

The Czech curriculum is defined by the Framework Educational Programme, which doesn't specify organization of biology disciplines or even school subjects (RVP G, 2007; RVP ZV, 2013). Biology can therefore be taught as a single subject or as a part of science. In any setting, it should include the crosscutting concepts. It is up to each school if they divide biology into individual disciplines such as zoology, botany or ecology, or embed the disciplines into a bigger picture – or within the "Big Ideas" put forward by Harlen et al. (2010). Despite this, most of the schools use the traditional approach of teaching biology as a separate subject with separate disciplines which are more or less isolated (Janštová & Jáč, 2015) in accordance with local tradition, and this is the majority approach in local textbooks (Rokos et al., 2019). Although it is commonly said teachers tend to teach the way they were taught, which can be seen as a drawback, research has shown they rather teach the way they would prefer to be taught (Cox, 2014) or by building on a diverse experience (Oleson & Hora, 2014).

On the other hand, teachers reflect and follow textbooks or manuals when preparing for their lessons (Janštová, 2015; Moulton, 1997). Only one current Czech biology textbook edition follows the 'ecological' approach (Rokos et al., 2019, p. 21). Such an arrangement of concepts to reflect the Big Ideas organization of biology in the common high school textbooks (Miller & Levine, 2010), has been shown to be more effective compared to topically organized textbooks when assessing students' performance in tests (Crawford & Carnine, 2000).

The current trend in teaching biology seems to be through the Big Ideas (Harlen et al., 2015), which is incorporated in national curriculums, including Estonia

(Government of the Republic of Estonia, 2014), Finland (*National Core Curriculum*, 2016), New Zealand (*The New Zealand Curriculum*, 2007), and Scotland (CfE, 2009); although our neighbour, Poland (MNE, 2018), organizes the biology curriculum according to phylogeny and traditional topics like human biology, zoology, botany and ecology.

The Czech curriculum has been changed (RVP G, 2007; RVP ZV, 2013), but the changes have been adapted only formally and the teachers have rarely changed their teaching practices (Dvořák et al., 2015; Janík, 2013; Štech, 2013). One of the reasons is probably not linking the general public and teachers together as partners for a discussion about the aim and realization of curricular reform. Methodology to support the reform was also missing (Straková, 2013). As a result, knowledge rather than problem solving, creativity and scientific literacy is required as is in some other countries (Koul & Fisher, 2002; Lindahl, 2003; Šorgo, 2012).

We took advantage of the fact that all students should have studied the basics of all biological disciplines at lower secondary school, and all the disciplines are repeated during upper secondary school biology according to the Czech curriculum. Therefore, the pupils know wha the disciplines are about, although they are not familiar with as many details. As mentioned above, the most widespread approach to teaching biology in Czechia is to organize a biology course through separately teaching the individual biological disciplines, not through big ideas or ecological approaches. Pupils' perception of biology disciplines is thus implicitly based on the more or less strictly divided disciplines, and it is possible to assume that their relations to individual disciplines are distinct enough to be measured. At the same time, it leads to the fact that Years 11 and 12 students have not yet substantially encountered or studied all the disciplines at the upper secondary school.

### 17.1.3 Perception of Usefulness

The term "usefulness" is seen as "usefulness for everyday life" inspired by the formulation: "Do middle school students see themselves using science?" (BouJaoude & Abd-El-Khalick, 1995). Students who perceive a discipline useful typically consider usefulness for themselves, e.g. future career, real life (Aschbacher et al., 2014) and usefulness for society. Perceiving a discipline useless and without connection to everyday life leads to perceiving it as boring (Rennie et al., 2001). Raved and Assaraf (2011) describe usefulness for life as a part of instrumental value within the Value of Science studies. Of course, usefulness for one's everyday life, and consequently to society, is only one of the possible reasons for including a topic or discipline into the curriculum. Importance, or in a broader sense relevancy, nevertheless is among the crucial ones (Dillon, 2009; Stuckey et al., 2013) although it may be vaguely defined (Stuckey et al., 2013).

## 17.2 Aims of the Study

We wanted to explore if perception of usefulness for life and difficulty of a specific science subject, here represented by biology, changes during (and due to) upper secondary school studies; and if it does – how? We were also interested in whether the school subject of biology is perceived by both boys and girls in a similar way. We decided to find out how students perceive all biological disciplines in terms of usefulness for life and difficulty, although we realize these are also only sections of a bigger picture.

Our objective was to find out how Czech upper secondary school students perceive individual biology disciplines in terms of usefulness for life and difficulty. We wanted to answer the following questions:

1. *Are there differences among individual biology disciplines in terms of usefulness for life and difficulty as perceived by upper secondary school students?*
2. *Is there a difference in the perception of usefulness for life and difficulty of the biology disciplines between the students who already studied the disciplines at upper secondary school and those who did not?*
3. *Is there a difference in the perception of usefulness for life and difficulty of the biology disciplines between boys and girls?*

## 17.3 Research Design and Method

### *17.3.1 Research Tool*

The questionnaire was inspired by Rusek (2013) who explored perception of different chemistry disciplines. The formulations in the first version of the questionnaire were changed (mostly shortened) after the first pilot testing with 15 students. After the second testing of the formulations with 21 students, no misunderstandings were found. Data obtained from these students were not included in the analysis. The questionnaire asked to mark the perception of usefulness for life and difficulty of individual biology disciplines on a 5-point Likert scale (by using the following questions: "How useful do you think these biological disciplines are for everyday life? and "How difficult do you think these biological disciplines are?". Biology was represented by the following disciplines (inspired by Janštová et al. (2015): cell biology, genetics, evolutionary biology, microbiology, protozoology, botany, zoology, mycology, human biology, physiology processes, ecology and environmental protection. Respondents also indicated if they had already studied the disciplines at upper secondary school or not, which was also confirmed by their teacher. As mentioned above, according to Czech biology curriculum, students should have studied the basics of all biological disciplines at lower secondary school so we assumed they all knew these disciplines.

### 17.3.2 Respondents

The questionnaire was distributed by one of the authors (NT) in nine classes of three upper secondary schools. Answering the questionnaire was voluntary and anonymous. Data from 192 students were analyzed (Table 17.1). All the respondents were taught biology as an individual biological discipline. Because teaching biology started by cell biology and microbiology at the beginning of the school year at all the schools in this study all the students had already studied these disciplines when taking the survey.

### 17.3.3 Data Analysis

Only complete questionnaire data were analyzed. Data was transcribed to Excel and analyzed using R (R Core Team, 2020). Perception of usefulness for life and difficulty of biology was calculated as a mean (and median in case of the graphs) value of the disciplines. A Mann-Whitney test was used to test the hypotheses that there are differences in perception of individual disciplines between boys and girls and between students who have and have not already studied the discipline at upper secondary school. Effect sizes (Vargha and Delaney's A) were calculated and evaluated according to Vargha and Delaney (2000).

## 17.4 Results

### 17.4.1 Perceived Usefulness for Life

Nearly all biology disciplines were perceived useful for one's life (having the mean and median above 3). Human biology was perceived as the most useful and protozoology as the least useful, although still close to a neutral value, as shown in Table 17.2 and Fig. 17.1.

Girls perceived the following disciplines more useful than did boys: cell biology ($p = 0.04$), genetics ($p = 0.004$), zoology ($p = 0.006$), physiology ($p = 0.005$) and

**Table 17.1** Respondents in the study

| Year of the study/gender | Boys | Girls | Total |
|---|---|---|---|
| 11 | 26 | 29 | 55 |
| 12 | 30 | 39 | 69 |
| 13 | 31 | 37 | 68 |
| Total | 87 | 105 | 192 |

**Table 17.2** Means of the perceived usefulness of biology disciplines for life

| Biology discipline | Usefulness for life – (means) | Perceived more useful by girls (effect sizes) | Perceived more useful if studied[a] (effect sizes) |
|---|---|---|---|
| Human biology | 4.86 | 0.45 | |
| Physiology | 4.68 | 0.41 | |
| Ecology and environmental protection | 4.41 | | |
| Genetics | 4.40 | 0.39 | |
| Zoology | 4.29 | 0.40 | |
| Microbiology | 3.92 | | |
| Evolutionary biology | 3.88 | | |
| Botany | 3.86 | | |
| Cell biology | 3.42 | 0.42 | |
| Mycology | 3.38 | | |
| Protozoology | 2.60 | | |

Note. The higher the mean, the higher the perceived usefulness for life. All the calculated effect sizes are medium

[a]At upper secondary school

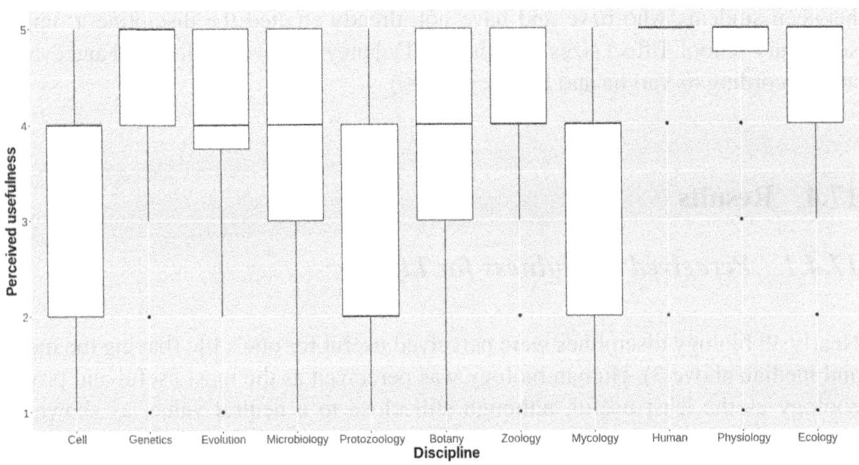


**Fig. 17.1** Medians of the perceived usefulness of biology disciplines for life
Note. The higher the mean, the higher the perceived usefulness for life. The box plot shows medians, 25th – 75th percentile, "minimum" (25th percentile – 1.5*interquartile range) and "maximum" (75th percentile + 1.5*interquartile range), the single dots represent outliers

human biology ($p = 0.02$). See Table 17.2 for effect sizes. On the other hand, boys did not perceive any biology discipline more useful compared to girls.

There was no difference in perceived usefulness for life between students who had already studied the given biology disciplines at upper secondary school (during years 11, 12 and the first months of year 13) and those who had not.

### 17.4.2 Perceived Difficulty

Genetics was seen as the most difficult; ecology and environmental protection was seen as easier than the rest except for zoology, see Table 17.3 and Fig. 17.2.

**Table 17.3** Means of the perceived difficulty of biology disciplines

| Biology discipline | Difficulty (means) | Perceived easier by girls (effect sizes) | Perceived easier if studied[a] (effect sizes) |
|---|---|---|---|
| Genetics | 2.40 | | |
| Cell biology | 2.68 | | |
| Microbiology | 2.78 | | |
| Protozoology | 2.79 | 0.41 | |
| Physiology | 2.79 | | 0.37 |
| Human biology | 2.96 | | 0.32 |
| Botany | 2.97 | 0.36 | |
| Mycology | 3.27 | | |
| Evolutionary biology | 3.46 | | |
| Zoology | 3.57 | 0.38 | |
| Ecology and environmental protection | 3.84 | 0.42 | |

Note. The lower the mean, the higher the perceived difficulty. All the calculated effect sizes are medium except for human biology, which is small
[a]At upper secondary school

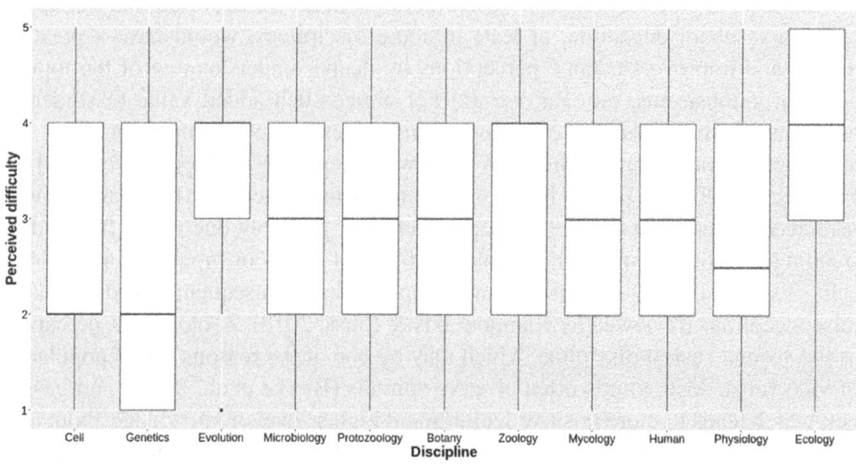


**Fig. 17.2** Medians of the perceived difficulty of biology discipline
Note. The lower the mean, the higher the perceived difficulty. The box plot shows medians, 25th – 75th percentile, "minimum" (25th percentile – 1.5*interquartile range) and "maximum" (75th percentile + 1.5*interquartile range), the single dots represent outliers

Girls, on contrary to boys, rated protozoology ($p = 0.03$), botany ($p < 0.001$), zoology ($p = 0.004$) and ecology and environmental protection ($p = 0.03$) as easier, see Table 17.3 for effect sizes. Boys did not perceive any biology discipline as easier than did girls. Students who already studied physiology and human biology at upper secondary school (during years 11, 12 and the first months of year 13) perceived these as easier than students who studied these disciplines only at lower secondary school ($p = 0.002$), see Table 17.3 for effect sizes.

## 17.5 Discussion and Conclusions

We are aware that this study has several limitations, one of which is the limited number of respondents and the fact they were only from three upper secondary schools. It would require a subsequent study possibly using a qualitative approach to describe the reasons why pupils find specific biology disciplines easier after studying it at upper secondary school. Also letting the pupils explain what they imagine as "usefulness" would shed more light on this problem. A longitudinal study following the perception of the same students over several years would probably be a more accurate approach to exploring the possible change of perception. Such a study would examine our findings more appropriately and would also allow us to examine how much the perception of biology disciplines is influenced also by lower secondary school.

All biology disciplines were perceived as useful, which provides a good starting point for biology educators, but the perception did not change during studying at upper secondary school. Our interpretation is that the teachers and school education do not adversely affect students' "usefulness" beliefs. We expected that the upper secondary school education, at least in some disciplines, would have a positive impact and improve students' perceptions by deeper understanding of the topics, personal enthusiasms, etc. But we did not notice such added value to students' beliefs about the usefulness of biology. This leaves space for improvement of the teaching techniques and approaches used while teaching biology. As reported by Prokop et al. (2007b), human biology was rated as most useful for life. Genetics was perceived as the most difficult biology discipline, probably due to the fact understanding genetics requires understanding different levels of organization (DNA – cell – tissue – organ – organism), and failing to do it consequently leads to many misconceptions (reviewed by Kampourakis & Reiss, 2018). Zoology was perceived as the second easiest discipline, which may be one of the reasons for its popularity or vice versa. Also, pupils often observe animals (Bjerke et al., 2001), and have a pet, which leads to more positive feelings and higher level of knowledge about animals (Prokop & Tunnicliffe, 2010). Moreover, the popularity of human zoology and biology corresponds to the popularity of these disciplines among Czech teachers (Bukáčková & Janštová, 2017).

Unlike usefulness, perceived difficulty of some biology disciplines changed during studying upper secondary school. Interestingly, the disciplines were perceived as easier when studying them. We have not found any discipline which the students perceived more difficult after studying it at upper secondary school than they expected. In-depth interviews with pupils are needed to find out the reasons and reasoning behind them.

Zoology and botany typically include learning phylogenetic relationships; ecology is typically focused on describing the individual concepts, with the practical work being represented by text-based work in Czech education (Janštová, 2015). These disciplines were perceived as easier by girls. The results of the comparison of boys and girls show that there is a difference between the perception of both monitored evaluations. Girls perceive biological disciplines within the Czech educational system as easier (especially those containing a lot of new concepts) and more useful (especially those connected to human health and reproduction). This finding can probably be partly explained by the generally better attitudes of girls towards biology than boys (Dawson, 2000), by girls choosing nursing professions more often than boys Uitto et al. (2006), or the influence of a different approach to learning can be considered. As memorizing has been shown to be used more by girls than boys (Tan, 2014), we hypothesize that girls use more memorizing to learn the content of the disciplines and their terminology. However, the list of disciplines in which girls perceive difficulty less than boys is relatively wide; the average effect size is also noticeable, especially taking into account the evaluation of effect size in the sense of Kraft (2020). Czech education has a high proportion of women teachers, according to OECD statistics for the last available year 2018 (OECD, 2018) the Czech Republic is the 5th most feminized education system in ISCED 1 (94% of teachers are women), and in ISCED 2 it is the 7th most feminized (78%). Specific data for biology teachers are not available. The concept of school-female stereotypes (Heyder & Kessels, 2013) presents us with a theoretical framework in which we can consider the influence of women's representation in teacher positions as a significant factor influencing pupils' attitudes and subsequent success. We therefore propose to further explore this area, i.e. the high degree of feminization of schools leads to an unbalanced concept of teaching biology, which is more suitable for girls. The Czech Republic can serve as an example of a school system with a high degree of feminization where girls perceive more than a third of biological disciplines as more useful and easier than do boys. Because the mentioned topics are similarly relevant to every citizen, the teachers should make them relevant for all pupils including boys.

## References

Aschbacher, P. R., Ing, M., & Tsai, S. M. (2014). Is science me? Exploring middle school students' STE-M career aspirations. *Journal of Science Education and Technology, 23*(6), 735–743.

Bjerke, T., Kaltenborn, B. P., & Ødegårdstuen, T. S. (2001). Animal-related activities and appreciation of animals among children and adolescents. *Anthrozoös, 14*(2), 86–94. https://doi.org/10.2752/089279301786999535

BouJaoude, S., & Abd-El-Khalick, F. (1995). *Lebanese middle school students' definitions of science and perceptions of its purpose and usage*. https://eric.ed.gov/?id=ED387328

Bukáčková, A., & Janštová, V. (2017). Methods of teaching organism recognition. How recommendation and practice differ. In M. Rusek, D. Stárková, & I. B. Metelková (Eds.), *Project-based Education in Science Education XIV* (Vol. 14, pp. 155–160) http://pages.pedf.cuni.cz/pvch/files/2011/11/PBE_XIV_final.pdf

CfE. (2009). Curriculum for excellence. https://education.gov.scot/education-scotland/scottish-education-system/policy-for-scottish-education/policy-drivers/cfe-building-from-the-statement-appendix-incl-btc1-5/experiences-and-outcomes

Cimer, A. (2012). What makes biology learning difficult and effective: Students' views. *Educational Research and Reviews, 7*(3), 61.

Cox, S. (2014). *Perceptions and influences behind teaching practices: Do teachers teach as they were taught?* Thesis. https://scholarsarchive.byu.edu/etd/5301

Crawford, D. B., & Carnine, D. (2000). Comparing the effects of textbooks in eighth-grade U. S. history: Does conceptual organization help? *Education and Treatment of Children, 23*(4), 387–422. JSTOR.

Dawson, C. (2000). Upper primary boys' and girls' interests in science: Have they changed since 1980? *International Journal of Science Education, 22*(6), 557–570. https://doi.org/10.1080/095006900289660

Dillon, J. (2009). On scientific literacy and curriculum reform. *International Journal of Environmental and Science Education, 4*(3), 201–213.

Dvořák, D., Starý, K., & Urbánek, P. (2015). Malá škola po pěti letech: Proměny školy v době reformy. *Pedagogická orientace, 25*(1), 9–31. https://doi.org/10.5817/PedOr2015-1-9

Goodrum, D., Rennie, L. J., & Hackling, M. W. (2001). *The status and quality of teaching and learning of science in Australian schools: A research report*. Department of Education, Training and Youth Affairs Canberra.

Government of the Republic of Estonia. (2014). *National curriculum for basic schools – Regulation (Last amendment 29 August 2014). Tallinn.*. https://www.hm.ee/sites/default/files/est_basic_school_nat_cur_2014_general_part_1.pdf

Harlen, W., Bell, D., Devés, R., Dyasi, H., Fernández de la Garza, G., Léna, P., Millar, R., Reiss, M., Rowell, P., & Yu, W. (2010). *Principles and big ideas of science education*. Association for Science Education College Lane, Hatfield, Herts. AL10 9AA. https://www.ase.org.uk/bigideas

Harlen, W., Bell, D., Devés, R., Dyasi, H., Fernández de la Garza, G., Léna, P., Millar, R., Reiss, M., Rowell, P., & Yu, W. (2015). *Working with big ideas of science education*. Science Education Programme (SEP) of IAP. https://www.ase.org.uk/bigideas

Heyder, A., & Kessels, U. (2013). Is school feminine? Implicit gender stereotyping of school as a predictor of academic achievement. *Sex Roles, 69*(11-12), 605–617.

Janík, T. (2013). Od reformy kurikula k produktivní kultuře vyučování a učení. *Pedagogická orientace, 23*(5), 634–663. https://doi.org/10.5817/PedOr2013-5-634

Janštová, V. (2015). What is actually taught in high school biology practical courses. *ICERI2015 Proceedings, 8*, 1501–1507.

Janštová, V., & Jáč, M. (2015). Výuka molekulární biologie na gymnáziích: Analýza současného stavu a možnosti její podpory. *Scientia in Educatione, 6*(1), 14–39.

Janštová, V., Jáč, M., & Dvořáková, R. (2015). Faktory motivující žáky středních škol k zájmu o obor biologie a účasti v předmětových soutěžích s biologickou tematikou. *e-Pedagogium, 15*(1), 56–71.

Janštová, V., Poláček, B., & Novotný, P. (2020). Geology field-trip improves students' knowledge better than computer work: A case study. In M. Rusek, M. Tóthová, & K. Vojíř (Eds.), *Project-based Education and other activating Strategies in Science Education XVII* (pp. 53–60).

Kampourakis, K., & Reiss, M. J. (2018). *Teaching biology in schools global research, issues, and trends*. Routledge.

Knippels, M. C. P. J. (2002). Coping with the abstract and complex nature of genetics in biology education: The yo-yo learning and teaching strategy. Dissertation. http://dspace.library.uu.nl/handle/1874/219

Koul, R. B., & Fisher, D. (2002). Science classroom learning environments in India. In *International educational research conference of the Australian Association for Research in Education (AARE)*.

Kraft, M. A. (2020). Interpreting effect sizes of education interventions. *Educational Researcher, 49*(4), 241–253. https://doi.org/10.3102/0013189X20912798

Lindahl, B. (2003). Pupils' responses to school science and technology? A longitudinal study of pathways to upper secondary school. *Göteborg Studies in Educational Sciences, 196*. http://www.researchgate.net/profile/Britt_Lindahl/publication/237722627_Pupils'_responses_to_school_science_and_technology_A_longitudinal_study_of_pathways_to_upper_secondary_school/links/02e7e533330220b96f000000.pdf

Miller, K. R., & Levine, J. S. (2010). *Miller & Levine biology: 2010 on-level, student edition (Student edition: 2010 edition)*. Prentice Hall.

MNE (Ministry of National Education of Poland). (2018). *Nowa Podstawa Programowa (Szkoła podstawowa IV-VIII)*. https://podstawaprogramowa.pl/Szkola-podstawowa-IV-VIII

Moulton, J. (1997). *How do teachers use textbooks*. A review of the research literature.

National Core Curriculum. (2016). https://www.oph.fi/en/statistics-and-publications/publications/new-national-core-curriculum-basic-education-focus-school

OECD (2018). Retrieved 11th March 2021. https://stats.oecd.org/Index.aspx?datasetcode=EAG_PERS_SHARE_AGE

Oleson, A., & Hora, M. T. (2014). Teaching the way they were taught? Revisiting the sources of teaching knowledge and the role of prior experience in shaping faculty teaching practices. *Higher Education, 68*(1), 29–45.

Osborne, J., Simon, S., & Collins, S. (2003). Attitudes towards science: A review of the literature and its implications. *International Journal of Science Education, 25*(9), 1049–1079. https://doi.org/10.1080/0950069032000032199

Prokop, P., Prokop, M., & Tunnicliffe, S. D. (2007a). Is biology boring? Student attitudes toward biology. *Journal of Biological Education, 42*(1), 36–39.

Prokop, P., Tuncer, G., & Chudá, J. (2007b). Slovakian students' attitudes toward biology. *Eurasia Journal of Mathematics, Science & Technology Education, 3*(4), 287–295.

Prokop, P., & Tunnicliffe, S. D. (2010). Effects of having pets at home on children's attitudes toward popular and unpopular animals. *Anthrozoös, 23*(1), 21–35. https://doi.org/10.2752/175303710X12627079939107

R Core Team. (2020). R: a language and environment for statistical computing (4.0.3) [Computer software]. R Foundation for Statistical Computing. https://www.R-project.org/

Raved, L., & Assaraf, O. B. Z. (2011). Attitudes towards science learning among 10th-grade students: A qualitative look. *International Journal of Science Education, 33*(9), 1219–1243.

Rennie, L. J., Goodrum, D., & Hackling, M. (2001). Science teaching and learning in australian schools: Results of a national study. *Research in Science Education, 31*(4), 455–498. https://doi.org/10.1023/A:1013171905815

Riess, F. (2000). Problems with German science education. *Science & Education, 9*(4), 327–331. https://doi.org/10.1023/A:1008712329753

Rokos, L., Holec, J., Daniš, P., Čiháková, K., Doubková, A., Janštová, V., Jáč, M., Kroufek, R., Pechoušková, R., Pražáková, M., Pražienka, M., & Vágnerová, P. (2019). Podkladová studie k revizi rámcových vzdělávacích programů v oblasti vzdělávání o živé a neživé přírodě: Jak budeme učit přírodopis, biologii a geologii v příštích letech? *Národní ústav pro vzdělávání.*. http://www.nuv.cz/file/3642/

Rusek, M. (2013). Vliv výuky na postoje žáků SOŠ k chemii. *Scientia in educatione, 4*(1), Article 1. https://doi.org/10.14712/18047106.43

RVP G (2007). *Rámcový vzdělávací program pro gymnázia*. Retreived 26.11.2021 from https://www.edu.cz/wp-content/uploads/2020/08/RVPG-2007-07_final.pdf

RVP ZV (2013). *Rámcový vzdělávací program pro základní vzdělávání*. Retreived 26.11.2021 from http://www.nuv.cz/file/433_1_1/

Salonen, A., Hartikainen-Ahia, A., Hense, J., Scheersoi, A., & Keinonen, T. (2017). Secondary school students' perceptions of working life skills in science-related careers. *International Journal of Science Education, 39*(10), 1339–1352. https://doi.org/10.1080/09500693.2017.1330575

Šorgo, A. (2012). Scientific creativity: The missing ingredient in Slovenian science education. *European Journal of Educational Research, 1*(2), 127–141.

Straková, J. (2013). Jak dál s kurikulární reformou. *Pedagogická orientace, 23*(5), 734–744. https://doi.org/10.5817/PedOr2013-5-734

Strgar, J. (2007). Increasing the interest of students in plants. *Journal of Biological Education, 42*(1), 19–23. https://doi.org/10.1080/00219266.2007.9656102

Stuckey, M., Hofstein, A., Mamlok-Naaman, R., & Eilks, I. (2013). The meaning of 'relevance' in science education and its implications for the science curriculum. *Studies in Science Education, 49*(1), 1–34. https://doi.org/10.1080/03057267.2013.802463

Štech, S. (2013). Když je kurikulární reforma evidence-less. *Pedagogická orientace, 23*(5), 615–633. https://doi.org/10.5817/PedOr2013-5-615

Tan, S. (2014). Gender differences in learning molecular biology using virtual learning environments. In Y. Cai & S. L. Goei (Eds.), *Simulations, serious games and their applications (s. 219–226)*. Springer. https://doi.org/10.1007/978-981-4560-32-0_14

Tytler, R., & Osborne, J. (2012). Student attitudes and aspirations towards science. In *In Second international handbook of science education* (pp. 597–625). Springer.

The New Zealand Curriculum. (2007). https://nzcurriculum.tki.org.nz/The-New-Zealand-Curriculum

Uitto, A., Juuti, K., Lavonen, J., & Meisalo, V. (2006). Students' interest in biology and their out-of-school experiences. *Journal of Biological Education, 40*(3), 124–129. https://doi.org/10.1080/00219266.2006.9656029

Vargha, A., & Delaney, H. D. (2000). A critique and improvement of the CL common language effect size statistics of McGraw and Wong. *Journal of Educational and Behavioral Statistics, 25*(2), 101–132.

# Chapter 18
# Analyzing the Relationships Between Pre-service Biology Teachers' Modelling Processes, Scientific Reasoning Competencies and General Cognitive Abilities

**Maximilian Göhner and Moritz Krell**

## 18.1 Introduction

Modelling is considered to be a core practice of reasoning in science (Giere et al., 2006; Passmore et al., 2014), notably within the specific discipline of biology (Laubichler & Müller, 2007). The introduction of authentic scientific reasoning and, therefore, modelling, into biology classrooms remains a common goal shared by many researchers (Haugwitz & Sandmann, 2010; Svoboda & Passmore, 2013; Upmeier zu Belzen et al., 2019). This goal has been emphasized in curriculum documents worldwide (e.g. KMK, 2005; NGSS Lead States, 2013; VCAA, 2016).

Earlier studies revealed that students and teachers mainly utilize models for representative purposes (i.e. models of phenomena). This is in contrast to scientists, who primarily use models as research tools (e.g. models for reasoning; Passmore et al., 2014). Specifically, the predictive power of models seems to be a challenging concept for students and teachers alike (Gouvea & Passmore, 2017; Krell & Krüger, 2016). Consequently, enacting a model-based curriculum and implementing a scientific perspective on modelling in classrooms poses a challenge for teachers (Windschitl et al., 2008). Hence, in order to support student learning, future biology teachers will need to develop modelling competencies and capabilities related to models and modelling in addition to pedagogical competencies during teacher education at the university level (Günther et al., 2019).

While the concept of modelling competencies has been the subject of intense research in science and biology education (Krell, 2013; Nicolaou & Constantinou, 2014; Upmeier zu Belzen et al., 2019), only a few studies published thus far have

M. Göhner (✉) · M. Krell
Department of Biology, Chemistry, Pharmacy, Freie Universität Berlin, Berlin, Germany
e-mail: maxgoe119@zedat.fu-berlin.de; moritz.krell@fu-berlin.de



K. Korfiatis, M. Grace (eds.), *Current Research in Biology Education*, Contributions from Biology Education Research,
https://doi.org/10.1007/978-3-030-89480-1_18

investigated the relationships between dimensions of modelling competencies and conceptually related constructs (Mierdel & Bogner, 2019; Ruppert et al., 2019). Studies such as these have the potential to provide valuable theoretical knowledge (Nehring & Schwichow, 2020). The present study adds to this body of research by providing an analysis of the relationships between the quality of modelling processes exhibited by pre-service biology teachers and their scientific reasoning competencies (SRC) and general cognitive abilities (GCA).

## 18.2 Modelling Competencies in Biology Education

Competencies are commonly conceptualized as complex latent constructs, which are necessary for problem-solving within specific domains (Klieme et al., 2008). Competencies encompass knowledge, skills, abilities, and further dispositions, for example motivation (Rychen & Salganik, 2003). An individual competence is defined as a system of dispositions that one needs in order to solve problems of a specific type. Hence, the extensionality of specific competencies (i.e. their constituent elements) are predefined by the type of problems to which they refer (Blömeke et al., 2015; Franke, 2005).

Modelling competencies are defined as the knowledge, skills, abilities, and motivational dispositions necessary to engage in the process of developing and using models for problem-solving in specific situations (Upmeier zu Belzen et al., 2019). Understanding the predictive power of models and having the capacity to use models as research tools are key elements of sophisticated modelling competencies (Gouvea & Passmore, 2017). Advanced modelling processes are, therefore, characterized by their complex and cyclic nature, including several modelling activities like a repeated testing of the model by deducing and evaluating model-based predictions (e.g. Giere et al., 2006; Schwarz et al., 2009). Wilmont et al. (2019) also suggest that switching between different modelling activities often and fluently, leading to homogenously distributed modelling activities, is a sign of sophisticated modelling processes, as this behaviour prevents modellers from getting stuck on problem features or missing out on relevant opportunities for sense-making.

Dimensions of modelling competencies include meta-modelling knowledge and the ability to engage in modelling processes (Nicolaou & Constantinou, 2014). Although there is a clear consensus on how one defines the construct of modelling competencies in science education (Upmeier zu Belzen et al., 2019), the relationships between modelling competencies and conceptually related constructs are not yet clearly established. Most published studies focus on meta-modelling knowledge or on the quality of modelling products rather than on the process of modelling itself. Therefore, qualitative insights into actual modelling processes as performed by both students and teachers are widely lacking (Nicolaou & Constantinou, 2014).

## 18.3 Relationship Between Modelling Competencies and Conceptually Related Constructs

Studies on the relationships between modelling competencies and conceptually related constructs can contribute to an improved understanding of the dimensions and the extensionality of specific constructs associated with modelling competencies (Nicolaou & Constantinou, 2014). For example, these studies can provide evidence for discriminant validity (Krell et al., 2019; Mayer et al., 2014). As competencies are complex constructs that encompass various aspects of the knowledge, skills, and abilities required for problem-solving (Klieme et al., 2008), these studies can provide valuable theoretical knowledge (Schwichow & Nehring, 2018). This theoretical knowledge can be used to guide further research, for example, related to the assessment of specific competencies (Nehring & Schwichow, 2020; Shavelson, 2013). Empirical studies that focus on the relationships between modelling competencies and potentially related constructs can also be used to test established assumptions about general abilities and prerequisites for effective modelling (e.g. creativity; Mierdel & Bogner, 2019). Furthermore, knowledge focused on abilities that positively influence the quality of student engagement in modelling processes might contribute to developing interventions and professional development, aimed at fostering modelling competencies among biology teachers (Schwichow & Nehring, 2018).

While theoretical considerations suggest a positive relationship between modelling competencies and general SRC, the specific relationship between these two constructs remains unclear (Göhner & Krell, 2020a). Some authors understand scientific reasoning as a modelling process that encompasses various other practices (Lehrer & Schauble, 2015), while others describe modelling as one of six distinct *styles of scientific reasoning* (Kind & Osborne, 2017).

GCA encompass analogical reasoning and spatial thinking, which were suggested to be among the prerequisites for effective engagement in modelling processes (Clement, 2008). In line with this observation, previous findings indicated a positive relationship between meta-modelling knowledge and GCA (Krell et al., 2014). Essential to the assessment of GCA are tasks that include spatial transformation (Liepmann et al., 2007), which was proposed as a critical ability for modelling (Nersessian, 2002).

In addition to SRC and GCA, domain-specific knowledge has been proposed as a factor contributing to modelling processes (Ruppert et al., 2019). Furthermore, creativity has also been presented as an important prerequisite for modelling (Clement, 2008). In one case study, Bailer-Jones (1999) described how creative strategies (e.g. visualization) were employed by scientists while modelling the complex phenomena associated with extended extragalactic radio sources. However, other studies found no significant associations between creativity and the quality of student and teacher engagement in hands-on modelling processes (Demirhan & Şahin, 2019; Mierdel & Bogner, 2019).

## 18.4 Aims of the Study and Research Questions

The present study contributes to research on modelling competencies in biology education by comparing SRC and GCA of pre-service biology teachers with their modelling processes. The data presented comes in part from previously published studies (Göhner & Krell, 2018, 2020a; Krell et al., 2019) and have been reanalyzed for use in the present study. We will address the following research question: what relationships can be identified when comparing SRC and GCA of pre-service biology teachers with their respective modelling processes? It is expected that pre-service biology teachers with higher SRC and/or GCA will conduct more sophisticated modelling processes, which are characterized by being rather homogeneous and complex (Schwarz et al., 2009; Wilmont et al., 2019) and include activities of deducing and evaluating model-based predictions (Gouvea & Passmore, 2017; Upmeier zu Belzen et al., 2019).

## 18.5 Research Design

The present study follows a mixed-method design, combining quantitative analyses of responses to two surveys assessing SRC and GCA with a qualitative analysis of modelling processes observed as participants engage in a black box modelling task.

### *18.5.1 Context of the Study*

Pre-service biology teachers from two German universities were asked to participate in the study based on a criterion-based sampling strategy, which is described below. Participants differ in their academic progress ranging from pre-service teachers in introductory biology courses during their bachelor's programme to pre-service teachers in advanced biology education courses at the end of their master's programme. At the end of their studies, pre-service biology teachers in Germany are expected to have developed sufficient knowledge and competencies regarding inquiry and reasoning in science, including modelling (KMK, 2019).

### *18.5.2 Assessment of SRC and GCA*

SRC and GCA of 202 participating pre-service biology teachers were assessed with established paper-and-pencil-based instruments. For the evaluation of SRC, a multiple-choice instrument was used that includes tasks related to two dimensions of reasoning in science: conducting scientific investigations and using scientific models (Krüger et al., 2020). GCA were assessed with a commercially available and standardized instrument with individual task formats that assessed verbal, numeric, and spatial intelligence as subscales (Liepmann et al., 2007).

### 18.5.3 Assessment of Modelling Processes

The results of the two paper-and-pencil-based assessments were used in a criterion-based sampling approach in order to increase the variance of the modelling processes (heterogeneous sample). Based on the expectations described in Sect. 18.4, pre-service biology teachers with SRC that were one standard deviation higher or lower than the mean scores of the respective norm sample (Krüger et al., 2020) and GCA that were a one-half standard deviation higher or lower than the mean scores of the respective norm sample (Liepmann et al., 2007) were invited to participate in the study ($N = 57$). Thirty-six pre-service biology teachers voluntarily agreed to participate and signed an informed consent form.

The participating pre-service biology teachers were individually presented with a black box (Fig. 18.1) that was previously shown to be an effective tool for encouraging modelling processes (Krell et al., 2019). This black box has a funnel for water input and an internal, hidden mechanism that determines water output. For example, repeated input of 400 ml yields a varied output pattern (e.g. 0 ml, 400 ml, 600 ml, 400 ml, 0 ml, 1000 ml, etc.) that cannot be readily explained. This feature of the black box facilitates the creation of varied analogies and models and motivates the participants toward further investigation (Krell et al., 2019). The participants were instructed to draw a model of the mechanism inside the black box on the chalkboard provided while thinking aloud (Ericsson & Simon, 1998). No further instructions or time constraints were provided. The participants were videotaped while participating in this activity.

**Fig. 18.1** Black box setting. Components are labelled as follows: (a) written task: participants were asked to develop a model of the internal mechanisms of the black box, (b) exemplarily labelled chalkboard, (c) video camera (1 of 3), (d) black box, (e) water reservoir, (f) beakers of different sizes, (g) volumetric flask

## 18.6 Data Analysis

### *18.6.1 Analysis of SRC and GCA*

For estimations of SRC, weighted likelihood estimates were computed in a one-dimensional "Rasch-model" (Krüger et al., 2020). The sum score of verbal, numerical and spatial intelligence was calculated to determine individual participants' GCA (Liepmann et al., 2007).

### *18.6.2 Qualitative Analysis of Modelling Processes*

Videos and verbalizations were fully transcribed and analyzed within the methodological framework of qualitative content analysis (Göhner & Krell, 2020b; Schreier, 2012). The modelling processes of the participants were coded using an established category system (Krell et al., 2019). This system facilitated a detailed breakdown of each participant's modelling process into 19 distinct modelling activities that are related to three phases of modelling: exploration (of the black box), model development, and use of models for predictions (Table 18.1). Cohen's Kappa indicated very high intra-rater agreement ($M_{\kappa} = 0.84$) and moderate to very high inter-rater agreement ($M_{\kappa} = 0.77$) for this analysis (Göhner & Krell, 2020a).

Based on the resulting frequencies and sequences of modelling activities, transition graphs were built for each participant's modelling process. Such graphs include knots, depicting the modelling activities, and edges, depicting the transitions between the respective modelling activities (Fig. 18.2). To test the expectations described in Sect. 18.4, the complexity and homogeneity of the resulting transition graphs (i.e. modelling processes) were estimated as described in Göhner & Krell (2020a): Complexity was estimated using the transition graph metric known as "communities", which involves counting all subgraphs within each transition graph. Participants may show a limited range of activities, resulting in transition graphs with more subgraphs, as some activities are not connected to others. To estimate complexity, the communities-score of each participant was subtracted from the maximum communities-score achieved in this study. To estimate homogeneity, the centrality of every transition graph was determined and reciprocally transformed. Centrality describes the dependence of a graph on a single node. To account for structural outliers, homogeneity was estimated using the sum of the three measures of centrality: closeness centrality, degree centrality, and betweenness centrality.

**Table 18.1** Modelling activities and their associations with the three phases of modelling. Detailed descriptions can be found in Göhner & Krell (2020a)

| Phases | Exploration | Model development | Model usage for predictions |
|---|---|---|---|
| Activities | 1. Perception of a phenomenon<br>2. Input/output (exploratory)<br>3. Summarizing/ describing observations<br>4. Input/output (pattern detection)<br>5. Recognizing patterns | 7. Graphically develop model<br>8. Change model to optimize consistency<br>9. Change model to optimize representation<br>10. Reject model due to poor consistency or representation<br>11. Evaluate consistency<br>12. Evaluate representation<br>13. Finding of consistency and representation | 14. Deduction of predictions<br>15. Input/output (i.e., test predictions)<br>16. Confirmation of prediction<br>17. Falsification of prediction<br>18. Change model due to falsified predictions<br>19. Reject model due to falsified predictions |

Activity 6, "activation of analogies and experiences" takes place throughout the modelling process. As such, this activity was not assigned to any one of the three phases listed above.

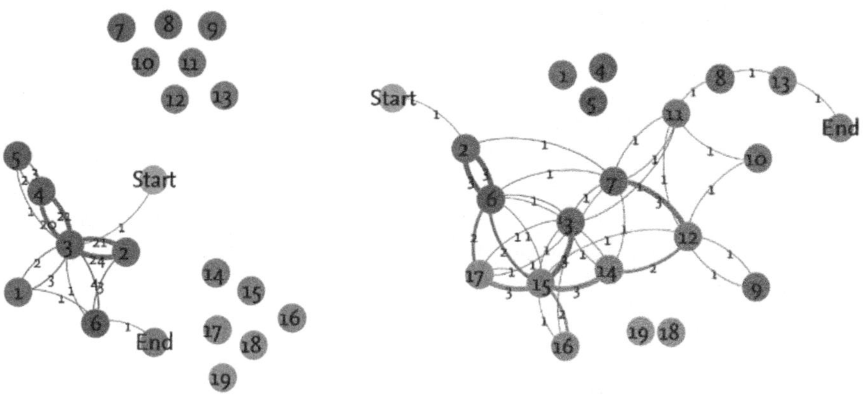


**Fig. 18.2** Transition graphs of "Carlo" (left) and "James" (right). Knots depict the modelling activities labelled with the same numbers as in Table 18.1. Additionally, all edges are labelled with the number of transitions, between the respective modelling activities. Unconnected knots have not been addressed by the participants

### 18.6.3 Analysis of the Relationships Between SRC, GCA, and Modelling Processes

To address the research question directly, Pearson correlational analyses were performed to identify relationships between the participants' SRC and GCA, their relative time of engagement in each of the modelling phases, and the transition graph metrics complexity and homogeneity. For time of engagement, in each modelling phase, the time in each phase was divided by the total time each participant was engaged in the entire task. Furthermore, SRC and GCA were compared between respondents who did and did not show activities associated with prediction in their modelling processes (*t*-test), as predictive activities were previously identified as a key element of sophisticated modelling competencies (Gouvea & Passmore, 2017).

## 18.7 Findings

Table 18.2 includes descriptive statistics for the relevant measures in this study. Based on the heterogeneous sampling approach described in Sect. 18.5.3, the participants were divided into four subsamples.

### 18.7.1 SRC and GCA

Overall, the groups' mean SRC were close to the norm sample's mean score, which is constrained to $M = 0$ (Krüger et al., 2020). With respect to the subscales used to evaluate GCA, the pre-service biology teachers achieved on average ~40 points, with a maximum value possible of 60 points (Liepmann et al., 2007). The mean scores of the subgroups reflect the heterogenous sampling approach, differing one (SRC) or one half (GCA) standard deviation from the respective norm samples for both variables, respectively.

### 18.7.2 Modelling Processes

Regarding the modelling processes, Table 18.2 shows that all participants predominantly invest time in activities assigned to the modelling phase of exploration. On average, the participants were engaged in exploratory activities for ~60% of the overall time on the task, while activities assigned to model development only account for ~20% of the time. Activities assigned to the modelling phase of prediction and the activity of finding analogies account for less than ~10% of the time each.

Figure 18.2 shows the transition graphs of Carlo and James (names are pseudonyms) as two examples. Carlo uses only exploratory modelling activities

**Table 18.2** Descriptive statistics for SRC and GCA; the relative (%) time of engagement in each modelling phase (Table 18.1); and the transition graph metrics, complexity, and homogeneity for each subsample (*SD* standard deviation)

| | Total sample | Subsamples | | | |
|---|---|---|---|---|---|
| | (N = 36) | High SRC, high GCA ($n$ = 17) | High SRC, low GCA ($n$ = 6) | Low SRC, high GCA ($n$ = 3) | Low SRC, low GCA ($n$ = 10) |
| Mean values regarding SRC and GCA | | | | | |
| Mean SRC (SD) | 0.21 (1.04) | 0.97 (0.23) | 0.91 (0.20) | −1.09 (0.20) | −1.13 (0.29) |
| Mean GCA (SD) | 114.56 (25.24) | 135.76 (9.14) | 90.33 (12.47) | 131.33 (6.11) | 88.00 (10.81) |
| Verbal | 38.50 (6.89) | 38.24 (7.30) | 39.00 (7.13) | 41.00 (3.61) | 37.90 (7.48) |
| Numeric | 39.03 (13.46) | 39.94 (12.44) | 38.50 (12.11) | 33.00 (25.00) | 39.60 (13.95) |
| Spatial | 37.03 (8.29) | 36.35 (9.99) | 40.83 (4.62) | 34.33 (8.39) | 36.70 (7.01) |
| Mean time (%) of engagement in modeling phases (SD) | | | | | |
| Exploration | 62 (18) | 65 (16) | 58 (21) | 66 (25) | 58 (18) |
| Model dev. | 21 (12) | 22 (12) | 18 (7) | 25 (25) | 21 (12) |
| Prediction | 3 (7) | 2 (4) | 6 (12) | 0 (0) | 4 (7) |
| Analogy | 5 (4) | 3 (3) | 7 (5) | 6 (1) | 6 (6) |
| Mean values of transition graph metrics (SD) | | | | | |
| Complexity | 4.89 (2.68) | 5.53 (2.96) | 4.83 (2.79) | 3.33 (0.58) | 4.30 (2.45) |
| Homogeneity | 1.52 (0.41) | 1.48 (0.34) | 1.55 (0.49) | 1.26 (.57) | 1.66 (0.45) |

(activities 1–5) and activity 6 "activation of analogies and experiences". These activities are often repeated, especially activities 2 (input/output (exploratory)), 3 (summarizing/describing observations), and 4 (input/output (pattern detection)) as illustrated by the high numbers on the respective edges. This leads to a transition graph with low complexity and homogeneity scores. James on the other hand uses more modelling activities and switches between them more fluently, leading to high complexity and homogeneity scores.

### *18.7.3 Relations Between SRC and GCA and the Modelling Processes*

To normalize the different scales for further statistical analyses, all data have been $z$-standardized. Table 18.3 presents Pearson correlation coefficients ($r$) between all measures included in this study. The correlation coefficients between SRC and GCA and the complexity and homogeneity measures were comparatively small ($|r|<.30$) and did not reach statistical significance ($p>.05$). Correlations between

**Table 18.3** Pearson correlation coefficients *r* and *p*-values for relationships between the SRC and GCA and its subscales, the time of engagement in each modelling phase, and the transition graph metrics complexity and homogeneity (n.s. = $p > .05$)

| Variable | SRC | | GCA | | verbal | | numeric | | spatial | |
|---|---|---|---|---|---|---|---|---|---|---|
| | R | P | r | p | r | p | R | p | r | p |
| Exploration | .06 | n.s. | .20 | n.s. | .09 | n.s. | −.01 | n.s. | −.11 | n.s. |
| Model development | .01 | n.s. | .07 | n.s. | −.03 | n.s. | −.04 | n.s. | −.04 | n.s. |
| Prediction | .02 | n.s. | −.21 | n.s. | −.03 | n.s. | .03 | n.s. | .19 | n.s. |
| Analogies | −.23 | n.s. | −.38 | <.05 | .12 | n.s. | .20 | n.s. | .29 | n.s. |
| Complexity | .21 | n.s. | .06 | n.s. | .25 | n.s. | .16 | n.s. | .00 | n.s. |
| Homogeneity | −.08 | n.s. | −.20 | n.s. | −.06 | n.s. | .14 | n.s. | −.05 | n.s. |

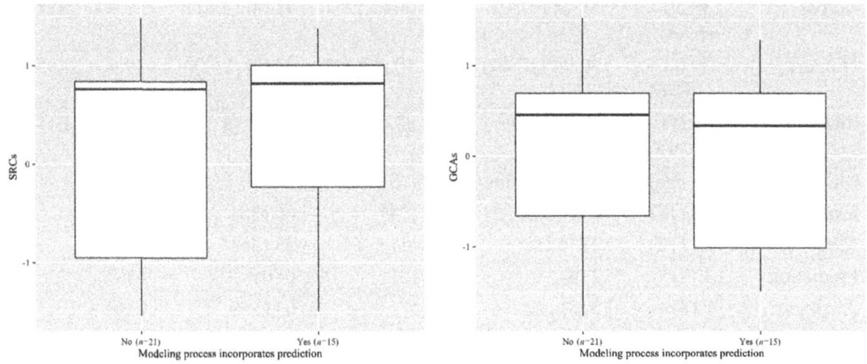


**Fig. 18.3** Boxplots with participants' SRC (left) and GCA (right) grouped into those that did and those that did not incorporate prediction-related activities in their modelling processes

SRC and GCA and the time of engagement in each modelling phase were also not statistically significant, save for the case of GCA and engagement in developing analogies, which revealed a significant negative correlation with medium effect size.

Figure 18.3 illustrates the participants' SRC and GCA grouped by the presence or absence of prediction activities associated with their respective modelling processes. No systematic differences in SRC and GCA between the two groups are apparent and no statistically significant differences could be found in test scores between the two groups (*t*-test).

## 18.8 Discussion

The aim of this study was to contribute to research on modelling competencies in biology education by evaluating the nature and quality of modelling processes exhibited by pre-service biology teachers with respect to their respective SRC and GCA. Hence, this study focuses specifically on modelling processes as one of the constituent dimensions of modelling competencies (Nicolaou & Constantinou, 2014). Contrary to our expectations, neither SRC nor GCA were systematically related to the modelling processes exhibited by the cohort of 36 selected pre-service biology teachers (Table 18.3 and Fig. 18.3). These unexpected findings are discussed while acknowledging the limitations of this study.

Naturally, the present study has limitations. Paper-and-pencil based instruments are limited by the nature of their task format. This type of instrument cannot capture complex performances (Martinez, 1999). Furthermore, the instrument used to assess SRC does not include tasks associated with all six styles of scientific reasoning (Kind & Osborne, 2017). Similarly, only verbal, numeric, and spatial intelligence are included in the test used to assess GCA that was applied in this study. This test did not include, for example, problem-solving abilities (Krell et al., 2019). Finally, the rather abstract task of investigating a black box was chosen to reduce the importance of content knowledge for engagement in the modelling process (Ruppert et al., 2019). However, this might be perceived as an inauthentic task compared to real-life modelling problems in biology. Furthermore, some cognitive processes and thoughts might not have been captured by the respondents' verbalizations (Ericsson & Simon, 1998).

First, and despite the limitations of this study as discussed above, the absence of a systematic relationship between SRC and modelling processes may stand in support of the recent conceptualization of modelling as a distinct style of scientific reasoning (Kind & Osborne, 2017). As such, these findings suggest a need for individual and explicit promotion of modelling in educational settings. Second, the absence of a systematic relationship between GCA and modelling processes challenges theoretical assumptions (e.g., Clement, 2008; Nersession, 2002) and also contradicts previously published observations (Krell et al., 2014). Specifically, it was surprising to find that there was no significant relationship between spatial intelligence and modelling processes (Table 18.3), as spatial transformation abilities have been proposed as one of the essential prerequisites for modelling (Nersessian, 2002). In general, our findings suggest that low GCA scores do not impose any constraints with respect to the complexity and homogeneity of modelling processes, at least among the 36 pre-service biology teachers who participated in this study. In fact, a statistically significant negative relationship between the time of engagement in analogy development and GCA was found (Table 18.3). However, this finding might suggest that participants with lower GCA require more time to develop appropriate analogies compared to participants with higher GCA.

Furthermore, the present results for both SRC and GCA taken together with previous results regarding creativity (Demirhan & Şahin, 2019; Mierdel & Bogner, 2019) might suggest that modelling processes are connected to entirely different abilities than those that have been theoretically conceptualized thus far. Earlier work has shown that creative strategies are important for scientists' ability to perform modelling (Bailer-Jones, 1999), although this was not found to be the case for students and teachers (Demirhan & Şahin, 2019; Mierdel & Bogner, 2019). This example might indicate that conceptually related constructs such as creativity and spatial intelligence could be important dispositions for modelling, depending on the modelling problem at hand. While complex modelling problems (e.g. expert modelling of extended extragalactic radio sources, Bailer-Jones, 1999) might require such dispositions, this might not be the case for simpler modelling problems (e.g. student modelling of DNA structure, Mierdel & Bogner, 2019). It is likely that the complexity of a problem and the amount of domain-specific knowledge required for a given modelling problem could be mediating factors in this case (Ruppert et al., 2019). Hence, the dispositions necessary for solving modelling problems might depend on the specific context. Context-dependency has been discussed previously as one feature of competencies in general (Rychen & Salganik, 2003).

Future studies might build on the findings of the present study with a focus on more authentic biological modelling problems. In such studies, modelling problems with different complexities should be considered as a means to test the possibility of context-dependency as suggested above. Furthermore, motivational and volitional effects might be considered, as the present black box investigation was an open and rather lengthy task for most of the participants (Nordheimer, 2019).

In conclusion, the present study demonstrates that established assumptions held by many science education researchers regarding factors that are positively associated with the capacity to perform scientific modelling might not be empirically valid. These results illustrate the need for critical reflection and for empirical proof of all assumptions, even those that are theoretically sound. Our results highlight the fact that educational interventions might be specifically designed with the aim of fostering modelling competencies, as the development of abilities that are theoretically related to this goal may not be fully sufficient.

**Acknowledgement** We thank the German Research Foundation (DFG) for funding the project TypMoL (project number 327507949). Any opinions, findings, conclusions, or recommendations expressed in this article are those of the authors and do not necessarily reflect the views of the DFG.

## References

Bailer-Jones, D. M. (1999). Creative strategies employed in modelling. *Foundations of Science, 4*, 375–388.

Blömeke, S., Gustafsson, J.-E., & Shavelson, R. J. (2015). Beyond dichotomies: Competence viewed as a continuum. *Zeitschrift für Psychologie, 223*, 3–13.

Clement, J. (2008). *Creative model construction in scientists and students*. Springer.

Demirhan, E., & Şahin, F. (2019). The effects of different kinds of hands-on modelling activities on the academic achievement, problem-solving skills, and scientific creativity of prospective science teachers. *Research in Science Education*. https://doi.org/10.1007/s11165-019-09874-0

Ericsson, K., & Simon, H. (1998). How to study thinking in everyday life. *Mind, Culture, and Activity, 5*, 178–186.

Franke, G. (2005). *Facetten der Kompetenzentwicklung* [Facets of competence development]. Bertelsmann.

Giere, R. N., Bickle, J., & Mauldin, R. (2006). *Understanding scientific reasoning*. Thomson.

Göhner, M., & Krell, M. (2018). Modellierungsprozesse von Lehramtsstudierenden der Biologie. [Modelling processes of pre-service biology teachers]. *Erkenntnisweg Biologiedidaktik, 17*, 45–63.

Göhner, M., & Krell, M. (2020a). Preservice science teachers' strategies in scientific reasoning: The case of modelling. *Research in Science Education.* https://doi.org/10.1007/s11165-020-09945-7

Göhner, M., & Krell, M. (2020b). Qualitative Inhaltsanalyse in naturwissenschaftsdidaktischer Forschung unter Berücksichtigung von Gütekriterien: Ein Review. [Qualitative content analysis in science education research under the consideration of quality criteria: A review]. *Zeitschrift Für Didaktik Der Naturwissenschaften., 26,* 207–225.

Gouvea, J., & Passmore, C. M. (2017). Models of' versus 'models for'. *Science & Education, 26*(1–2), 49–63.

Günther, S. L., Fleige, J., Upmeier zu Belzen, A., & Krüger, D. (2019). Using the case method to foster preservice biology teachers' content knowledge and pedagogical content knowledge related to models and modelling. *Journal of Science Teacher Education, 30*(4), 321–343.

Haugwitz, M., & Sandmann, A. (2010). Collaborative modelling of the vascular system. *Journal of Biological Education, 44*(3), 136–140.

Kind, P., & Osborne, J. (2017). Styles of scientific reasoning. *Science Education, 101*(1), 8–31.

Klieme, E., Hartig, J., & Rauch, D. (2008). The concept of competence in educational contexts. In J. Hartig, E. Klieme, & D. Leutner (Eds.), *Assessment of competencies in educational contexts* (pp. 3–22). Hogrefe.

KMK. (2005). *Bildungsstandards im Fach Biologie für den Mittleren Schulabschluss* [Biology education standards for the Mittlere Schulabschluss]. Wolters Kluwer.

KMK. (2019). *Ländergemeinsame inhaltliche Anforderungen für die Fachwissenschaften und Fachdidaktiken in der Lehrerbildung* [Common federal content requirements for disciplines and didactics in teacher education]. Sekretariat der Kultusministerkonferenz

Krell, M. (2013). *Wie Schülerinnen und Schüler biologische Modelle verstehen* [How students understand biological models]. Logos.

Krell, M., & Krüger, D. (2016). Testing models: A key aspect to promote teaching activities related to models and modelling in biology lessons? *Journal of Biological Education, 50*(2), 160–173.

Krell, M., Upmeier zu Belzen, A., & Krüger, D. (2014). Students' levels of understanding models and modelling in biology: Global or aspect-dependent? *Research in Science Education, 44,* 109–132.

Krell, M., Walzer, C., Hergert, S., & Krüger, D. (2019). Development and application of a category system to describe pre-service science teachers' activities in the process of scientific modelling. *Research in Science Education, 49*(5), 1319–1345.

Krüger, D., Hartmann, S., Nordmeier, V., & Upmeier zu Belzen, A. (2020). Measuring scientific reasoning competencies: Multiple aspects of validity. In O. Zlatkin-Troitschanskaia, H. A. Pant, M. Toepper, & C. Lautenbach (Eds.), *Student learning in German higher education* (pp. 261–280). Springer VS.

Laubichler, M. D., & Müller, G. B. (2007). *Modelling biology.* MIT Press.

Lehrer, R., & Schauble, L. (2015). The development of scientific thinking. In R. M. Lerner (Ed.), *Handbook of child psychology and developmental science* (pp. 671–715). John Wiley & Sons.

Liepmann, D., Beauducel, A., Brocke, B., & Amthauer, R. (2007). *Intelligenz-Struktur-Test 2000 R: IST 2000 R.* Hogrefe.

Martinez, M. E. (1999). Cognition and the question of test item format. *Educational Psychologist, 34*(4), 207–218.

Mayer, D., Sodian, B., Koerber, S., & Schwippert, K. (2014). Scientific reasoning in elementary school children. *Learning and Instruction, 29,* 43–55.

Mierdel, J., & Bogner, F. (2019). Is creativity, hands-on modelling and cognitive learning gender-dependent? *Thinking Skills and Creativity, 31,* 91–102.

Nehring, A., & Schwichow, M. (2020). Was ist Wissen, was ist Können? Deutungen des Kompetenzbegriffs und deren psychometrische Konsequenzen im Kontext von Fachwissen und Variablenkontrollstrategie [What is knowledge, what is ability? Interpreting the notion of competence and investigating the psychometrical consequences in the context of content knowledge and the control-of-variables-strategy]. *Zeitschrift Für Didaktik Der Naturwissenschaften.* https://doi.org/10.1007/s40573-020-00113-y

Nersessian, N. (2002). The cognitive basis of model-based reasoning in science. In P. Carruthers, S. Stich, & M. Siegal (Eds.), *The cognitive basis of science* (pp. 133–153). Cambridge University Press.

NGSS Lead States. (2013). Next generation science standards: For states, by states: Appendix f – science and engineering practices in the ngss. In NGSS Lead States (Ed.), *Next generation science standards: For states, by states*. The National Academies Press.

Nicolaou, C. T., & Constantinou, C. P. (2014). Assessment of the modelling competence: A systematic review and synthesis of empirical research. *Educational Research Review, 13*, 52–73.

Nordheimer, R. (2019). *Volition in Modellierungsprozessen von Biologie-Lehramtsstudierenden* [Volition in modelling processes of pre-service biology teachers]. Unpublished master's thesis. Freie Universität Berlin.

Passmore, C. M., Gouvea, J. S., & Giere, R. (2014). Models in science and in learning science: Focusing scientific practice on sense-making. In M. R. Matthews (Ed.), *International handbook of research in history, philosophy and science teaching* (pp. 1171–1202). Springer.

Ruppert, J., Duncan, R. G., & Chinn, C. A. (2019). Disentangling the role of domain-specific knowledge in student modelling. *Research in Science Education, 49*(3), 921–948.

Rychen, D., & Salganik, L. (2003). A holistic model of competence. In D. Rychen & L. Salganik (Eds.), *Key competencies for a successful life and a well-functioning society* (pp. 41–62). Hogrefe & Huber.

Schreier, M. (2012). *Qualitative content analysis in practice*. SAGE.

Schwarz, C. V., Reiser, B., Davis, E., Kenyon, L., Acher, A., Fortus, D., … Krajcik, J. (2009). Developing a learning progression for scientific modelling: Making scientific modelling accessible and meaningful for learners. *Journal of Research in Science Teaching, 46*(6), 632–654.

Schwichow, M., & Nehring, A. (2018). Variablenkontrolle beim Experimentieren in Biologie, Chemie und Physik: Höhere Kompetenzausprägungen bei der Anwendung der Variablenkontrollstrategie durch höheres Fachwissen? [Controlling variables in biology, chemistry and physics. Higher competences through higher content knowledge?]. *Zeitschrift für Didaktik der Naturwissenschaften, 24*, 217–233.

Shavelson, R. J. (2013). On an approach to testing and modelling competence. *Educational Psychologist, 48*, 73–86.

Svoboda, J., & Passmore, C. M. (2013). The strategies of modelling in biology education. *Science & Education, 22*(1), 119–142.

Upmeier zu Belzen, A., van Driel, J., & Krüger, D. (2019). Introducing a framework for modelling competence. In A. Upmeier zu Belzen, D. Krüger, & J. van Driel (Eds.), *Models and modelling in science education. Towards a competence-based view on models and modelling in science education* (pp. 3–19). Springer.

Victorian Curriculum and Assessment Authority (VCAA). (2016). *Victorian Curriculum: F-10*. Victorian Curriculum and Assessment Authority. https://victoriancurriculum.vcaa.vic.edu.au/science/curriculum/f-10

Windschitl, M., Thompson, J., & Braaten, M. (2008). How novice science teachers appropriate epistemic discourses around model-based inquiry for use in classrooms. *Cognition and Instruction, 26*(3), 310–378.

Wilmont, I., Barendsen, E., & Hoppenbrouwers, S. (2019). A case study of executive functions in real process modelling sessions. In H. A. Proper & J. Stirna (Eds.), *Lecture notes in business information processing. Advanced information systems engineering workshops* (Vol. 349, pp. 17–28). Springer.

# Chapter 19
# The Pitfalls of Using Presentation Technology in the Biology Classroom

**Andrej Šorgo and Vida Lang**

## 19.1 Introduction

The introduction of digital instructional technologies into classrooms has impacted biology education in many ways (Bognar, 2016; Kirkup, & Kirkwood, 2005; Pevzner, & Shamir, 2009), resulting in a mix of traditional and digital technologies used in the classroom (Livingstone, 2012). Slovenian schools are well equipped with both digital and traditional technologies, and all schools and most classrooms have access to the Internet. In addition, almost every lower secondary and general upper secondary school has a specialized biology classroom, most of which are equipped with a stationary or mobile computer along with a projector or i-board. Many schools are also equipped with data loggers along with traditional equipment for laboratory work, which is mandatory (Šorgo, & Špernjak, 2012). However, even if a classroom is fully equipped with digital technology, it is at the discretion and autonomy of the teacher to use it or not (Šorgo et al., 2010; Šorgo, & Špernjak, 2012).

Because PowerPoint (PPT) allows for the compilation of a range of multimedia formats into a single file, it has become the gold standard for presentations from elementary school to university (Adams, 2006; Baker et al., 2018) and has transformed not only the format but also the pedagogy of presentations (Adams, 2010; Stoner, 2007). Almost immediately after the introduction of PPT in 1987, it was already being praised by some and heavily criticized by others (Kernbach et al.,

A. Šorgo (✉)
Faculty of Natural Sciences and Mathematics, University of Maribor, Maribor, Slovenia
e-mail: andrej.sorgo@um.si

V. Lang
Faculty of Natural Sciences and Mathematics, University of Maribor, Maribor, Slovenia
e-mail: vida.lang1@um.si



K. Korfiatis, M. Grace (eds.), *Current Research in Biology Education*, Contributions from Biology Education Research,
https://doi.org/10.1007/978-3-030-89480-1_19

2015; Stoner, 2007). For example, Adams (2006) recognized PPT as a "killer app" that displaced several other teaching practices. Szabo, & Hastings (2000) wrote:

> It is suggested that the efficacy of PowerPoint lecturing may be case specific rather than universal (p.175).

In the field of biology education, studies on the use of PPT are under-represented below the undergraduate level, even though biology relies heavily on visualization (O'Day, 2006). The results of comparative studies between PPT-based biology instruction and more traditional instruction provided mixed results and appear to be context dependent. Adebanjo (2020) reported a positive impact of using PPT, while Prokop et al. (2016) reported negative results. As far as we know, there are no Slovenian studies on the effectiveness of PPT in biology classes.

## 19.2 Aims and Scope of the Research

As with any other technology or teaching method, Slovenian teachers are completely free to use PPT (or any other presentation technology) or not. Since presentations can be prepared at home, which is a great incentive for the introduction and use of an ICT application in school (Šorgo et al., 2010), PPT has recently been used regularly by most, if not all, biology teachers in Slovenia (Špernjak, 2014). However, studies from the students' perspective on PPT in biology classrooms are completely lacking.

The research was conducted as an exploratory study with the aim of determining the status of ICT-based multimedia use (PPT) in lower secondary school biology classes and its position among traditional (e.g. wallcharts and living organisms) and computer-based technologies and applications such as computer games. The intention was to use the results as a basis for possible evidence-based interventions in the classroom by matching teachers' use of multimedia with their students' views of multimedia use. We hypothesize that better use of presentations based on student feedback can lead to a better learning environment and better learning outcomes.

The research questions were as follows:

RQ1: How often are students exposed to traditional and computer-based instructional technologies in the biology classroom?
RQ2: How often are different multimedia formats used in the biology lessons?
RQ3: What are teachers' classroom practices when using PPT or similar applications in biology classrooms?
RQ4: What are students' opinions about the format and use of PPT in biology classes?
RQ5: From the students' point of view, what can be done to make biology more interesting?

## 19.3 Research Methodology

### 19.3.1 Instruments

Data were collected in the form of an online survey distributed to lower secondary students. The scales and items of interest are listed in the tables in the Findings section. When students were asked about frequency, the response format was a 5-point scale (1 = never; 2 = less than ¼ of lessons; 3 = between ¼ and ½ of lessons; 4 = between ½ and ¾ of lessons; 5 = in more than ¾ of lessons). When students were asked their opinion, the response format was a 7-point scale with anchors at 1 = completely agree; and 7 = completely disagree, with 4 as the midpoint of the scale.

### 19.3.2 Sample and Sampling

The sample population were lower secondary school students from the last two years of Slovenian 9-year compulsory primary and lower secondary unified school (about 40,000 students). We collected responses from 643 respondents (51% boys), and 610 (94/9%) answered all questions. The questionnaire was compiled as an online application. A request for assistance was sent to all Slovenian biology teachers. We received positive responses from 26 teachers who allowed their students to complete the questionnaire or provided them with a link to the questionnaire.

### 19.3.3 Data Analysis

The analysis followed the well-known methodology of descriptive statistics (Field, 2009). The most frequent responses (mode), means, and Standard Deviations are given as measures of central tendencies.

## 19.4 Findings

### 19.4.1 Frequency of Use of Instructional Technologies and Multimedia Formats in Biology Lessons

Students were instructed to indicate a frequency for the use of traditional and computer-assisted instructional technologies in the biology lessons on a scale ranging from never (1) to in more than ¾ of the lessons (5) (Tables 19.1 and 19.2).

Using the measures of central tendencies calculated from the student reports (Table 19.1), it can be seen that a projector connected to a computer has become the

**Table 19.1** Student-reported frequency of use of traditional and computer-based instructional technologies in biology lessons

| Instructional technology | Mode | Mean | SD |
|---|---|---|---|
| Projector connected to computer | 5 | 4.50 | 0.94 |
| Green- or white-board | 5 | 3.59 | 1.30 |
| Wallcharts or posters | 2 | 2.84 | 1.16 |
| 3-D models | 2 | 2.42 | 1.04 |
| Living organisms or their parts | 2 | 2.35 | 1.14 |
| i-board | 1 | 2.55 | 1.73 |
| Camera, connected to the computer | 1 | 1.62 | 1.16 |
| Overhead projector | 1 | 1.61 | 0.87 |
| Computer with data-logger | 1 | 1.51 | 0.86 |
| Tablet | 1 | 1.37 | 0.88 |
| Smart phone | 1 | 1.32 | 0.84 |

Note. Results are given by decreasing modes

**Table 19.2** Student-reported frequency of use of digital multimedia formats in biology lessons

| A teacher in biology lessons uses | Mode | Mean | SD |
|---|---|---|---|
| PowerPoint or similar program | 5 | 4.05 | 1.10 |
| e-materials / e-textbooks | 5 | 3.51 | 1.40 |
| videos / movies | 3 | 3.43 | 1.07 |
| On-line quizzes or tasks | 2 | 2.17 | 1.01 |
| Computer games | 1 | 1.35 | 0.77 |

Note. Results are given by decreasing means

dominant experience of instructional technology in biology classes for the majority of students. Green or white boards are becoming less important but are still frequently used, while all other technologies are used only occasionally or even never in most cases.

From the measures of central tendencies in Table 19.2, it can be seen that PPT has become the dominant multimedia instructional format in biology classes, along with e-textbooks. Computer games are almost never used, while videos and quizzes are used only occasionally.

### 19.4.2 Instructional Practices Among Teachers When Using PPT in Biology Lessons

Students were asked about the practices they experienced from teachers when using PPT on a one-to-five frequency scale ranging from "never" (1) to "in more than ¾ of lessons" (5). It was stressed to the students that they should provide answers related only to Biology. Results are presented in Table 19.3.

**Table 19.3** Measures of central tendencies about students' experiences with PPT use in biology lessons

| Statements | Mode | Mean | SD |
|---|---|---|---|
| The teacher provides enough time to transcribe content from the presentation. | 5 | 3.97 | 1.30 |
| The teacher uses PPT for presenting the content of a new topic. | 5 | 3.86 | 1.20 |
| The teacher expects us to transcribe text from the slides. | 5 | 3.56 | 1.26 |
| The teacher gives us presentations in electronic form. | 5 | 3.11 | 1.48 |
| Students present homework using PPT. | 2 | 2.69 | 1.31 |
| The teacher replaces practical laboratory work with a PPT presentation. | 2 | 2.55 | 1.24 |
| The teacher uses PPT for knowledge assessment | 2 | 2.40 | 1.18 |
| The teacher repeatedly has problems when using computer technology. | 2 | 2.40 | 1.29 |
| There is too much text in presentations. | 2 | 2.19 | 1.15 |
| The teacher uses PPT for quizzes, assessment or games. | 1 | 2.28 | 1.19 |
| Figures and text are too small | 1 | 1.81 | 1.05 |
| The teacher gives us printed presentations. | 1 | 1.74 | 1.00 |

Note. Results are given by decreasing means

From the measures of the central tendencies (Table 19.3) of the student reports, it is evident that in the first three places are practices that allow students to copy content into their notebooks, which is also given to students in electronic (but not printed) form by more than half of the teachers. It is evident that the majority of teachers have learned about formats of illustrations and text, but some of them still put too much text on slides, and some of them still struggle with computer technology.

### 19.4.3 Students' Opinions About Their Preferred Format and Use of PPT in Biology Lessons

Students were asked about their preferred format and use of PPT. The response format was a 7-point scale *(1 – completely agree; 7 – completely disagree)*. Results are presented in Table 19.4. It can be seen (Table 19.4) that the majority of students would prefer presentations that are rich in multimedia elements and would prefer videos over pictures and both over pure verbal text. They would also prefer active applications over static formats. A large majority of students disagree with the statement that: "I'd prefer to see a teacher writing on the board than using a PPT presentation" (Mode = 7); this can be identified as a drive for teachers to use PPT even more intensively. A negative side effect of more frequent use and another warning sign of increasing student passivity is the finding that nearly 45% agree with the following statement, "When a teacher gives a PowerPoint presentation, I just listen because I don't have to take notes." Knowing that note-taking can positively

**Table 19.4** Opinions about preferred format and use of PPT in biology lessons

| Statements | Mode | Mean | SD |
|---|---|---|---|
| PowerPoint presentations that include animation and videos of the learning topic are more interesting than an explanation where only figures, photos, and text are included. | 1 | 3.50 | 2.27 |
| PowerPoint presentations that include photos and sketches of the learning topic are more interesting than an explanation in front of the board. | 1 | 3.60 | 2.17 |
| I would remember more in class if more animation, photos, sound, and movies were included in the presentation. | 1 | 3.60 | 2.24 |
| I would be happier to study towards testing my knowledge if it took place as quizzes and games. | 1 | 3.60 | 2.14 |
| PowerPoint is useful and interesting because it makes it easier to follow lessons. | 1 | 3.64 | 2.12 |
| I am very happy when a teacher uses PowerPoint instead of writing on the board. | 1 | 3.66 | 2.20 |
| Teacher's PowerPoints are more useful than a textbook, as PowerPoint also includes clips and animations. | 1 | 3.70 | 2.15 |
| When a teacher has a PowerPoint presentation, I just listen because I don't have to take notes. | 4 | 3.84 | 2.00 |
| I'd rather have a teacher write on a whiteboard than use a PowerPoint presentation. | 7 | 4.28 | 2.29 |
| PowerPoint presentations that include only text without photos and animation are more interesting, because I can more easily follow the lesson. | 7 | 4.50 | 2.40 |

Note: Results are sorted by increasing modes

influence learning, teachers should design PPT-based lessons to encourage active note-taking strategies and limit word-for-word copying from slides, which is common practice (see Table 19.2).

### *19.4.4 Students' Suggestions of What Should Be Done to Make Biology More Interesting?*

Looking for solutions that would lead to a more interesting biology class, students were asked, "What should the teacher do to make biology more interesting?" They responded on a scale of 1 to 7, with numbers closer to 1 indicating greater agreement, and mode indicate the most frequently provided answer.

From the results in Table 19.5, it can be seen that there is a wide variation in views on what can make teaching more interesting. According to the modes and clustering, it is obvious that experimental work and videos/films are at the top of the wish list. PPT is still on the "positive" side as a medium of interest and slightly behind 3-D models (skeleton, heart, etc.), animation, fieldwork, and live materials, but ahead of i-table and traditional multimedia formats and presentation tools. It also seems that adding explanations is not high on the wish list for many students.

**Table 19.5** Results of descriptive analysis of students' suggestions how to make biology more interesting

| Instructional technology | Mode | Mean | SD |
|---|---|---|---|
| 3-D models (skeleton, heart, etc.) | 1 | 3.37 | 2.31 |
| Animations | 1 | 3.43 | 2.45 |
| Videos / movies | 1 | 3.48 | 2.60 |
| Experimental work | 1 | 3.50 | 2.60 |
| Field work | 1 | 3.52 | 2.40 |
| Live materials | 1 | 3.58 | 2.34 |
| **PowerPoint** | 1 | 3.64 | 2.31 |
| i – table | 1 | 3.84 | 2.26 |
| Posters | 4 | 3.80 | 1.82 |
| More explanation | 4 | 3.87 | 1.96 |
| Green- or white-board | 4 | 4.04 | 2.01 |
| Overhead projector | 7 | 4.01 | 2.22 |

Note. Results are ordered by increasing mode

## 19.5 Conclusions

Combining the results from Tables 19.1 and 19.2, it appears that a computer equipped with a projector that enables the presentation of predominantly PPT-based presentations, along with a whiteboard/greenboard, are students' dominant experiences with instructional technology. The results confirm previous findings (Šorgo et al., 2010) that essentially all biology teachers use digital technology, most often for presentations, word processing, web browsing, and email. However, even if teachers use digital technology on a daily basis, this does not mean that they allow all available digital tools in the classroom. The evidence is found at the bottom of the column, where digital devices and apps that are never or rarely used, such as smartphones, tablets, and data loggers, are positioned. Even an outdated overhead projector is used more often than Moodle or computer games. It seems that traditional tools such as models, living organisms or their parts and wallcharts have lost the battle with PPT. The limited use of models and organisms or their parts is particularly concerning because, after all, biology is a science that deals with living things, and they are, by default, three-dimensional. After recognizing that laboratory and experimental work is key to science and science education, and that computers equipped with data loggers may have advantages in some types of biology laboratories (Šorgo et al., 2010; Špernjak & Šorgo, 2018), the use of data loggers is below acceptable levels.

Inspecting Table 19.3, it is easy to see that teachers replaced black (green, white) boards and transparencies with PPT transparencies. We cannot say whether in some or most lessons PPTs are accompanied by lively classroom discussion, but the figures tell us that students are most often allowed (or even expected) to transcribe content from presentations into their notebooks. Transcribing text or tracing numbers is hardly recognizable as anything other than simple memorization. A dangerous sign is that teachers at least occasionally transcribe lab work into PPT format. It seems that most teachers have at least occasional problems with ICT but have learned the lesson that numbers and fonts in presentation should be large enough; however, they still forget not to put too much text on a slide. Assessment, quizzes, and games are not commonly used in biology lessons. It was interesting to see that at least with some teachers and at least occasionally PPT is already used as a presentation tool for homework and in student presentations. This finding raises some questions about where and when students learned to use PPT, since the acquisition of computer-based presentation skills is not part of the official curricula at this school level.

Several conclusions can be drawn from the results presented in the tables in the Findings section. The first is that PPT has become the ubiquitous program used in almost every biology class. Since the presentation of PPT can be anything from simple text to rich multimedia presentations, this finding in itself cannot necessarily be seen as a bad thing (Babik, & Luther, 2020; O'Day, 2006) as long as we recognize that PPT is well on its way to becoming a "killer app" (Adams, 2006). Among students, it is an option preferred over traditional blackboard or whiteboard instruction. However, if students are asked, they would prefer active methods with real objects over multimedia, which is used by most teachers as a delivery medium. As it turns out, PPTs are used by teachers as a summary of what is to be copied and memorized, which contradicts the suggestions for active and engaging lectures (Moravec et al., 2010).

It is impossible to suppress PPT in the classroom, and there is no reason to try. It can be considered an effective tool that allows combinations of different formats in one file that can be easily transferred between home and school. It also allows the use of presentations offered by publishers and accompanying textbooks, or even articles with wonderful slides, clips and movies, which makes the lessons more interesting and livelier (Moravec et al., 2010; O'Day, 2006). On the other hand, PPT can also encourage misuse of the format to promote dull, lifeless, and boring biology lessons (Tranter, 2004).

From this it can be deduced for practice that there is no reason to suppress the use of PPT, however not as a central medium for transcribing the material to be learned, but as a supporting tool for the graphic representation of images of objects and processes that cannot be presented in the classroom for practical reasons.

The main limitation of the study was participation, which was voluntary for teachers as promoters and students as respondents. Thus, although complete anonymity was assured, teachers were gatekeepers between the researchers and the students. While the sample of student opinions can be considered representative, an unknown number of students provided data on the behaviour of the same teacher.

Another problem that can reduce the usefulness of comparative studies from different years is the term PPT itself. A presentation can be anything from a string of simple text to a highly complex structure consisting of numbers, charts, voice messages, video clips, animations, etc. (Bartsch, & Cobern, 2003), with new capabilities being added to the toolbox with each new version (Baker et al., 2018). Due to the dynamic nature of technology, many findings have only short-term value due to rapid development (e.g. Mehlinger, 1996). Therefore, the utility and impact of different modes and options for PPT presentations should be tested regularly, not only as large-scale studies, but also as part of practitioners' small-scale instructional research that promotes better evidence-based instructional outcomes (Šorgo, & Heric, 2020).

**Acknowledgements** We would like to thank former students Anja Jug, Sabina Gaberšek, Jasmina Heric, and Tadeja Gašparič from Faculty of Natural Sciences and Mathematics who participate in data collection.

## References

Adams, C. (2006). PowerPoint, habits of mind, and classroom culture. *Journal of Curriculum Studies, 38*(4), 389–411.

Adams, C. A. (2010). Teachers building dwelling thinking with slideware. *Indo-Pacific Journal of Phenomenology, 10*(1), 1–12.

Adebanjo, A. A. (2020). Effects of lecture method supplemented with powerpoint presentation on students' academic achievement in biology. *KIU Journal of Humanities, 5*(2), 215–222.

Babik, J. M., & Luther, V. P. (2020). Creating and presenting an effective lecture. *Journal of Continuing Education in the Health Professions, 40*(1), 36–41.

Baker, J. P., Goodboy, A. K., Bowman, N. D., & Wright, A. A. (2018). Does teaching with PowerPoint increase students' learning? A meta-analysis. *Computers & Education, 126*, 376–387.

Bartsch, R. A., & Cobern, K. M. (2003). Effectiveness of PowerPoint presentations in lectures. *Computers & Education, 41*(1), 77–86.

Bognar, B. (2016). Theoretical backgrounds of e-learning. *Croatian Journal of Education: Hrvatski časopis za odgoj i obrazovanje, 18*(1), 225–256.

Field, A. (2009). *Discovering statistics using SPSS*. Sage.

Kernbach, S., Bresciani, S., & Eppler, M. J. (2015). Slip-sliding-away: A review of the literature on the constraining qualities of PowerPoint. *Business and Professional Communication Quarterly, 78*(3), 292–313.

Kirkup, G., & Kirkwood, A. (2005). Information and communications technologies (ICT) in higher education teaching—A tale of gradualism rather than revolution. *Learning, media and technology, 30*(2), 185–199.

Livingstone, S. (2012). Critical reflections on the benefits of ICT in education. *Oxford Review of Education, 38*(1), 9–24.

Mehlinger, H. D. (1996). School reform in the information age. *Phi Delta Kappan, 77*(6), 400–407.

Moravec, M., Williams, A., Aguilar-Roca, N., & O'Dowd, D. K. (2010). Learn before lecture: A strategy that improves learning outcomes in a large introductory biology class. *CBE—Life Sciences Education, 9*(4), 473–481.

O'Day, D. H. (2006). Animated cell biology: A quick and easy method for making effective, high-quality teaching animations. *CBE—Life Sciences. Education, 5*(3), 255–263.

Pevzner, P., & Shamir, R. (2009). Computing has changed biology—Biology education must catch up. *Science, 325*(5940), 541–542.

Prokop, P., Majerčíková, D., & Vyoralová, Z. (2016). The use of realia versus powerpoint presentations on botany lessons. *Journal of Baltic Science Education, 15*(1), 18–27.

Šorgo, A., Verčkovnik, T., & Kocijančič, S. (2010). Information and communication technologies (ICT) in biology teaching in Slovenian secondary schools. *Eurasia Journal of Mathematics, Science and Technology Education, 6*(1), 37–46.

Šorgo, A., & Špernjak, A. (2012). Practical work in biology, chemistry and physics at lower secondary and general upper secondary schools in Slovenia. *Eurasia Journal of Mathematics, Science and Technology Education., 8*(1), 11–19.

Šorgo, A., & Heric, J. (2020). Motivational and demotivational factors affecting a teacher's decision on whether to do research. *Center for Educational Policy Studies Journal, 10*(3), 77–97.

Špernjak, A. (2014). Usefulness of Prezi and PowerPoint presentation. In: Biljanović, P. (Ed.). MIPRO 2014: proceedings. MIPRO 2014, 37th International Convention, Opatija, Croatia. 880–882.

Špernjak, A., & Šorgo, A. (2018). Differences in acquired knowledge and attitudes achieved with traditional, computer-supported and virtual laboratory biology laboratory exercises. *Journal of Biological Education, 52*(2), 206–220.

Stoner, M. R. (2007). PowerPoint in a new key. *Communication Education, 56*(3), 354–381.

Szabo, A., & Hastings, N. (2000). Using IT in the undergraduate classroom: Should we replace the blackboard with PowerPoint? *Computers & education, 35*(3), 175–187.

Tranter, J. (2004). Biology: Dull, lifeless and boring? *Journal of Biological Education, 38*(3), 104–105.

# Part IV
# Textbook Analysis

# Chapter 20
# Human Reproduction in Greek Secondary Education Textbooks (1870s to Present)

**Georgios Ampatzidis and Anastasia Armeni**

## 20.1 Introduction

Most children have already built a basic understanding of human reproduction by the age of 11 years (Driver et al., 1994). However, research with students of different ages shows that they demonstrate limited knowledge in regards to certain aspects of sex and reproduction such as the reproductive physiology and anatomy (Ampatzidis et al., 2019), the menstrual cycle (Yip, 1998) and sexual health (Bolshakova et al., 2020; Westwood & Mullan, 2006). Poor knowledge concerning human reproduction has been linked with risks such as sexually transmitted diseases and unintended pregnancies (Donati et al., 2000; Sydsjö et al., 2006), while a strong understanding of sex and reproduction issues seems to delay the age when students become sexually active. Moreover, when students become sexually active, they are more likely to adopt practices that protect their sexual health (Walker & Milton, 2006).

Several informal sources are considered to help students build their understanding and develop the attitudes that would help them make informed decisions concerning their sexual lives - these include their parents, other family members, their peers, churches and media (Milton, 2003; Valois et al., 1995). Informal sources seem to be important; Sydsjö et al. (2006), investigating students in Sweden aged 13-25 years old, report that among different sources of information (namely school,

G. Ampatzidis (✉)
Department of Chemistry, University of Patras, Patras, Greece
e-mail: ampatzidis@upatras.gr

A. Armeni
Department of Obstetrics and Gynecology, Medical School, National and Kapodistrian University of Athens, Athens, Greece
e-mail: anarmeni@med.uoa.gr



K. Korfiatis, M. Grace (eds.), *Current Research in Biology Education*, Contributions from Biology Education Research,
https://doi.org/10.1007/978-3-030-89480-1_20

television, friends, parents, siblings, healthcare personnel, books, internet, newspapers and magazines) their respondents were most pleased with answers provided by their friends when it came to sexual and reproductive matters. However, the authors also report that the most frequently used source of information among their respondents was school (Sydsjö et al., 2006).

It is argued that however helpful an informal source of information may be, most young people build understanding on sex and reproduction mainly through school education (Westwood & Mullan, 2006). In many cases sex education topics are introduced already in primary school (McKay et al., 2017; Milton, 2003) and they are considered a valued part of school education (Milton et al., 2001). On the other hand, school education largely relies on the use of textbooks since they play a central role in teaching and learning (Gerouki, 2008). Weiss et al. (2001) mention that textbooks are used by over 90% of secondary science teachers in USA in order to support them in planning and delivering instruction while Abd-El-Khalick et al. (2008) note that textbooks may determine, to a degree larger than is wanted by educators, what is taught and learned in the classroom.

Thus, it seems that school education, which relies considerably on the use of textbooks, plays a crucial role in advancing students' understanding of human reproduction. Considering also the fact that abortion rates are considerably high in Greece (Ioannidi-Kapolou, 2004), while only a small percentage of people 16–45 years old appear well-informed concerning contraception (Tountas et al., 2004), we decided to explore the units of Greek secondary education textbooks that deal with human reproduction with regard to: (a) reproductive anatomy, (b) reproductive physiology, (c) reproductive health and (d) sexual behaviour. Our research covers a period from the late nineteenth century to present, in order to detect possible differences across decades related, for instance, to (i) scientific advancements, and (ii) ideas about sexuality. We investigate how changes in scientific knowledge and ideas about sex and reproduction are reflected in relevant Greek secondary education textbooks from the 1870s to present in order to discuss how textbooks have contributed in the building of a meaningful understanding of reproduction.

## 20.2 Methods

Access to old textbooks was achieved through the historic collection of the Greek Institute of Educational Policy. The institute's website includes a database of digitalized versions of Greek textbooks from the 1820s to the 1980s, which can be browsed based on specific criteria (i.e. date, language, education level and course). We performed a query for biology and anthropology textbooks of secondary education written in Greek, regardless of the date. 'Anthropology' was used as a name of courses used from the nineteenth century till the 1990s that dealt mostly with human biology (i.e. different organs and systems of the human body). Our query was performed in September 2019 and it resulted in 107 e-books, which we reviewed along with 6 biology books in use during the school year 2019–2020, in search of units of human reproduction.

Some books included information concerning mammal, or more generally, animal reproduction without treating human reproduction as a special subunit; such books were not analyzed. For example, in 'General Biology Lessons for the 3rd grade of Lyceum' (Krimpas & Kalopisis, 1977) there is a unit about reproduction which deals with sexual reproduction (through the example of plants) and asexual reproduction, but human reproduction is not treated as a special subunit. Humans are mentioned in some cases – for instance, there is a picture of a human egg being fertilized by human sperm – but the authors generally refer to groups of organisms such as mammals, vertebrates or 'organisms that reproduce sexually' to describe sperm and egg production, meiosis and fertilization. In 'Genetics for the 1st grade of Technical Vocational School' (Yfoulis, 1978) there is a unit about reproduction which contains a subunit entitled 'Fertilization and life cycles in humans and animals'; again, humans are mentioned as part of a bigger group of organisms and there is not a special mention of human reproduction. Text similar to the two examples above was not part of the analysis of this study.

Moreover, some of the e-books we reviewed were different editions of the same book; in such cases we used for our analysis the version which included most information relevant to our study. For example, among the different editions of 'Anthropology for the 2nd grade of Lyceum' that we reviewed (1969, 1970, 1971, 1972, 1974, 1975, 1976, 1977, 1978, 1979, 1980, 1981) we used the 1977 edition (Aspiotis, 1977) for our analysis that has two extra paragraphs comparing to previous editions; the text does not change in later editions. Finally, some books made no reference to the reproductive system, even when other systems were described in detail – as it was the case, for instance, in 'Human Somatology' (Sperantzas, 1966) – and they were of course not analyzed for our research.

Taking into consideration the above, our analysis concerned six textbooks:

1. Somatological and Psychological Anthropology (Chortakis, 1872) –K7 to K12.
2. Anthropology for the 2nd grade of Lyceum (Aspiotis, 1977) – K11.
3. Anthropology for the 2nd grade of Gymnasium (Aspiotis, 1980) – K8.
4. Anthropology for the 2nd grade of Gymnasium (Argyris & Kavouras, 1981) – K8.
5. Biology for the 1st grade of Lyceum (Kastorinis et al., 2011) – K10.
6. Biology for the 1st grade of Gymnasium (Mavrikaki et al., 2017) – K7.

The analysis of the units referring to human reproduction was driven by the objectives to present and discuss:

- textbooks' representations of the reproductive system;
- how questions of human reproduction are dealt with in the textbooks;
- how questions of reproductive health are dealt in the textbooks;
- how questions of sexual behaviour are dealt in the textbooks.

The textual corpus of our study was treated using a coding scheme informed by one of the instruments constructed and validated within the project 'Biology, health and environmental education for better citizenship', for comparing school textbooks of European and African countries (Carvalho & Clément, 2007). More specifically, for our coding scheme we draw on the instrument constructed for textbook analysis on the topic 'Human reproduction and sex education'. This is a grid constituted by 19 categories divided in several subcategories such as reproductive anatomy, reproduction (reproductive cells, reproductive hormones, conception, embryonic development, etc.), STDs and sexual behaviours (see for instance Bernard et al., 2007). Table 20.1 shows our coding scheme after removing empty categories (i.e. only categories represented at least in one of the textbooks are shown) and merging/rearranging categories with little content (for instance, 'sexuality' and 'contraception methods' were put together under the 'sexual behaviour' category). Each unit of the textbooks studied was determined as a unit of analysis and each unit could be coded in more than one category. Both authors coded independently textbook-5, which is the one including the most information on human reproduction, and the inter-rater agreement was over 90%. The cases of disagreement were reviewed and discussed by the two coders and the rest of the analysis was carried out by the first author.

We acknowledge the limitations of our study: we argue that by reviewing 113 books we have illustrated the trends of dealing (or failing to deal) with reproduction in Greek textbooks used in biology and anthropology classes from the 1870s to present. However, our research may not be considered an exhaustive account of how reproduction is represented in Greek secondary education textbooks for all ages since – apart from the textbooks currently in use – we have not included textbooks that are not part of the historic collection of the Greek Institute of Educational Policy.

**Table 20.1** The coding scheme

| Representation of reproductive system | Reproduction stages | Hormones | Reproductive health | Sexual behaviour |
|---|---|---|---|---|
| Male/female reproductive organs (MRO/FRO)<br>• Internal (I)<br>• External (E) | Sperm production (SP)<br>Egg production (EP)<br>Fertilization (F)<br>Embryonic development (ED) | Testosterone (T)<br>Estradiol (E)<br>Progesterone (P)<br>Luteinizing hormone (LH)<br>Follicle stimulating hormone (FSH)<br>Human chorionic gonadotropin (HCG)<br>Prolactin (Pl) | Sexually transmitted diseases (STD)<br>• Gonorrhea (G)<br>• Syphilis (S)<br>• HIV<br>• Hepatitis (H)<br>• Chlamydia (C)<br>• HPV infection (HPV)<br>• Herpes (Hp)<br>• Fungal infection (F)<br>• Pthiriasis (P)<br>Healthy pregnancy (HP)<br>Infertility (I)<br>Prenatal screening (PS) | Sexuality (S)<br>Contraception methods (CM) |

## 20.3 Results

As shown in Table 20.2, all textbooks refer at least to some of the organs that constitute the male and female reproductive system. It is worth noticing that female breasts are considered organs of the female reproductive system in textbook-1:

> Breasts are also considered as female genitalia, which are mainly composed of glands consisting of lobes. Blood vessels and milk secretory pores terminate at the ends of the lobes (Chortakis, 1872, p. 32).

Furthermore, female external genitalia are mentioned only in textbook-5:

> The vulva includes mons pubis, labia majora, labia minora and clitoris. (Kastorinis et al., 2011, p. 205).

Concerning the different stages of reproduction, sperm production is the only event described in all six textbooks. Egg production and fertilization are outlined in all textbooks apart from textbook-1. Finally, four textbooks include information in relation to the embryonic development as well. More specifically:

**Table 20.2** Analysis of six textbooks with regard to reproduction content

| Textbooks | Representation of reproductive system | Reproduction stages | Hormones | Reproductive health | Sexual behaviour |
|---|---|---|---|---|---|
| Textbook-1 (1872) | MRO (I, E)<br>FRO (I) | SP | | | |
| Textbook-2 (1977) | MRO (I, E)<br>FRO (I) | SP<br>EP<br>F | T<br>E | STD (G, S) | S |
| Textbook-3 (1980) | MRO (I, E)<br>FRO (I) | SP<br>EP<br>F<br>ED | T<br>E | STD (G, S) | S |
| Textbook-4 (1981) | MRO (I, E)<br>FRO (I) | SP<br>EP<br>F<br>ED | T<br>E<br>P | | |
| Textbook-5 (2011) | MRO (I, E)<br>FRO (I, E) | SP<br>EP<br>F<br>ED | T<br>E<br>P<br>LH<br>FSH<br>HCG<br>Pl | HP<br>I<br>PS<br>STD (G, S, HIV, H, C, HPV, Hp, F, P) | CM |
| Textbook-6 (2017) | MRO (I, E)<br>FRO (I) | SP<br>EP<br>F<br>ED | | STD (HIV, H)<br>HP | CM |

- In textbook-3 there is information concerning the embryos' nurture through the umbilical cord and placenta, the origin of human tissues from the three germ layers (ectoderm, mesoderm and endoderm), the embryos' features in different developmental stages and the birth.
- In textbook-4 there is information concerning the embryos' features in different developmental stages, the embryos' nurture through the umbilical cord and placenta and the birth.
- In textbook-5 there is information concerning the implantation, the formation of the placenta and the embryos' nurture through the umbilical cord, the embryos' features in different developmental stages and the birth.
- In textbook-6 there is information concerning the embryos' nurture through the umbilical cord and placenta and the birth.

Most textbooks mention testosterone and estradiol as the hormones which induce the development of secondary sex characteristics. For example, in textbook-2 we read:

> Testicles and ovaries not only produce sperm and ova respectively; they also secrete hormones. Thus, testicles mainly secrete testosterone and ovaries mainly secrete estradiol. Sex characteristics also depend on these hormones. (Aspiotis, 1977, p. 151).

In textbook-4, progesterone also appears:

> Testicles are the male genitalia. They are the primary genitals, which produce the male germ cells that compose the sperm and secrete male, androgenic hormones, such as testosterone. (…). The main function of the ovaries is the production of ova, which are female germ cells as well as the production of the sex hormones, namely estradiol and progesterone. (Argyris & Kavouras, 1981, p. 111).

Textbook-5 mentions prolactin, progesterone, luteinizing hormone, follicle stimulating hormone and human chorionic gonadotropin as well, since it includes a detailed account of hormone level fluctuation during the menstrual cycle and pregnancy.

Concerning reproductive health, there are references to STDs in four out of six textbooks. Textbook-2 and textbook-3 name gonorrhea and syphilis:

> Terrible diseases of the genital system always threaten human. We note only two: gonorrhea and syphilis, ultimate complications of which could be blindness and even insanity. (Aspiotis, 1977, p. 151).

Textbook-5 refers to a larger list of STDs, while it should be noted that textbook-6 actually includes a homework assignment for students to search information about HIV and hepatitis and ways to prevent them, but it cites no information on the diseases or their prevention.

Textbook-5 and textbook-6 refer to factors that contribute to a healthy pregnancy. For instance, in textbook-6 we read:

> During pregnancy, development of the fetus and the health of a pregnant woman are affected by: 1. Environmental factors: air pollution, pesticides, chemical additives in food, etc. 2. Factors related to lifestyle during pregnancy: the pregnant woman should not smoke, consume alcoholic beverages and medicines not recommended by the doctor. A balanced diet

> and exercise contribute to normal development of the fetus and the good health of the pregnant woman. (Mavrikaki et al., 2017, p. 127).

Moreover, textbook-5 makes reference to infertility problems and in vitro fertilization as a procedure for infertile couples to have children. Lastly, textbook-5 is the only textbook to discuss birth control and various contraception methods as ways of family planning, also making connections to STD prevention.

With regard to sexual behaviour, our last axis of analysis, we note that sexuality is not directly addressed in any of the textbooks investigated. However it is implied in textbook-2 and textbook-3 in which right after introducing STDs, the authors discuss how anything different than 'normal' related to sexual activity may have a negative effect on a person's education and social status. What follows is the relevant quote from textbook-2:

> The genital system exerts a great impact on human psyche. Non-normal tendencies and any perversion associated with the genital system are accompanied by severe consequences involving an individual's proper education, social status and advance. (Aspiotis, 1977, p. 151).

## 20.4 Discussion

Reviewing 113 books used in biology and anthropology courses we noticed that human reproduction is rarely part of their content. Several books describe sexual reproduction through the example of plants or by referring to groups of organisms such as organisms that reproduce sexually, mammals and vertebrates (e.g. Krimpas & Kalopisis, 1977). In other cases, reproduction is absent: for instance, in 'Fundamental Anthropology' (Eleftheropoulos, 1883) there is a chapter about the fluids of the human body, in which the author mentions blood, saliva, bile, tears, mucus, earwax, fat, bolus, urine and sweat but he makes no reference to sperm or vaginal fluid. Other examples are 'Human Somatology' (Sperantzas, 1966) which, as already mentioned, makes no reference to the reproductive system, although other systems are described in detail (skeletal system, muscular system, digestive system, respiratory system, circulatory system, nervous system), and 'Compact Anthropology' (Kasimis, 1886) where again other systems and organs of the human body are mentioned but there are no references to human reproduction.

The reluctance of textbooks' authors to mention human reproduction may be associated with the period of time the textbooks were written, related with taboos and restrictions of earlier times (Curran et al., 2009). Actually, the omission of the reproductive system concerns mostly the older books we reviewed, with the exception of textbook-1. On the other hand, the practice of discussing reproduction through references to other organisms or big groups of organisms may be related to Allen's (2004) argument that a discourse of desire and erotics has been missing from sex education. She claims that textbooks often emphasize the internal parts of the reproductive system and ignore the body's ability for pleasure; for example, the clitoris is often not mentioned and penis is rarely illustrated erect in relevant graphics (Allen, 2004). A missing discourse of erotics seems to be also the case for the textbooks we analyzed - in only one of them female external genitalia are mentioned and they all focus on structure and events of reproduction ignoring the pleasure involved.

Several authors suggest that pleasure should be included in sex/reproduction classes. For example, Wood et al. (2019) argue that an emphasis of sex classes on pleasure may lead youngsters to adopt safe sex practices and feel more in control when it comes to sexual encounters. Scholer (2002) suggests that describing arousal and orgasm is a good way to review the reproductive system anatomy at the end of a relevant class. Moreover, she claims that orgasm and arousal should anyway be included in classes about reproduction because of their role in procreation: males release sperm through orgasm and female orgasm increases the speed of sperm transport through the vagina and fallopian tubes through muscular contractions of the female reproductive tract (Scholer, 2002).

Concerning the different stages of reproduction, we note that sperm production, egg production, fertilization and embryonic development are described in most textbooks (four out of six). The embryonic development is outlined in different ways but the embryo's nurture through the umbilical cord, the role of placenta and the delivery are mentioned in all four textbooks. It is noteworthy that while sperm production is outlined in textbook-1 there is no information with regard to egg production. Textbook-1 also omits information concerning fertilization and embryonic development. We should, though, underline the fact that in the period textbook-1 was written (1872) the understanding of complex procedures such as egg production, embryogenesis and fertilization was limited; valuable descriptions of fertilization, egg maturation and cleavage were produced in the late nineteenth century (Lopata, 2009).

Four out of the six textbooks that we investigated mention some of the hormones involved in reproduction. Textbook-1 makes no reference to hormones, as expected, since the concept of hormones was introduced in early 1900s (Eaton, 2005). Textbook-4 and textbook-5 present testosterone as restricted to men, estradiol as restricted to women and both as having sex-restricted functions. These ideas are not in accordance with current scientific knowledge but they are accordant with the 'sex-dualistic conceptualization of steroid hormones' which prevailed in early twentieth century and seem to be well-established in school science (Nehm & Young, 2008). Textbook-2 and textbook-3 argue that mainly testicles release testosterone and mainly ovaries release estradiol in an effort to avoid presenting them as sex-restricted; however, they do not elaborate on the role of testosterone in females and the role of estradiol in males.

Regarding reproductive health, we note that four out of six textbooks mention STDs. Actually, textbook-6 contains no information on STDs but its authors engage students in researching HIV, hepatitis and prevention methods as a homework. Textbook-2 and textbook-3 introduce gonorrhea and syphilis, arguably two of the most important STDs at the time they were written. Textbook-5 lists a thorough account of STDs mentioning the pathogens and relevant symptoms of men and women while it discusses contraception as a way to prevent STDs and also as a means of family planning. Contraception is considered vital for a healthy and risk free sexual life and there are many studies exploring different aspects of it, such as attitudes towards contraception (e.g. Marshall et al., 2020), the connection between practicing contraception and following relevant school subjects (e.g. Bourke et al., 2014), and developing teaching interventions to support students understanding on the usage of contraception methods (Rosenthal, 2010). Nevertheless, only textbook-5 includes information on different methods of contraception and textbook-6 suggests students should do their own research on contraception by asking them to explore methods of prevention of hepatitis and HIV as homework. Ioannidi-Kapolou (2004) argues that Greek education in the early 2000s lacked scientifically accurate information on contraception and this had important repercussions for women's lives resulting in unintended pregnancies in many cases. The lack of relevant information from education may also be associated with what Tountas et al., (2004) report, namely that only a small percentage of people 16–45 years old appear well informed concerning contraception.

Finally, sexuality is considerably absent from the textbooks investigated. Apart from a paragraph in textbook-2 and textbook-3 introducing the idea of 'normal' in sexual activity, no other textbook attempts to contribute to such a discussion. On the one hand, we may again argue that a discourse of desire and erotics is missing from the textbooks we researched. On the other hand, the introduction of the idea of 'normal' may be linked to what Curran et al. (2009) define as compulsory heterosexuality. Compulsory heterosexuality is not associated only with conceptions and acts related to sexuality. It is rather considered a foundational societal and cultural structure (Herz & Johansson, 2015) and it is strongly linked to heteronormativity, i.e. the idea that heterosexuality is the norm in sexuality assuming the status of 'true' sexuality on the grounds of male-female biological binary and reproduction (Robinson, 2005). It is argued that heteronormative school environments – approving heterosexual practices and presuming that students identify themselves as heterosexual – contribute to the vulnerability of non-heterosexual students and they potentially reduce interaction between teachers and students, impacting the building and application of new knowledge (Dinkins & Englert, 2015). Our analysis shows that heteronormativity seems to inhere in all textbooks that we explored.

## 20.5 Conclusion

In our study we focus on textbooks as a reflection of what is taught in the classroom. The discussion of our findings suggests certain actions to be taken concerning textbooks currently in use in Greek secondary education, in order to enhance teaching and learning of human reproduction. For example, when dealing with sex/reproduction issues, textbooks should discuss the pleasure involved through introducing the physiology of arousal and orgasm. Moreover, a discussion of sexuality should be involved; textbooks seem to assume heterosexuality as the norm, reinforcing its dominance in students' learning environments. Heteronormativity should be avoided in order for a more meaningful, inclusive understanding of sex and reproduction to be promoted. As a final remark, we note that however important textbooks may be for teaching and learning (Abd-El-Khalick et al., 2008), teachers could supplement limited text and, as Elliott (2003) argues, they may need to when it comes to reproduction.

## References


Abd-El-Khalick, F., Waters, M., & Le, A.-P. (2008). Representations of nature of science in high school chemistry textbooks over the past four decades. *Journal of Research in Science Teaching, 45*(7), 835–855.

Allen, L. (2004). Beyond the birds and the bees: Constituting a discourse of erotics in sexuality education. *Gender and Education, 16*(2), 151–167.

Ampatzidis, G., Georgakopoulou, D., & Kapsi, G. (2019). Clitoris, the unknown: What do postgraduate students of educational sciences know about reproductive physiology and anatomy? *Journal of Biological Education*. Advanced online publication. https://doi.org/10.1080/00219266.2019.1679658

Argyris, I., & Kavouras, A. (1981). *Anthropology for the 2nd grade of gymnasium*. School Book Publishing Organization (OEDB).

Aspiotis, N. (1977). *Anthropology for the 2nd grade of lyceum*. School Book Publishing Organization (OEDB).

Aspiotis, N. (1980). *Anthropology for the 2nd grade of gymnasium*. School Book Publishing Organization (OEDB).

Bernard, S., Clément, G., & Carvalho, G. S. (2007, August). *Prevention of sexual transmitted diseases in biology textbooks from different countries*. Paper presented at the ESERA 2007 Conference, Malmö, Sweden.

Bolshakova, M., Galimov, A., Unger, J. B., Rohrbach, L. A., & Sussman, S. (2020). Russian adolescent sexual behaviour and contraceptive knowledge. *Sex Education, 20*(5), 568–582.

Bourke, A., Boduszek, D., Kelleher, C., McBride, O., & Morgan, K. (2014). Sex education, first sex and sexual health outcomes in adulthood: Findings from a nationally representative sexual health survey. *Sex Education, 14*(3), 299–309.

Carvalho, G. S., & Clément, P. (2007, August). *Construction and validation of the instruments to compare teachers' conceptions and school textbooks of 19 countries – The European Biohead-Citizen project*. Paper presented at the AREF 2007 International Conference, Strasbourg, France.

Chortakis, N. (1872). *Somatological and psychological anthropology*. Depasta Brothers.

Curran, G., Chiarolli, S., & Pallotta-Chiarolli, M. (2009). 'The C words': Clitorises, childhood and challenging compulsory heterosexuality discourses with pre-service primary teachers. *Sex Education, 9*(2), 155–168.

Dinkins, E. G., & Englert, P. (2015). LGBTQ literature in middle school classrooms: Possibilities for challenging heteronormative environments. *Sex Education, 15*(4), 392–405.

Donati, S., Medda, E., Spinelli, A., & Grandolfo, M. E. (2000). Sex education in secondary schools: An Italian experience. *Journal of Adolescent Health, 26*(4), 303–308.

Driver, R., Squires, A., Rushworth, P., & Wood-Robinson, V. (1994). *Making Sense of Secondary Science: Research into children's ideas*. Routledge.

Eaton, L. (2005). College looks back to discovery of hormones. *BMJ, 330*(7506), 1466.

Eleftheropoulos, A. (1883). *Fundamental anthropology*. Ermis.

Elliott, K. J. (2003). The Hostile Vagina: Reading vaginal discourse in a school health text. *Sex Education, 3*(2), 133–144.
Gerouki, M. (2008). Pushed to the margins – Sex and relationships in Greek primary textbooks. *Sex Education, 8*(3), 329–343.
Herz, M., & Johansson, T. (2015). The normativity of the concept of heteronormativity. *Journal of Homosexuality, 62*(8), 1009–1020.
Ioannidi-Kapolou, E. (2004). Use of contraception and abortion in Greece: A review. *Reproductive Health Matters, 12*, 174–183.
Kasimis, P. (1886). *Compact anthropology*. S. K. Vlastos.
Kastorinis, A., Kostaki-Apostolopoulou, M., Mparona-Mamali, F., Peraki, V., & Pialoglou, P. (2011). *Biology for the 1st grade of lyceum*. CTI Diophantus.
Krimpas, K., & Kalopisis, I. (1977). *General biology lessons for the 3rd grade of lyceum*. School Book Publishing Organization (OEDB).
Lopata, A. (2009). History of the egg in embryology. *Journal of Mammalian Ova Research, 26*(1), 2–9.
Marshall, S. A., Driver, N., & Allison, M. K. (2020). Attitudes towards contraception: Focus groups with Arkansas teenagers and parents. *Sex Education*. Advanced online publication. https://doi.org/10.1080/14681811.2020.1759526
Mavrikaki, E., Gouvra, M., & Kampouri, A. (2017). *Biology for the 1st grade of gymnasium*. CTI Diophantus.
McKay, E., Vlazny, C., & Cumming, S. (2017). Relationships and sexuality education topics taught in Western Australian secondary schools during 2014. *Sex Education, 17*(4), 454–470.
Milton, J. (2003). Primary school sex education programs: Views and experiences of teachers in four primary schools in Sydney, Australia. *Sex Education, 3*(3), 241–256.
Milton, J., Berne, L., Peppard, J., Patton, W., Hunt, L., & Wright, S. (2001). Teaching sexuality education in high schools: What qualities do Australian teachers value? *Sex Education, 1*(2), 175–186.
Nehm, R. H., & Young, R. (2008). "Sex hormones" in secondary school biology textbooks. *Science & Education, 17*(10), 1175–1190.
Robinson, K. H. (2005). 'Queerying' gender: heteronormativity in early childhood education. *Australasian Journal of Early Childhood, 30*(2), 19–28.
Rosenthal, M. S. (2010). Name that contraceptive! A game for the human sexuality classroom. *American Journal of Sexuality Education, 5*(2), 189–199.
Scholer, A.-M. (2002). Sexuality in the science classroom: One teacher's methods in a college biology course. *Sex Education, 2*(1), 75–86.
Sperantzas, S. (1966). *Human somatology*. School Book Publishing Organization (OEDB).
Sydsjö, G., Selling, K. E., Nyström, K., Oscarsson, C., & Kjellberg, S. (2006). Knowledge of reproduction in teenagers and young adults in Sweden. *The European Journal of Contraception & Reproductive Health Care, 11*(2), 117–125.
Tountas, Y., Creatsas, G., Dimitrakaki, C., Antoniou, A., & Boulamatsis, D. (2004). Information sources and level of knowledge of contraception issues among Greek women and men in the reproductive age: A country-wide survey. *The European Journal of Contraception & Reproductive Health Care, 9*(1), 1–10.
Valois, R. F., Roth, N. L., Montgomery, E., & Waring, K. A. (1995). Sex education content of ninth grade health education textbooks: A rhetorical analysis. *Journal of Sex Education and Therapy, 21*(3), 192–209.
Walker, J., & Milton, J. (2006). Teachers' and parents' roles in the sexuality education of primary school children: A comparison of experiences in Leeds, UK and in Sydney, Australia. *Sex Education, 6*(4), 415–428.
Weiss, I. R., Banilower, E. R., McMahon, K. C., & Smith, P. S. (2001). *Report of the 2000 national survey of science and mathematics education*. Horizon Research.

Westwood, J., & Mullan, B. (2006). Knowledge of secondary school pupils regarding sexual health education. *Sex Education, 6*(2), 151–162.
Wood, R., Hirst, J., Wilson, L., & Burns-O'Connell, G. (2019). The pleasure imperative? Reflecting on sexual pleasure's inclusion in sex education and sexual health. *Sex Education, 19*(1), 1–14.
Yfoulis, A. (1978). *Genetics for the 1st grade of technical vocational lyceum*. Eugenides Foundation.
Yip, D. Y. (1998). Children's misconceptions on reproduction and implications for teaching. *Journal of Biological Education, 33*(1), 21–26.

# Chapter 21
# Exploring Logico-Semantic Relations of Historical Gene Models in Similar Genetic Content in Greek Biology School Textbooks

**Akrivi Christidou, Penelope Papadopoulou, and Anastasia G. Stamou**

## 21.1 Introduction

Genetics has been assigned its own subject matter and development of a complex role of holding multiple concepts with complicated relations (Flodin, 2009). The didactic transposition of knowledge derived from the genetic research context to the genetic educational context is documented as difficult, and leads to both teaching and learning obstacles (Flodin, 2009, 2017; Gericke et al., 2014). Often, it has been stated in international bibliography that students share severe alternative conceptions about genes, at undergraduate and graduate level, with social impact in their informed future decisions as citizens (Albuquerque et al., 2008; Flodin, 2009; Gericke et al., 2014). Since textbooks constitute crucial teaching aids and sources for students, the inadequacy or inconsistency of textbook discourse about genes contributes to difficulties in understanding genetics rather than clarifying the issue (Liu & Khine, 2016; Gericke et al., 2014).

A well-known strand of international research in secondary school and university science textbook analysis of genetic topics concerns the representations of the gene concept and its functions by the five historical gene models: The Mendelian model, the Classical model, the Biochemical-Classical model, the Neoclassical model and the Modern model. The gene concept, particularly how it was perceived by the

A. Christidou (✉) · P. Papadopoulou
Department of Early Childhood Education, University of Western Macedonia, Florina, Greece
e-mail: akchrist@hotmail.gr; ppapadopoulou@uowm.gr

A. G. Stamou
Department of Linguistics and Language Teaching, Aristotle University of Thessaloniki, Thessaloniki, Greece
e-mail: anstamou@del.auth.gr



K. Korfiatis, M. Grace (eds.), *Current Research in Biology Education*, Contributions from Biology Education Research,
https://doi.org/10.1007/978-3-030-89480-1_21

scientists during the development of the biology research field, has been described as: (1) a trait with physical place (locus), in the Mendelian model and used in Mendelian genetics; (2) a nucleotide sequence that provides instructions, is expressed and regulated, in the Classical model and usually found in molecular biology; (3) a factor or DNA that interacts, moves or duplicates, in the Biochemical-Classical model and encountered in genomics; (4) a regulator, thus DNA controls, directs and defines a pattern, in the Neoclassical model and used in developmental biology; and (5) a marker that can be fixed or added, in the Modern model and is mostly useful in population genetics (Gericke & Hagberg, 2007; Gericke & Hagberg, 2010a, b). Actually, the genotype gradually separated from the phenotype and the environmental factors were progressively taken into account with regard to gene function, among the historical gene models (Gericke & Hagberg, 2007; Gericke & Hagberg, 2010a, b).

The historical models have been grouped by Gericke and Hagberg (2007) and were implemented by those and also other researchers in secondary and university biology and chemistry textbooks from different countries, to identify the types of gene models used as well as the way they can affect students' understanding and learning (Aivelo & Uitto, 2015; Gericke & Hagberg, 2010a, b; Santos et al., 2012). In particular, it has been reported that in Swedish and Brazilian biology textbooks the Neoclassical model prevails, while in Finnish biology textbooks the Mendelian model is the most common. On the other hand, the Biochemical-Classical model dominates in international biology textbooks. In all cases, the elements of the Modern model are limited or even absent (Aivelo & Uitto, 2015; Gericke & Hagberg, 2010a, b; Santos et al., 2012).

Our research focuses on all seven current Greek biology textbooks (Christidou & Papadopoulou, 2020). Six of them are being taught in Greek secondary school and one textbook is being used in the last grade of Greek primary school. The material contains not only direct but also indirect references to the gene concept and its functions, and addresses students aged 13–18. The textbook analysis was based on seven predefined groups of epistemological features composing the five historical gene models. These mainly describe the relationship between the structure and the function of the gene, the organizational level and the definition of gene function, the genotype and the phenotype, the genetic and the environmental factors. Every single epistemological feature variant from each group corresponds to one or more of the historical models, but there are also some unclassified features often called non-historical (Gericke & Hagberg, 2010a, b; Santos et al., 2012). Some feature variants emerged also from our analysis expressing points not mentioned before, which are: "The gene and its functions are related to DNA", "The gene is present in frequencies", "There is a distinction between genotype and phenotype, with chromosomal explanation" (Christidou & Papadopoulou, 2020).

Our previous content analysis has revealed the simultaneous presence of multiple gene models in each one of the seven Greek biology textbooks (Christidou & Papadopoulou, 2017a, b). In general, the epistemological feature variants of the Biochemical-Classical model have been found to dominate (25.9%), followed by those of the Classical model (24.1%), the Neoclassical model (15.9%) and the Mendelian model (15.5%). Finally, the presence of the epistemological feature variants of the Modern model are limited (9%) (Christidou & Papadopoulou, 2020). Additionally, the percentage of appearance of the gene models seemed to be dependent upon the different genetic content. For instance, in topics such as Genetics/ Molecular biology and Biotechnology/Bioethics, the gene concept is presented mostly through the characteristics of the Biochemical-Classical model, while in Evolution chapters, the Neoclassical model and the Modern model are prevalent for describing the gene functions. In glossaries that contain genetic definitions, the Classical model prevails (Christidou & Papadopoulou, 2017a, b; Christidou & Papadopoulou, 2020).

As follows, there is conceptual variation describing the gene concept and its functions in Greek biology textbooks. The conceptual variation has been ascribed for describing different meanings of a concept, as is usually the case with the term 'gene', that is the bearer of a large diversity of explanations which relate to each biological field (Gericke et al., 2014). Equally important to the above findings, a deeper analysis in the composition of the gene models found in the seven Greek biology textbooks showed that the conceptual variation of the gene concept and its functions is mainly due to the Classical model and the Neoclassical model, which appear in the textbooks through their unique epistemological feature variants with an equal total percentage of occurrence of 30%. Besides, the prevalent presence of the Biochemical-Classical model is in fact mostly based on the epistemological feature variants shared with other historical models (approximately 94%). Concerning the Mendelian model and the Modern model, they appear almost only through their non-exclusive epistemological features (about 90%, each), (Christidou & Papadopoulou, 2019a).

There are some interesting key points concluded from the outcomes of the categorical content analysis in the seven Greek biology textbooks. Some serious concerns arise about the possibility of reinforcing the formation of alternative conceptions and misunderstandings by the Greek students. This speculation results from the predominance of the Biochemical-Classical model in the analysis material that, as known from literature, contains internal inconsistencies concerning the gene function and the explanatory reduction (Gericke & Hagberg, 2007). It should also be noticed that the conceptual variation found in describing the gene function is observed only implicitly in the context of the biology textbooks that may affect students' learning procedure (Gericke et al., 2014). The degraded presence of the Modern model in Greek biology textbooks should result in limiting students' knowledge to more outdated conceptions about genes (Aivelo & Uitto, 2015). Moreover, deeper analysis in the composition of the present gene models revealed the absence of environmental elements in the effect of gene function that could direct students to strong deterministic views (Gericke et al., 2014).

Therefore, there is a further need of research in the Greek biology textbooks from another perspective regarding the emphasis on the presence of multiple gene models, their cohesion and their possible contribution to didactic problems. For this purpose, the genetic content was analyzed at the level of lexicogrammar and then the outcomes from both aspects of study were combined. In this study, critical discourse analysis is applied to specific texts with similar genetic content of the Greek biology textbooks. The aim of the study is to examine how the parallel presence of multiple gene models already found in Greek biology textbooks, which is a potential source of alternative conceptions formed by students, is actually reflected in the cohesion and coherence of the genetic meanings, and how it is differentiated among textbooks in junior high school and high school.

## 21.2 Methodology

In order to examine the influence of the multiple gene models in Greek biology textbooks on the cohesion and the coherence of the genetic content, the relations between the paragraphs as written by the authors and the succession of the present gene models were initially investigated in three selective and similar subjects regarding genetics in junior high school and high school biology textbooks. In other words, the application of the critical discourse analysis attempted to reveal the type of transition of information from one paragraph to the next.

Hence, the chosen content analysis material includes three representative chapters of similar individual topics regarding genetics among the biology textbooks used in the 2nd – 3rd grade of junior high school, indicated from now on as Textbook 1, and in the 3rd grade of high school, indicated from now on as Textbook 2. The latter is part of studying for university entry exams. As shown in Table 21.1, the analyzed chapters in this study cover three subjects that concern gene function and regulation, Mendelian inheritance and mutations. Both textbooks deal extensively with genetics.

Specifically, the analyzed chapters from Textbook 1 are:

Chapter 5.2 – The flow of genetic information, covering 3 pages, Chapter 5.5 – Inheritance, covering 3 pages, and
Chapter 5.6 – Mutations, covering 1 page.

The chapters from Textbook 2 are:

Chapter 2 – Expression of genetic information & Gene regulation: Control of gene regulation, covering 5 pages,
Chapter 5 – Mendelian inheritance, covering 14 pages, and
Chapter 6 – Mutations, covering 13 pages.

**Table 21.1** The chapters from the biology textbooks for the 2nd – 3rd grade of junior high school (Textbook 1), and for the 3rd grade of high school (Textbook 2), which were chosen as the content analysis material

| Textbook 1 | Textbook 2 |
|---|---|
| Chapter 5.2. The flow of genetic information | Chapter 2. Expression of genetic information & Gene regulation: Contro of gene regulation |
| Chapter 5.5. Inheritance | Chapter 5. Mendelian inheritance |
| Chapter 5.6. Mutations | Chapter 6. Mutations |

Each paragraph can be an autonomous, meaningful piece of text and has been attributed some epistemological feature variants as shown in the analysis above. The paragraphs are in the book chapters and were used as the unit of analysis.

The critical discourse analysis adopted in our research, is based on the theory of systemic functional grammar developed by Halliday and focuses on textual metafunction of meaning on the level of lexicogrammar (Knain, 2001). On this level, there are three metafunctions of meaning, namely ideational, interpersonal and textual meaning. Those are realized in transitivity, mood system/structure and theme-rheme, respectively, in which clauses are used as representation, exchanging and message (Ngongo, 2018). Critical discourse analysis is a critical, text-centric analysis tool (Stamou, 2014). This tool of descriptive qualitative analysis is indeed a social semiotic approach to language showing how it functions in the context (Knain, 2001).

We used the logico-semantic relations that refer to the expansion of the meaning of a primary clause. Expansion can be expressed as: 1. an **extension**, by adding a new idea or stating an exception to a primary clause; 2. an **elaboration**, by repeating, commenting, simplifying or determining in detail the primary clause; and 3. an **enhancement**, by adding an idea due to circumstances such as time, place, cause, condition or even result (Ngongo, 2018). Appropriate transitional words or phrases may be used in each case and can be divided into the above types of relations (Campbell et al., 1994; Halliday & Hasan, 1976). All of the following transitional words and their respective logico-semantic relations that are represented, if they were present in the text, should be traced at the beginning of every paragraph of the analysis material from Textbook 1 and 2, showing the transitions. To ensure the reliability of the results, analysis was carried out independently by three researchers, and in case of disagreement they discussed it until they reached an agreement.

For the extension of a meaning there are additive words or phrases expressing **addition** such as: and, and also, not only, either...(or), but also, as well, or, moreover, furthermore, in addition, besides, in fact, alternatively. There are adversative words or phrases expressing **contrast** (but, while, whilst, yet, still, however, whereas, although), **concession** (nevertheless, despite, albeit, regardless, for all that), **dismissal** (either way), and **replacement** (instead, on the contrary, at least). Alternatively, there are sequential words or phrases signaling logical sequence such as **continuation** (next, subsequently, afterwards, eventually, previously, before) (Halliday & Hasan, 1976).

For the elaboration of a meaning there are additive words or phrases expressing **introduction** (such as, particularly, for example, for instance, especially, notably), **reference** (considering, regarding, concerning), **similarity** to other ideas (similarly, equally, in the same way), **identification** (that is, namely, specifically, thus) and **clarification** (that is to say, in other words). There are adversative words or phrases signaling **emphasis** (even more, above all, indeed), causal words or phrases showing **respective** (in this respect, here, apart from this) and sequential words or phrases representing **resumption** (to resume, to get back to the point), **conclusion** (to conclude, eventually, finally, at last), **digression** (to change the topic, incidentally, by the way) and **summation** (in summary, all in all, overall, on the whole, in conclusion, ultimately, altogether) (Halliday & Hasan, 1976).

Hereafter, for the enhancement of a meaning there are causal words or phrases expressing **cause/reason** (because, as, since, for, due to, to the extent that), **condition** (on condition that, if, provided that, even if), **effect /result** (as a result, consequently, hence, thus, therefore, accordingly), **purpose** (so that, in order to, so as to, for the purpose of, with this intention), and **consequence** (then, in that case, otherwise, under the circumstances). Further, there are sequential words or phrases signaling chronological sequence such as **ordinal** (initially, in the first place, to start with, originally, at first, secondly) (Halliday & Hasan, 1976).

## 21.3 Results

In this study, when critical discourse analysis was applied to three dominant and representative chapters from Textbook 1 and Textbook 2, all with similar genetic topics, we found that there was some variety in logico-semantic relations among the paragraphs and their gene models. However, the most common transition found is the *extension*. The outcomes of critical discourse analysis are presented selectively in Tables 21.2, 21.3 and 21.4 only for the exact corresponding paragraphs of these chapters, facilitating, in this way, the comparison of the resulting relations.

For the topic "The flow of genetic information", there are four corresponding paragraphs in the two textbooks: Genetic information and replication, Stages of transcription, Definition of the gene concept, and The central dogma of biology. As shown in Table 21.2, three extensions including two additions and one opposition, in addition to one elaboration of the primary meaning expressed as summation, were found in Textbook 1. These logico-semantic relations depict the transition from one paragraph to the following one, as well as the kind of sequence of the gene models that are present. The corresponding paragraphs of Textbook 2, and the gene models found to describe the flow of genetic information, are linked with three elaborations of the primary meaning expressed as exemplification and summation, two extensions as additions and one temporal enhancement. The exact matching among the two textbooks is shown in Table 2, which is divided into two sections respectively. Each section has three columns. The first column of each section presents the excerpt from the beginning of each paragraph in which the logico-semantic relations are examined. In the second column of each section, the logico-semantic relations creating the transitions among the paragraphs are found. In the third column, the historical gene models that represent the gene function in the respective paragraph, are shown.

Regarding the topic of "Inheritance", there are four corresponding paragraphs in the two textbooks: Mendel studied inheritance in peas, Definitions of genotype and phenotype, Cross-pollination of pea plants, and the 3:1 ratio. In Table 21.3, the sequence of the logico-semantic relations among these paragraphs and the gene models that represent the topic of inheritance are shown with the exact matching

**Table 21.2** Comparative analysis among the paragraphs of similar content from the chapter: "The flow of genetic information" in the two Greek biology textbooks

| "The flow of genetic information" | | | | | |
|---|---|---|---|---|---|
| Textbook 1: Chapter 5.2 | | | Textbook 2: Chapter 2 | | |
| Excerpt | Logico-semantic relations | Historical models | Excerpt | Logico-semantic relations | Historical models |
| **Paragraph: *Genetic information and replication*** | | | | | |
| *"DNA of every cell contains genetic information that is necessary for its structure and function..."* | Extension (and) | Neoclassical model | *"DNA of an organism is the molecular "hard disk drive" that contains stored precise instructions, which determine the structure and function of the organism ..."* | Elaboration (exemplification) | Neoclassical model |
| **Paragraph: *Stages of transcription*** | | | | | |
| *"The mRNA that is synthesized must contain the genetic information about the amino acid sequence of that particular protein... "* | Extension (and) | Biochemical model Neoclassical model Modem model | *"All cells of a multicellular organism have the same DNA..."* | Extension (and) | Mendelian model Biochemical model Neoclassical model Modem model |
| | | | *"At the initiation of transcription of a gene, RNA polymerase binds to the promoter and causes local unwinding of the double-stranded DNA..."* | Temporal enhancement | Neoclassical model Modem model |
| **Paragraph: *Definition of the gene concept*** | | | | | |
| *"But transcription does not only produce mRNA..."* | Extension (but) | Mendelian model Biochemical model Neoclassical model Modern model | *"The haploid human genome has $3\times10^9$ base pairs. Of this, a small percentage is transcribed into DNA, that is genes."* | Extension (and) | Neoclassical model |

(continued)

**Table 21.2** (continued)

| "The flow of genetic information" | | | | | |
|---|---|---|---|---|---|
| Textbook 1: Chapter 5.2 | | | Textbook 2: Chapter 2 | | |
| Excerpt | Logico-semantic relations | Historical models | Excerpt | Logico-semantic relations | Historical models |
| **Paragraph:** ***The central dogma of biology*** | | | | | |
| *"The processes of copying, transcription and translation are done with the help of special enzymes. These processes are summarized in the central dogma of biology..."* | Elaboration (summation) | Biochemical model<br>Neoclassical model | *"This relationship is summarized in the following figure, where the arrows indicate the direction of transfer of genetic information..."* | Elaboration summation) | Neoclassical model |
| | | | *"This pattern is the central dogma of molecular biology as it was named by F. Crick (1958) ..."* | Elaboration (exemplification) | Mendelian model<br>Biochemical model<br>Neoclassical model<br>Modem model |

among the two textbooks, in the same way as explained above for Table 21.2. In Textbook 1, there are two elaborations expressed as exemplification and summation, one extension expressed as addition and one enhancement transition expressed as consequence. Moreover, in Textbook 2, two extensions with additional meaning and two temporal enhancements are detected.

For the topic "Mutations", there are three corresponding paragraphs in the two textbooks: Definition of mutations, Gene and chromosomal mutations and Genetic diversity. Similarly, the comparative analysis of transitions among the paragraphs in the two Greek biology textbooks and their gene models that represent the gene function in this topic is indicated in Table 21.4. In Textbook 1, one additional extension and one oppositional extension are found as well as a consequential transition of enhancement and an exemplification transition of elaboration. Furthermore, the analysis of Textbook 2 showed that the transitions occur as two additional and one oppositional extensions, and also one elaboration with summation.

**Table 21.3** Comparative analysis among the paragraphs of similar content from the chapter: "Inheritance" in the two Greek biology textbooks

| "Inheritance" | | | | | |
|---|---|---|---|---|---|
| Textbook 1: Chapter 5.5 | | | Textbook 2: Chapter 5 | | |
| Excerpt | Logico-semantic relations | Historical models | Excerpt | Logico-semantic relations | Historical models |
| **Paragraph: *Mendel studied inheritance in peas*** | | | | | |
| *"Mendel studied extensively the way in which the characteristics of organisms are inherited..."* | Elaboration (summation) | Mendelian model | *"When Mendel crossed pure short plants with pure tall plants, he found that all plants of the F 1 generation were tall ..."* | Temporal enhancement | Mendelian model |
| **Paragraph: *Definitions of genotype and phenotype*** | | | | | |
| *"Our characteristics are determined by genes found on homologous chromosomes..."* | Extension (and) | Mendelian model Classical model Biochemical model | *"A person with the same alleles of a certain gene for ci particular trait is homozygous for that trait..."* | Extension (and) | Mendelian model Classical model Biochemical model |
| **Paragraph: *Cross-pollination of pea plants*** | | | | | |
| *"If we denote by the predominant allele for brown eyes and by m the recessive allele for blue, then: ..."* | Elaboration (exemplification) | Mendelian model Biochemical model Neoclassical model Modem model | *"Mendel crossed short (ΨΨ) plants with tall (ΨΨ) plants..."* | Extension (and) | Classical model |
| **Paragraph 3:*1 ratio*** | | | | | |
| *"Therefore, the probability of giving birth to a child with blue eyes is 25% (1/4), while with brown eyes 75% (3/4), end this will apply every time a zygote is created..."* | Enhancement (consequence) | Classical model | *"Then, he crossed the F1 generation' plants with each other ..."* | Temporal enhancement | Mendelian model Classical model Biochemical model Neoclassical model |

**Table 21.4** Comparative analysis among the paragraphs of similar content from the chapter: "Mutations" in the two Greek biology textbooks

| "Mutations" | | | | | |
|---|---|---|---|---|---|
| Textbook 1: Chapter 5.6 | | | Textbook 2: Chapter 6 | | |
| Excerpt | Logico-semantic relations | Historical models | Excerpt | Logico-semantic relations | Historical models |
| **Paragraph: *Definition of mutations*** | | | | | |
| *"Organisms sometimes show new characteristics, which are due to changes in their DNA ..."* | Extension (and) | non-historical features | *"The genetic material can be filtered in many different ways ..."* | Extension (and) | Mendelian model<br>Biochemical model<br>Neoclassical model<br>Modem model |
| **Paragraph: *Gene and chromosomal mutations*** | | | | | |
| *"In humans, a mutation in a gene has a consequent not to produce a pigment called melanin..."* | Enhancement (consequence) | Mendelian model<br>Classical model<br>Biochemical model<br>Neoclassical model<br>Modem model | *"Geneticists classify mutations into two broad categories: genetic and chromosomal..."* | Elaboration (summation) | Neoclassical model |
| *"In another case, the change in the genetic material concerns a change in the number of chromosomes ..."* | Elaboration (exemplification) | Mendelian model<br>Classical model<br>Biochemical model | | | |
| **Paragraph: *Genetic diversity*** | | | | | |
| *"The mutations do not always cause disease..."* | Extension (but) | Mendelian model<br>Biochemical model<br>Neoclassical model<br>Modem model | *"Mutations contribute to genetic diversity in the Population and are responsible for many inherited diseases ..."* | Extension (and) | Biochemical model<br>Neoclassical model |
| | | | *"Although most mutations lead to results that are not favorable for the organism, some of them exhibit advantages ..."* | Extension (but) | Mendelian model<br>Classical model<br>Biochemical model<br>Neoclassical model<br>Modem model |

## 21.4 Conclusions

From the critical discourse analysis applied in the whole of Textbook 1, it seems that the text flows without any obvious problems in the use of speech according to transitions among the paragraphs, as in most cases the new information is added sequentially – mostly by extensions. The addition or removal of the gene models in the text indeed happens without particular contradictions, since there are only sporadic contrasts in transitions. It seems that the gene models and all the information are added cumulatively to students' knowledge. This is also shown in the individual outcomes presented above (Tables 21.2, 21.3, and 21.4), particularly for Textbook 1. In Textbook 2, the critical discourse analysis applied to the whole text of the textbook proved a greater variety of logico-semantic relations, reflecting the more complex scientific information transformed into school knowledge in higher school grades (Christidou & Papadopoulou, 2019b). The sequence of the gene models in the specific chapters referred to previously (Tables 21.2, 21.3 and 21.4), as generally in Textbook 2, also seems complementary in their genetic meanings. What is meant by 'complementary' is that the gene models contribute to the sequence and the addition of knowledge included in the paragraphs of these biology textbooks.

A different perspective on the presence and connection of the gene models is reflected in this study. At the level of lexicogrammar, the critical discourse analysis in the genetic content of the two selected Greek biology textbooks, shows that the genetic meanings are coherent and follow one another in a natural way. Accordingly, the multiple gene models found in these texts are connected to each other through logico-semantic relations, basically without any contrast relations being used. Consequently, triggering students' alternative conceptions is less likely. Moreover, the research outcomes reveal that the authors might have been unaware of the historical gene models, which are only implicitly presented in Greek biology textbooks.

It is remarkable that the various gene concepts could be useful in different biological research fields, fulfilling several suitable purposes such as treating the gene as a trait, an information-structure, an actor in the cell, a regulator in embryonic development, or as a marker for evolutionary change (Flodin, 2009). Thus, genetic conceptual variation is necessary and desirable for researchers, especially when different gene concepts act in a complementary rather than an incompatible way (Gericke et al., 2014). In genetic education, the conceptual variation originates from the presence of multiple gene models, as detected in biology school textbooks of many countries (Aivelo & Uitto, 2015; Christidou & Papadopoulou, 2017a, b, 2019a, 2020; Gericke & Hagberg, 2010a, b; Santos et al., 2012). This could bring logical inconsistencies, self-contradictions and conceptual incoherence, causing concept confusion to students or even unidentifiable alternative conceptions (Gericke et al., 2014). In this case, it is proposed that genetic conceptual variation should be explicitly addressed in Greek biology textbooks, so both learners and teachers identify and realize that gene function could be presented in different ways depending on the biological framework (Gericke et al., 2014). It should be obviously indicated and clearly explained in Greek biology textbooks that the gene concept might differentiate according to the historical model used in each topic, in order to transform and convey the scientific information to the readers (Albuquerque et al., 2008). The textbook reader's notion of the gene models could affect their understanding and promote knowledge. Another prerequisite could be the presentation of the historical development of the gene concept through capturing the history of science in Greek biology textbooks. Consequently, when implemented properly, conceptual variation also turns out to be a useful tool for teachers and students (Gericke et al., 2014).

Nevertheless, in this study when analysis deepened into logico-semantic relations, the conceptual variation of the gene concept in the two specific Greek biology textbooks, did not appear to include notable contradictions regarding the sequence and cohesion of the present gene models. This contrasts with the speculation of the former findings in Greek biology textbooks, and to the conclusions of Gericke et al., 2014 for biology textbooks from six other countries. The use of multiple gene models mostly seems to have an explanatory and complementary role in the text, and presumably does not trigger alternative conceptions. This might have happened either because the authors of the biology textbooks do not distinguish between the differences of gene models, or they are unaware of them. As a consequence, the information in textbooks is added mostly as complementary. In a different case, the authors may be familiar with the gene models, but for the sake of the didactic transformation they do not choose to refer to them; therefore, the pieces of information are being added successively. As the Greek secondary biology textbooks had been written before international research introduced the historical gene models in the literature, it is not possible that the second case happened.

In conclusion, the findings of this study could be taken into consideration by the authors of the Greek biology textbooks for future editions. Future perspectives to help clarify the impact of the genetic conceptual variation, could refer to the research on conceptions of genetics by Greek students who have previously been exposed to the genetic content material of Greek biology textbooks.

**Acknowledgements** This research is part of a PhD being funded by the Greek General Secretariat for Research and Technology (GSRT) and the Hellenic Foundation for Research and Innovation (HFRI).

## References

Aivelo, T., & Uitto, A. (2015). Genetic determinism in the Finnish upper secondary school biology textbooks. *Nordic Studies in Science Education, 11*(2), 139–152.

Albuquerque, P. M., de Almeida, A. M. R., & El-Hani, N. C. (2008). Gene concepts in higher education cell and molecular biology textbooks. *Science Education International, 19*(2), 219–234.

Campbell, G., Buckhoff, M., & Dowell, J. A. from MSU – Michigan State University. (1994). https://www.msu.edu/user/jdowell/135/transw.html

Christidou, A., & Papadopoulou, P. (2020). Representations of gene models in Greek secondary school biology textbooks. In B. Puig, P. B. Anaya, M. J. G. Quílez, & M. Grace (Eds.) *Biology education research. Contemporary topics and directions.* A selection of papers presented at the XIIth conference of European Researchers in Didactics of Biology (ERIDOB). (pp. 225–236). University of Zaragoza. ISBN: 978-84-16723-97-3.

Christidou A., & Papadopoulou P. (2019a, 19–21 April). *The compilation of gene models in biology textbooks of Greek education.* 11th Pan-Hellenic conference on science education and new technologies in education: Redefining the teaching and learning of science and technology in the 21st century, University of Western Macedonia, Florina, Greece.

Christidou, A., & Papadopoulou, P. (2019b). Succession and connection of gene models through logico-semantic relations in Greek biology textbooks. In A. Polyzos & M. Georgiou (Eds.), *Proceedings of the 5th Pan-Hellenic conference on biology education* (pp. 218–226). Panhellenic Association of Bioscientists. ISBN: 978-618-81159-6-5.

Christidou, A., & Papadopoulou, P. (2017a). Representations of historical gene models in Greek biology school textbooks: The case of textbook of the 3rd grade of junior high school. In D. Stavrou, A. Michailidi, & A. Kokolaki (Eds.), *Proceedings of the 10th Pan-Hellenic conference on science education and new technologies in education: Bridging the gap between natural sciences, society and educational practice* (pp. 446–454). Pedagogical Department of Primary Education, Laboratory of Teaching Science, University of Crete. ISBN 978-960-86978-3-6.

Christidou, A., & Papadopoulou, P. (2017b). Representations of historical gene models in Greek biology secondary school textbooks: A Comparison. In A. Polyzos & L. Anthis (Eds.), *Proceedings of the 4th Pan-Hellenic conference on biology education* (pp. 133–142). Panhellenic Association of Bioscientists. ISBN: 978-618-81159-5-8.

Flodin, V. S. (2017). Characterisation of the context-dependence of the gene concept in research articles. *Science and Education, 26*, 141–170.

Flodin, V. S. (2009). The necessity of making visible concepts with multiple meanings in science education: The use of the gene concept in a biology textbook. *Science and Education, 18(1)*, 73–94.

Gericke, N. M., Hagberg, M., Santos, V. C., Joaquim, L. M., & El-Hani, C. N. (2014). Conceptual variation or incoherence? Textbook discourse on genes in six countries. *Science & Education, 23*(2), 381–416.

Gericke, N. M., & Hagberg, M. (2010a). Conceptual incoherence as a result of the use of multiple historical models in school textbooks. *Research in Science Education, 40*(4), 605–623.

Gericke, N. M., & Hagberg, M. (2010b). Conceptual variation in the depiction of gene function in upper secondary school textbooks. *Science and Education, 19*(10), 963–994.

Gericke, N. M., & Hagberg, M. (2007). Definition of historical models of gene function and their relation to students' understanding of genetics. *Science and Education, 16*, 849–881.

Halliday, M. A. K., & Hasan, R. (1976). *Cohesion in English.* Longman Group Limited.

Knain, E. (2001). Ideologies in school science textbooks. *International Journal of Science Education, 23*(3), 319–329.

Liu, Y., & Khine, M. S. (2016). Content analysis of the diagrammatic representations of primary science textbooks. *Eurasia Journal of Mathematics, Science and Technology Education, 12(8)*, 1937–1951.

Ngongo, M. (2018). Taxis and logico – Semantic relation in undergraduate students' English theses writing text: A systemic functional linguistics approach. *Journal of Arts, Science and Commerce, 9*(2), 146–152.

Santos, V. C., Joaquim, L. M., & El-Hani, C. N. (2012). Hybrid deterministic views about genes in biology textbooks: A key problem in genetics teaching. *Science and Education, 21*(4), 543–578.

Stamou, A. (2014). Critical discourse analysis: Studying the ideological role of language. In M. Georgalidou, M. Sifianou, & B. Tsakona (Eds.), *Discourse analysis: Theory and applications* (pp. 149–187). Nissos. ISBN: 978-960-9535-85-4.

# Part V
# Outdoor and Environmental Education

# Chapter 22
# Children's Interests and Orientations Towards Nature: Views from Young Children in England

**Richard Sheldrake and Michael Reiss**

## 22.1 Introduction

Supporting and protecting nature are increasing concerns in many countries (United Nations, 2015). Science education, biology education, and environmental education aim to inspire interest, curiosity, and understanding about the natural world (Royal Society of Biology, 2019), which may also help future generations to protect the natural environment. Previous research indicates that supporting and protecting nature associates with people holding orientations and affinities towards nature (often referred to as 'nature connection') focused around enjoyment and appreciation of nature (e.g. Mackay & Schmitt, 2019). However, most of the research has only been undertaken with adults. In order to gain new insights, the research presented here surveyed primary school children (aged between 7 and 10 years) in England, and considered what might associate with their enjoyment and appreciation of nature, sense of oneness with and responsibility towards nature (reflecting orientations towards supporting/protecting nature), and other views about nature.

### *22.1.1 Appreciating, Supporting, and Protecting Nature*

Children and young people often value nature, recognise and appreciate the diverse experiences that can be possible in nature, and associate natural places with positive feelings and/or relaxation (Bonnett & Williams, 1998; Wiens et al., 2016). Primary school children in England have particularly conveyed their appreciation of animals

R. Sheldrake (✉) · M. Reiss
UCL Institute of Education, London, United Kingdom
e-mail: r.sheldrake@ucl.ac.uk; m.reiss@ucl.ac.uk



K. Korfiatis, M. Grace (eds.), *Current Research in Biology Education*, Contributions from Biology Education Research,
https://doi.org/10.1007/978-3-030-89480-1_22

and plants (Harvey et al., 2020). Primary school children in England have also expressed concern with the welfare of animals, emphasised that people need the environment to live, and conveyed their awareness of environmental concerns (Bonnett & Williams, 1998). Children across various other countries have frequently conveyed empathy and sympathy for nature, recognition of dependency between people and nature, and recognition of the impacts that people have on nature (Aaron & Witt, 2011; Chawla, 2020; Collado et al., 2016; Rios & Menezes, 2017; Strife, 2012).

Considered from a wider perspective, children across England typically express positive affinities and orientations towards nature (e.g. Hughes et al., 2019). These affinities/orientations are often referred to as 'nature connection' (or equivalent terms such as 'nature relatedness'), and have been considered as aggregations of numerous aspects including: inherently valuing experiences of nature and enjoying being in nature; feeling in harmony and connected with nature; having affinities towards and appreciation of wildlife; feeling a perceived responsibility and sympathy for nature; and/or recognising the importance or value of nature as an aspect of someone's personal identity (e.g. Tam, 2013). Measures of nature connection have been found to associate with positive attitudes towards the environment (Mackay & Schmitt, 2019) and with actions and behaviours that support/protect the environment (Diessner et al., 2018).

Children's nature connection has been found to associate with visiting and/or otherwise engaging with nature (e.g. Szczytko et al., 2020). In children and adults, nature connection has linked with other activities such as watching wildlife, reading books about the natural world, and watching nature-related media (e.g. Hunt et al., 2017). Children's reported nature connection has also been found to associate with their perceptions of their family's values towards nature (Cheng & Monroe, 2012). Having and/or perceiving more local nature near their home, or otherwise living close to nature, has also been found to associate positively with nature connection for children and adults (Shanahan et al., 2017), which highlights the contextual importance of having access and/or opportunity to engage with nature. Girls often express higher nature connection than boys, and younger children often express higher nature connection than older children, although it remains unclear why (Hughes et al., 2019; Richardson et al., 2019). It remains unclear whether differences in children's views about nature can be completely explained by differences in their engagement with nature, and/or whether these difference link with other aspects of their lives such as their education.

### *22.1.2 Interests and Learning About Nature*

Research has often focused on orientations towards appreciating, supporting, and protecting nature, and less (but some) research has focused on children's interests towards learning about nature. Engaging in informal and/or extra-curricular activities, such as reading books and watching media about science, links with children's general interests in science (Bonnette et al., 2019), and children's interests in biology have been found to relate to their experiences in nature and engagement with nature-related media (Uitto et al., 2006). Children's connections to nature (encompassing their enjoyment in experiencing nature), previous experiences in nature, and family values towards nature have all been found to associate with their interest in participating in nature-based activities and their interest in environmentally-friendly practices (Cheng & Monroe, 2012). Children's interest in learning about aspects of nature has often been found to have been fostered by outdoor learning experiences (Hinds, 2011) and by environmental education programmes (Ballantyne et al., 2001). Experiences of outdoor learning have also been found to help develop children's interests and motivations towards specific areas of their studies, such as natural history (Stern et al., 2008). Outdoor learning can also provide memorable experiences (Dierking & Falk, 1997; Liddicoat & Krasny, 2014), which can sometimes help foster children's interest in engaging with nature and increase their environmental awareness and behaviours to protect/support nature (Liddicoat & Krasny, 2014). Nevertheless, more insights would benefit education and also wider understanding; for example, it remains unclear how children's views about nature, including their interest in learning about nature, relate to their wider views about their education, such as their confidence in undertaking their schoolwork and their sense of school belonging.

### *22.1.3 Research Aims*

Biology education aims to facilitate children's interests towards learning about the natural world, while environmental education and wider policies across England aim to help children to support and protect nature (e.g. Department for Environment, Food and Rural Affairs, 2018; Knapp, 2000). Nevertheless, it remains less clear, as the above review indicates, what might help foster these various views and behaviours. In order to gain new insights, the research presented here surveyed primary school children in England.

The research considered the children's self-reported engagement with various activities in daily life and their views about nature and learning. The analysis aimed to reveal similarities and/or differences in views and/or reported activities across boys and girls. Prior research has highlighted links between gender and views about nature; hence the importance of considering this area (Hughes et al., 2019; Richardson et al., 2019).

The analysis then focused on revealing independent predictors of the children's enjoyment/appreciation of nature, empathy/affinity for animals, sense of oneness/responsibility for nature, and interest in learning about nature. Prior research has not concurrently considered children's views about learning and about nature, so it remained unclear whether and/or which views about learning (or other reports from children) might associate with children's views about nature and especially children's interest in learning about nature. This analytical approach also helped to consider whether any observed differences in views across boys and girls might be explained by differences in their wider views and/or activities.

## 22.2 Methods

The research was approved by the ethics committee of the host institution. Children in schools and children attending outdoor learning activities at nature reserves and wildlife centres in England were invited to participate; parental information sheets and consent materials were disseminated and collected via the children's schools and/or by reserve/centre staff. Questionnaires were completed before any activities were undertaken, in order to gain insights into children's general views.

### *22.2.1 Participants*

The research considered the questionnaire responses of 679 children: 356 (52.6%) identified as girls and 321 (47.4%) identified as boys. The children were aged between 7 and 10 years old (on average, 8 years old), and were based within various regions of England.

### *22.2.2 Questionnaires*

A questionnaire was designed to cover a range of areas, encompassing validated items and also new items in order to explore under-researched issues. For example, the 'Connection to Nature Index' questionnaire items (Cheng & Monroe, 2012) were included to maximise comparability and contextualisation with prior research; other existing (and new) items/questions are detailed in the following sections.

For most items on the questionnaire, children expressed their agreement or disagreement against various statements, with response categories of 'Strongly disagree' (scored as 1), 'Disagree' (2), 'Neither agree nor disagree' (3), 'Agree' (4), and 'Strongly agree' (5).

Some items on the questionnaire were aggregated to provide single indicators ('factors' or 'scales'). These indicators were calculated as the arithmetic mean of the relevant items. Factor analysis showed that the relevant items could be aggregated into their respective indicators, and acceptable reliability (internal consistency across the items) was observed via Cronbach's alpha coefficients (as reported below).

#### 22.2.2.1 Children's Characteristics and Contexts

The questionnaire asked for children's gender, age, and whether either of their parents (or guardians) went to university. The questionnaire also considered the children's views about their education and school context through indicators of:

- *Enjoying and being confident in learning* ('I usually do well in school work', 'I enjoy learning at school', and 'I can do most things at school if I try'; 3 items, Cronbach's alpha = 0.705);
- *Sense of belonging at school* ('I feel like I belong at my school', 'I get on well with my classmates', and 'I get on well with my teachers'; 3 items, Cronbach's alpha = 0.697);
- *Life/learning aspirations* ('Doing well in school will help me in the future', 'I have goals and plans for the future', and 'I think I will be successful when I grow up'; 3 items, Cronbach's alpha = 0.636).

These areas have been similarly measured in other research with children (e.g. Lereya et al., 2016). Children's interests and motivations towards their studies have often been found to associate with numerous aspects of their education including their confidence and their sense of school belonging (e.g. Allen et al., 2018). These areas have not been considered alongside children's views about nature within prior research and so might offer new insights.

#### 22.2.2.2 Children's Activities and Engagement with Nature

The questionnaire measured children's engagement with various nature-related activities and aspects of life through specific items considering:

- *'I watch nature and wildlife programmes or videos'*;
- *'I read books about nature and wildlife'*;
- *'My parents encourage me to spend time outdoors in nature'*;
- *'I spend time outdoors in nature'*.

These were measured as engagement frequencies from 'Never or almost never' (scored as 1), 'A few times a year' (2), 'A few times a month' (3), 'A few times a week' (4), to 'Every day or almost every day' (5). Additionally, levels of agreement from 'Strongly disagree' (scored as 1), 'Disagree' (2), 'Neither agree nor disagree' (3), 'Agree' (4), to 'Strongly agree' (5) were used to measure:

- *'I live near nature, such as a park, some woods, or the countryside'*.

Previous research with children and adults has highlighted the potential relevance of some of these activities/engagement (e.g. Hunt et al., 2017; Uitto et al., 2006).

#### 22.2.2.3 Children's Views About Nature

The questionnaire considered the children's views about nature through indicators of:

- *Enjoyment/appreciation of nature* (e.g. 'Being outdoors makes me happy', 'Being in the natural environment makes me feel peaceful'; 7 items, Cronbach's alpha = 0.814);
- *Empathy/affinity for animals and wildlife* (e.g. 'I feel sad when wild animals are hurt', 'I enjoy touching animals and plants'; 4 items, Cronbach's alpha = 0.756);
- *Sense of oneness/responsibility for nature* (e.g. 'Humans are part of the natural world', 'People cannot live without plants and animals', 'My actions will make the natural world different', 'Picking up litter on the ground can help the environment'; 5 items, Cronbach's alpha = 0.708).

These items/indicators were from the 'Connection to Nature Index', which was designed for use with children (Cheng & Monroe, 2012) and has been previously applied in England and the United Kingdom (e.g. Kerr, 2015). This maximised comparability and contextualisation with prior research.

The questionnaire also measured:

- *Interest in learning about nature* ('I like learning about nature', 'I like learning about plants and animals', and 'I would like to learn more about nature in school'; 3 items, Cronbach's alpha = 0.866).

These were new items, developed for this questionnaire, in order to extend understanding and gain new insights. Interest in learning about nature may reflect an aspect of children's contextualised ecological/environmental orientation (Larson et al., 2011), but has been conceptualised and/or measured in varying ways within some research (Rowland et al., 2019) and has not been considered within other research (which has often focused on links between enjoyment/appreciation of nature and time spent outdoors). Children's interest in learning about particular topics are especially important during their education and for their wider trajectories towards (or away from) further studies and/or careers (Eccles, 2009).

### 22.2.3 Analytical Approaches

Statistically significant results are indicated through p-values less than 0.05. Magnitudes of difference between averages are quantified through Cohen's D-values (Cohen, 1988). Associations between the children's responses were explored through predictive modelling, which reveals the independent association between a predictor and the outcome while accounting for all of the other predictors. Predictive modelling was undertaken via linear ordinary-least-squares (OLS) estimation (linear regression). The models showed acceptable fit and the various residual histograms and plots highlighted that the underlying assumptions were met. Predictive (independent) associations were quantified as standardised coefficients (β-values).

## 22.3 Results

On average (Table 22.1), the children expressed positive views concerning nature and their learning, around the 'Agree' response category (values of 4 on the 1-5 scale from strong disagreement to strong agreement). The children also expressed somewhat frequent engagement with the various nature-related activities, around the 'A few times a month' response category (values of 3 on the 1-5 frequency scale). Compared to boys, on average, girls reported more frequent reading of books about nature/wildlife and more positive views for confidence/enjoyment in learning, sense of belonging in school, enjoyment/appreciation of nature, empathy/affinity for animals, oneness/responsibility for nature, and interest in learning about nature.

**Table 22.1** Sample summary and gender differences

| | All | | Gender: Girls | | Gender: Boys | | Gender difference | |
|---|---|---|---|---|---|---|---|---|
| Indicator (1–5 scales unless otherwise shown) | M | SD | M | SD | M | SD | Cohen's D | Sig. (p) |
| Gender (0 = girls, 1 = boys) | .47 | .50 | – | – | – | – | – | – |
| Parents/guardians went to university (0 = no, 1 = yes) | .65 | .48 | .64 | .48 | .66 | .47 | .037 | .664 |
| Age (years) | 8.36 | .86 | 8.44 | .88 | 8.27 | .81 | **.207** | **.007** |
| Confidence/enjoyment in learning | 4.30 | .75 | 4.39 | .68 | 4.19 | .81 | **.279** | **<.001** |
| Aspirations in life/learning | 4.51 | .66 | 4.54 | .62 | 4.48 | .70 | .093 | .242 |
| Belonging in school | 4.37 | .76 | 4.45 | .75 | 4.28 | .76 | **.232** | **.004** |
| 'I spend time outdoors in nature' | 3.66 | 1.32 | 3.72 | 1.29 | 3.58 | 1.34 | .104 | .191 |
| 'I watch nature and wildlife programmes or videos' | 2.99 | 1.49 | 3.05 | 1.45 | 2.91 | 1.53 | .095 | .240 |
| 'I read books about nature and wildlife' | 2.84 | 1.44 | 2.95 | 1.45 | 2.71 | 1.42 | **.166** | **.041** |
| 'My parents encourage me to spend time outdoors in nature' | 3.38 | 1.52 | 3.42 | 1.52 | 3.33 | 1.52 | .060 | .442 |
| 'I live near nature, such as a park, some woods, or the countryside' | 4.13 | 1.21 | 4.13 | 1.22 | 4.13 | 1.20 | .004 | .964 |
| Enjoyment/appreciation of nature | 3.97 | .83 | 4.14 | .72 | 3.79 | .91 | **.431** | **<.001** |
| Empathy/affinity for animals | 4.45 | .72 | 4.52 | .67 | 4.38 | .77 | **.192** | **.016** |
| Oneness/responsibility for nature | 4.33 | .72 | 4.39 | .67 | 4.26 | .76 | **.185** | **.020** |
| Interest in learning about nature | 4.24 | .95 | 4.37 | .88 | 4.10 | 1.00 | **.294** | **<.001** |

Notes: Means ('M'), standard deviations ('SD'), and the magnitude ('D'; Cohen's D) and statistical significance ('Sig. (p)'; *p*-value) of the differences between girls and boys are reported. Significant p-values ($p < .05$) and the associated magnitudes are highlighted in bold for clarity

### 22.3.1 Predictive Associations

Predictive modelling considered the children's characteristics and contextual experiences and views as predictors of the children's enjoyment/appreciation of nature, empathy/affinity for animals, oneness/responsibility for nature, and interest in learning about nature (Table 22.2). This highlighted that higher aspirations in life/learning and more frequent watching of nature/wildlife media and reading of books about nature/wildlife positively predicted enjoyment/appreciation of nature, empathy/affinity for animals, oneness/responsibility for nature, and interest in learning about nature.

#### 22.3.1.1 Enjoyment/Appreciation of Nature

The children's enjoyment/appreciation of nature was most strongly predicted by 'I spend time outdoors in nature', together with the other nature-related activities, confidence/enjoyment in learning, aspirations in life/learning, but was negatively predicted by being a boy (compared to being a girl).

#### 22.3.1.2 Empathy/Affinity for Animals

The children's empathy/affinity for animals was positively predicted by their wider educational views (aspirations in life/learning, and sense of belonging in school), by the various nature-related activities, by living near to nature, but was negatively predicted by being a boy (compared to being a girl).

#### 22.3.1.3 Oneness/Responsibility for Nature

The children's sense of oneness/responsibility for nature was positively predicted by their sense of school belonging, living near to nature, watching nature/wildlife programmes/videos, reading books about nature/wildlife, encouragement from parents to spent time outdoors, and aspirations in life/learning. When accounting for these aspects of life, 'I spend time outdoors in nature' had no association with the children's sense of oneness/responsibility for nature.

#### 22.3.1.4 Interest in Learning About Nature

The children's interest in learning about nature was positively predicted by aspirations in life/learning, watching nature/wildlife programmes/videos, sense of school belonging, reading books about nature/wildlife, spending time outdoors, and confidence/enjoyment in learning, but was negatively predicted by being a boy (compared to being a girl).

**Table 22.2** Predictive models

| | Enjoyment/ appreciation of nature | | Empathy/affinity for animals | | Oneness/responsibility for nature | | Interest in learning about nature | |
|---|---|---|---|---|---|---|---|---|
| Predictor | β | Sig. (p) | β | Sig. (p) | β | Sig. (p) | β | Sig. (p) |
| Intercept/constant | **NA** | **<.001** | **NA** | **<.001** | **NA** | **<.001** | **NA** | **.011** |
| Age (years) | −.049 | .120 | −.044 | .228 | .043 | .216 | −.053 | .106 |
| Gender (0 = girls, 1 = boys) | **−.145** | **<.001** | **−.085** | **.020** | −.036 | .293 | **−.100** | **.002** |
| Parents/guardians went to university (1 = yes, compared to no) | −.067 | .069 | .028 | .517 | −.026 | .523 | −.065 | .091 |
| Parents/guardians went to university (1= missing, compared to no) | −.036 | .329 | −.007 | .866 | .000 | .999 | .022 | .565 |
| Confidence/enjoyment in learning | **.127** | **.004** | −.092 | .073 | .042 | .384 | **.135** | **.003** |
| Aspirations in life/learning | **.105** | **.007** | **.165** | **<.001** | **.096** | **.025** | **.180** | **<.001** |
| Belonging in school | .071 | .084 | **.177** | **<.001** | **.224** | **<.001** | **.165** | **<.001** |
| 'I spend time outdoors in nature' | **.216** | **<.001** | **.145** | **.001** | .058 | .149 | **.156** | **<.001** |
| 'I watch nature and wildlife programmes or videos' | **.136** | **<.001** | **.153** | **<.001** | **.146** | **<.001** | **.170** | **<.001** |
| 'I read books about nature and wildlife' | **.197** | **<.001** | **.106** | **.013** | **.117** | **.003** | **.158** | **<.001** |
| 'My parents encourage me to spend time outdoors in nature' | **.176** | **<.001** | .042 | .317 | **.139** | **<.001** | .051 | .168 |
| 'I live near nature, such as a park, some woods, or the countryside' | .020 | .549 | **.128** | **.001** | **.170** | **<.001** | .003 | .930 |
| Explained variance (via adjusted R-squared) | 47.6% | | 27.9% | | 37.2% | | 43.5% | |

Notes: Standardised predictive coefficients ('β') and their statistical significance ('Sig. (p)'; p-value) are reported. Significant p-values ($p < .05$) and the associated coefficients are highlighted in bold for clarity

## 22.4 Discussion

The results presented here provide a number of insights for education and policy. On average, children expressed positive enjoyment/appreciation of nature, empathy/affinity for animals, oneness/responsibility for nature, and interest in learning about nature, with girls conveying more positive views than boys. The importance of nature-related activities was highlighted: children's enjoyment/appreciation of nature, empathy/affinity for animals, oneness/responsibility for nature, and interest in learning about nature were all positively predicted by more frequent watching of nature/wildlife media and reading of books about nature/wildlife. Enjoyment/appreciation of nature, empathy/affinity for animals, and interest in learning about nature were also positively predicted by spending time outdoors in nature. The importance of children's wider educational contexts and views was also highlighted: children's enjoyment/appreciation of nature and interest in learning about nature were both positively predicted by their confidence/enjoyment in their learning at school and by their aspirations in life/learning. Children's empathy/affinity for animals and oneness/responsibility for nature were also both positively predicted by their sense of belonging within school. Essentially, there may be many ways to promote and foster children's views.

These results affirm and extend earlier research, which has linked children's affinities/orientations towards nature ('nature connection') with visiting and/or otherwise engaging with nature outdoors (e.g. Szczytko et al., 2020), and with other activities such as watching wildlife, reading books about the natural world, and watching nature-related media (e.g. Hunt et al., 2017). Previous research has often considered nature connection through aggregating enjoyment/appreciation of nature, empathy/affinity for animals, oneness/responsibility for nature, and/or other views. The results presented here show the importance of considering these aspects separately, given that different arrays of predictors were relevant for each outcome. Children's empathy/affinities towards animals may be an important area for educators to consider. Recent research highlights that children appreciate animals and plants (Harvey et al., 2020), and that children often characterise nature as being inseparable from living things such as wild animals and feel empathy and concern with the welfare of many aspects of nature including animals (Bonnett & Williams, 1998).

These results also revealed other new insights. Higher confidence/enjoyment in learning, aspirations in life/learning, and sense of school belonging independently and positively predicted the children's interest in learning about nature, while accounting for their personal characteristics and engagement with nature-related activities. Confidence/enjoyment in learning and sense of belonging in school have been found to link to various beneficial outcomes within education such as attainment (e.g. Allen et al., 2018). Essentially, children's views about nature may not necessarily link only with nature-related activities. Ensuing that children can feel supported and confident at school may help provide freedom to develop more specific learning interests.

These results showed that girls conveyed more positive views about nature than boys, which has been seen in prior research (Hughes et al., 2019; Richardson et al., 2019). The predictive models then highlighted an important new insight: gender associated with the children's enjoyment/appreciation of nature, empathy/affinity for animals, and interest in learning about nature, even when accounting for their other reported characteristics and views. Essentially, the gender differences in children's views about nature (such as girls expressing higher interest in learning about nature than boys) were not explained by differences in their other reported views and activities (such as girls reporting more frequent reading of books about nature than boys). Ultimately, the gender differences in children's views about nature can be inferred to follow from other aspects of life that were not measured by the questionnaire. This highlights that more research must be undertaken in order to understand why different children hold different views about nature. Applying different approaches or methods may also be beneficial. This study applied questionnaires, similarly to most prior research (Cheng & Monroe, 2012; Hughes et al., 2019; Richardson et al., 2019). Undertaking detailed interviews with children may help reveal more insights.

### *22.4.1 Educational and Wider Implications*

The findings suggest that various nature-related activities could be encouraged, facilitated, and/or integrated within education. The results presented here show that children's reports of more frequently watching nature/wildlife programmes/videos positively predicted their interest in learning about nature and sense of oneness/responsibility for nature. Children's reports of more frequently spending time outdoors in nature also positively predicted their enjoyment/appreciation of nature, empathy/affinity for animals, and interest in learning about nature (although had no predictive association with oneness/responsibility for nature when accounting for the other predictors). Outdoor learning activities and excursions may be beneficial within education, as these have often helped foster children's interests and appreciation of wildlife (e.g. Lindemann-Matthies, 2005) and children's interests in learning about other aspects of nature (e.g. Hinds, 2011). Nevertheless, further research would need to consider associations in more detail; for example, spending time in nature might foster enjoyment/appreciation of nature and empathy/affinity for animals, which might then foster oneness/responsibility for nature.

Children, on average within this sample, manifested positive enjoyment/appreciation of nature, empathy/affinity for animals, oneness/responsibility for nature, and interest in learning about nature. Girls tended to convey more positive views about nature than boys, especially for their enjoyment/appreciation of nature, which affirms previous research (e.g. Richardson et al., 2019). We would argue that it may be beneficial to support children to find their own personally enjoyable ways to experience nature, which might involve different activities for different children; children differ from one another and it should not be assumed that what interests and engages one child interests and engages another. Future research may need to focus on considering why different children express different views; the presented results highlighted that gender differences in views could not be explained by differences in the children's contexts or engagement in particular nature-related activities. Children's local and/or wider context (such as living or studying in rural or urban contexts) may also be relevant, and could facilitate, limit, and/or otherwise influence how nature could or should be perceived and/or engaged with.

The findings also suggest wider implications for biology education and environmental education. There may be many ways to promote and foster children's views, but different aspects of life may be more or less relevant for different outcomes. This is especially relevant when education has multiple aims or outcomes, such as fostering interests in learning about nature and also orientations towards supporting/protecting nature (considered as oneness/responsibility for nature within the research presented here). For example, the presented results highlighted that children's reports of spending time outdoors in nature positively associated with their reported interest in learning about nature but had no association (when accounting for the other predictors) with their reported sense of oneness/responsibility for nature. Educators will still need to convey and contextualise how and why nature could and/or should be supported and protected, including the positive and negative impacts that people can have on the natural world. More generally, applying diverse and multiple approaches with education may be beneficial. Attempting to find one single nature-related activity that might foster any and every nature-related view in children may be difficult.

### 22.4.2 Limitations

The findings may not necessarily be generalisable to children of different ages and/or children in different contexts (such as in different areas of England and/or in different countries). The analysis considered children's questionnaire responses at only one time point, and so cannot definitively establish whether some views or aspects of life influence or entail other views or aspects of life. Essentially, the results do not show and cannot prove causality. For example, the presented results revealed that the children's reports of spending more time in nature (and other activities and views) positively associated with their enjoyment/appreciation of nature, empathy/affinity for animals, and interest in learning about nature. This only shows one perspective onto potentially complex circumstances. Prior research has highlighted that enjoying and appreciating nature may be a motivation to spend time in nature (Flowers et al., 2016; Lin et al., 2014), and also that enjoying and appreciating nature may follow from spending time outdoors in nature (Nisbet, 2015; Richardson et al., 2016). It is possible to infer that positive views about nature and spending time outdoors are reciprocal, and may form a positive feedback cycle. Similarly, reading books about nature might foster an interest in learning about nature, and an interest in learning about nature might motivate someone to read books about nature. The direction of any associations remains an important area to consider and clarify through further empirical research.

Additionally, many aspects of life may influence children's views, and the questionnaire could only consider a limited number of aspects within a reasonable length. Social-cognitive perspectives on learning and motivation highlight the relevance of someone's context and circumstances (including social contexts and norms) and someone's emerging identity, which link with their various attitudes and self-beliefs (including self-confidence and interest/enjoyment in various activities), which then link with their intentions/aspirations and actions (Eccles, 2009). This suggests further areas to explore when considering children's interest in learning about nature.

## References


Aaron, R., & Witt, P. (2011). Urban students' definitions and perceptions of nature. *Children Youth and Environments, 21*(2), 145–167.

Allen, K., Kern, M., Vella-Brodrick, D., Hattie, J., & Waters, L. (2018). What schools need to know about fostering school belonging: A meta-analysis. *Educational Psychology Review, 30*(1), 1–34.

Ballantyne, R., Fien, J., & Packer, J. (2001). Program effectiveness in facilitating intergenerational influence in environmental education: Lessons from the field. *Journal of Environmental Education, 32*(4), 8–15.

Bonnett, M., & Williams, J. (1998). Environmental education and primary children's attitudes towards nature and the environment. *Cambridge Journal of Education, 28*(2), 159–174.

Bonnette, R., Crowley, K., & Schunn, C. (2019). Falling in love and staying in love with science: Ongoing informal science experiences support fascination for all children. *International Journal of Science Education, 41*(12), 1626–1643.

Chawla, L. (2020). Childhood nature connection and constructive hope: A review of research on connecting with nature and coping. *People and Nature, 2*(3), 619–642.

Cheng, J. C.-H., & Monroe, M. (2012). Connection to nature: Children's affective attitude toward nature. *Environment and Behavior, 44*(1), 31–49.

Cohen, J. (1988). *Statistical power analysis for the behavioral sciences* (2nd ed.). Lawrence Earlbaum Associates.

Collado, S., Íñiguez-Rueda, L., & Corraliza, J. (2016). Experiencing nature and children's conceptualizations of the natural world. *Children's Geographies, 14*(6), 716–730.

Department for Environment, Food and Rural Affairs. (2018). *A green future: Our 25 year plan to improve the environment*. Department for Environment, Food and Rural Affairs.

Dierking, L., & Falk, J. (1997). School field trips: Assessing their long-term impact. *Curator, 40*(3), 211–218.

Diessner, R., Genthôs, R., Praest, K., & Pohling, R. (2018). Identifying with nature mediates the influence of valuing nature's beauty on proenvironmental behaviors. *Ecopsychology, 10*(2), 97–105.

Eccles, J. (2009). Who am I and what am I going to do with my life? Personal and collective identities as motivators of action. *Educational Psychologist, 44*(2), 78–89.

Flowers, E., Freeman, P., & Gladwell, V. (2016). A cross-sectional study examining predictors of visit frequency to local green space and the impact this has on physical activity levels. *BMC Public Health, 16*(420), 1–8.

Harvey, C., Hallam, J., Richardson, M., & Wells, R. (2020). The good things children notice in nature: An extended framework for reconnecting children with nature. *Urban Forestry & Urban Greening, 49*(126573), 1–8.

Hinds, J. (2011). Woodland adventure for marginalized adolescents: Environmental attitudes, identity and competence. *Applied Environmental Education & Communication, 10*(4), 228–237.

Hughes, J., Rogerson, M., Barton, J., & Bragg, R. (2019). Age and connection to nature: When is engagement critical? *Frontiers in Ecology and the Environment, 17*(5), 265–269.

Hunt, A., Stewart, D., Richardson, M. J. H., Bragg, R., White, M., & Burt, J. (2017). *Monitor of engagement with the natural environment: Developing a method to measure nature connection across the English population (adults and children)*. Natural England.

Kerr, K. (2015). *Report for the Royal Society for the Protection of Birds (RSPB): Connection to Nature questionnaire on the Northern Ireland kids life and times survey*. Queen's University Belfast.

Knapp, D. (2000). The Thessaloniki declaration: A wake-up call for environmental education? *Journal of Environmental Education, 31*(3), 32–39.

Larson, L., Green, G., & Castleberry, S. (2011). Construction and validation of an instrument to measure environmental orientations in a diverse group of children. *Environment and Behavior, 43*(1), 72–89.

Lereya, S. T., Humphrey, N., Patalay, P., Wolpert, M., Böhnke, J., Macdougall, A., & Deighton, J. (2016). The student resilience survey: Psychometric validation and associations with mental health. *Child and Adolescent Psychiatry and Mental Health, 10*(44), 1–15.

Liddicoat, K., & Krasny, M. (2014). Memories as useful outcomes of residential outdoor environmental education. *Journal of Environmental Education, 45*(3), 178–193.

Lin, B., Fuller, R., Bush, R., Gaston, K., & Shanahan, D. (2014). Opportunity or orientation? Who uses urban parks and why. *PLoS ONE, 9*(1), e87422.

Lindemann-Matthies, P. (2005). 'Loveable' mammals and 'lifeless' plants: How children's interest in common local organisms can be enhanced through observation of nature. *International Journal of Science Education, 27*(6), 655–677.

Mackay, C., & Schmitt, M. (2019). Do people who feel connected to nature do more to protect it? *A meta-analysis. Journal of Environmental Psychology, 65*(101323), 1–9.

Nisbet, E. (2015). *Answering nature's call: Commitment to nature contact increases well-being: Results of the 2015 David Suzuki Foundation's 30×30 Nature Challenge*. David Suzuki Foundation.

Richardson, M., Cormack, A., McRobert, L., & Underhill, R. (2016). 30 Days wild: Development and evaluation of a large-scale nature engagement campaign to improve well-Being. *PLoS ONE, 11*(2), e0149777.

Richardson, M., Hunt, A., Hinds, J., Bragg, R., Fido, D., Petronzi, D., & White, M. (2019). A measure of nature connectedness for children and adults: Validation, performance, and insights. *Sustainability, 11*(3250), 1–16.

Rios, C., & Menezes, I. (2017). 'I saw a magical garden with flowers that people could not damage!': Children's visions of nature and of learning about nature in and out of school. *Environmental Education Research, 23*(10), 1402–1413.

Rowland, A., Knekta, E., Eddy, S., & Corwin, L. (2019). Defining and measuring students' interest in biology: An analysis of the biology education literature. *CBE—Life Sciences Education, 18*(3), 1–14.

Royal Society of Biology. (2019). *Biology changing the world: Royal Society of Biology strategic plan 2019–2021*. Royal Society of Biology.

Shanahan, D., Cox, D., Fuller, R., Hancock, S., Lin, B., Anderson, K., & Gaston, K. (2017). Variation in experiences of nature across gradients of tree cover in compact and sprawling cities. *Landscape and Urban Planning, 157*, 231–238.

Stern, M., Powell, R., & Ardoin, N. (2008). What difference does it make? Assessing outcomes from participation in a residential environmental education program. *Journal of Environmental Education, 39*(4), 31–43.

Strife, S. J. (2012). Children's environmental concerns: Expressing ecophobia. *Journal of Environmental Education, 43*(1), 37–54.

Szczytko, R., Stevenson, K. T., Peterson, M. N., & Bondell, H. (2020). How combinations of recreational activities predict connection to nature among youth. *Journal of Environmental Education*, 1–15. https://doi.org/10.1080/00958964.2020.1787313

Tam, K.-P. (2013). Concepts and measures related to connection to nature: Similarities and differences. *Journal of Environmental Psychology, 34*, 64–78.

Uitto, A., Juuti, K., Lavonen, J., & Meisalo, V. (2006). Students' interest in biology and their out-of-school experiences. *Journal of Biological Education, 40*(3), 124–129.

United Nations. (2015). *Transforming our world: The 2030 agenda for sustainable development*. United Nations.

Wiens, V., Kyngäs, H., & Pölkki, T. (2016). The meaning of seasonal changes, nature, and animals for adolescent girls' wellbeing in northern Finland: A qualitative descriptive study. *International Journal of Qualitative Studies on Health and Well-Being, 11*(30160), 1–15.

# Chapter 23
# Picturing the Urban Environment: Using Photovoice to Bring University Students' Views and Voices into Urban Environmental Education

**Marianna Kalaitsidaki, Katerina Mantadaki, and Aris Tsantiropoulos**

## 23.1 Introduction

Urban environmental education is environmental education (EE) that occurs in cities (Russ & Krasny, 2015). It is a contemporary trend in the field of EE especially in the USA. Today, half of the world's population live in cities (Muller & Werner, 2010). Cities face a range of familiar environmental problems such as pollution, traffic related issues, and lack of green and open spaces, as well as social problems such as poverty and unemployment (Leou & Kalaitsidaki, 2017). Urban EE can contribute to human wellbeing and environmental integrity in cities in different ways: it can facilitate learning in science and ecology, address urban environmental, social and cultural issues, contribute to positive youth development, develop an understanding of cities as social-cultural-ecological systems, and allow students to explore aspects of cities like art and history and engage them in actions with environmental, cultural and social outcomes (Russ & Krasny, 2015). Cities were once considered the antithesis of nature, but in the context of urban EE, cities are redefined as places where nature can be found and ecosystem services are provided. They are places were learners can readily observe how ecosystem and social processes are intertwined (Russ & Krasny, 2017). One framework for urban EE is place-based education, an approach described by Sobel (2004) that uses the local community and environment for teaching and learning with the goal to connect students to their place. Schools have become detached from both the communities

M. Kalaitsidaki (✉) · K. Mantadaki
Department of Primary Education, University of Crete, Crete, Greece
e-mail: mkalaitz@edc.uoc.gr

A. Tsantiropoulos
Department of Philosophy and Social Studies, University of Crete, Crete, Greece
e-mail: aris@phl.uoc.gr



K. Korfiatis, M. Grace (eds.), *Current Research in Biology Education*, Contributions from Biology Education Research,
https://doi.org/10.1007/978-3-030-89480-1_23

in which they are situated and the lived experiences of their students (Sobel, 2004). Sense of place, refers to the level of connectedness individuals feel to a specific place and the meanings associated with this place (Withrow-Clark et al., 2015). Different people ascribe different meanings to a particular place that reflect the physical, natural, social, cultural, economic and other aspects of places (Russ, 2015). Many environmental education programmes have as a goal to nurture ecological place meaning, defined as appreciating nature-related phenomena including ecosystems, as symbols of a place (Russ & Krasny, 2017) Developing sense of place through urban EE may promote pro-environmental behaviour. By making environmental education more local and providing youth with tools to investigate their environment, young people can develop a critical lens to view their surroundings, identify strengths and weakness and then act to bring change (Bellino & Adams, 2014).

Local environments can be examined and documented through *photovoice*, a participatory qualitative research method developed by Wang and Burris (1997), by which under-represented or marginalized groups identify, represent and finally enhance their community through a specific photographic technique that involves photography and group dialogue and reflection. Specifically, participants record (with photographs), reflect and critique personal or community issues that are stated as research questions. Visual images are accompanied by stories shared by the participants (Wang, 1999). In this way, in addition to the photographic representation of the participant experiences, critical reflections and group dialogues provide greater context and deeper meaning for the photos. Photovoice is a creative way to explore the human experience through the lenses of people directly impacted by the experience or the circumstance, rather than through researcher interpretation (Farley et al., 2017). In this research method participants are not passive research objects but have the position of an active participant whereby they are emancipated to act to implement change, bottom-up (Wang, 1999). Earlier photovoice applications focused on a range of social issues including mental and physical illness, and gender discrimination (reviewed in Woodgate et al., 2017). Applications were then followed with young people, both in out-of-school and educational settings; for example with secondary students to empower teachers and students in literacy engagements (Adams et al., 2014), and with youth with disabilities for an inclusive campus (Agarwal et al., 2015).

It this framework we chose photovoice as a methodological tool to understand perceptions of the urban environment in the context of our environmental education courses at the University of Crete in the city of Rethymno. We present a trial application of photovoice with Primary Education student teachers in order to identify how they perceive the urban environment in the city they study in. The research questions are:

1. Which components of the local urban environment did these students value and for what reason?
2. Do they ascribe any ecological meaning to the city and appreciate urban biodiversity and urban ecosystem services? If they do, what type of ecological meaning does that have?

To the best of our knowledge, this is the first application of photovoice in our country (Greece), so this pilot study will enable us to identify advantages and challenges of using it at the university level that may prove useful for future applications for us and other researchers in the field.

## 23.2 Research Methodology

### *23.2.1 Context and Participants*

The Department of Primary Education of the University of Crete, is located 4.5 km from the centre of Rethymno, a city with a population of over 32,000 situated by the sea, on the island of Crete. It accepts approximately 200 students every year; the programme of study is 4 years, after which students will eventually work as primary school teachers.

Students come to Rethymno to study from various parts of the country. Female students outnumber males (78% female, 22% males in 2015). Environmental courses (Biology, Ecology, Environmental Education) are included in the programme of study taught by the first author. In this study we used a convenience sample, seeking volunteers to take part in a study that involved photography among students attending daily classes. Finding students to take part in the study proved quite difficult. Students either did not respond or showed initial interest and then withdrew later. This resulted in three rounds of photovoice with three small groups of students (A, B and C) with a total number of 11 participants. By chance, students were in different semesters of study (see Table 23.1). Moreover, ten students came to Rethymno to study from other parts of the country, a variable that could have an impact on their perceptions. It should be noted that seven students showed initial interest but did not show up later, while five more gave up later. Consequently group B ended up with one participant.

**Table 23.1** The profile of the participants

| Participants | Year of study | Semester | Group | Place of origin | Population of place of origin |
|---|---|---|---|---|---|
| Student 1 | 1 | 2 | A | Tripoli | 29,297 |
| Student 2 | 1 | 2 | A | Chios | 26,361 |
| Student 3 | 1 | 2 | A | Piraeus | 163,688 |
| Student 4 | 4 | 8 | B | Lamia | 52,000 |
| Student 5 | 1 | 1 | C | Athens | 3,753,000 |
| Student 6 | 1 | 1 | C | Aridaia | 7000 |
| Student 7 | 1 | 1 | C | Heraklion | 151,324 |
| Student 8 | 1 | 1 | C | Athens | 3,753,000 |
| Student 9 | 1 | 1 | C | Rethymno | 32,468 |
| Student 10 | 1 | 1 | C | Athens | 3,753,000 |
| Student 11 | 1 | 1 | C | Chania | 53,910 |

### 23.2.2 Photovoice Procedure and Data Analysis

As stated in the previous section, the difficulty with recruiting participants and the pending end of the academic semester required us to implement three rounds of photovoice. The first round in early May 2017 was with students 1, 2 and 3, the second round at end of May with student 4, and the third round the following semester with students 5-11. Following Wang & Burris (1997), each group was given digital cameras, attended a seminar on digital photography with hands-on practice and they were asked to take 10 photographs according to the question "*What I like/What I don't like in Rethymno's environment*" over the course of two weeks. Then participants in each group presented their photographs while the first two authors acted as facilitators, asking the questions: "What do we see here?", "Why did you take this photograph?". Sessions were recorded and transcribed. Thematic analysis was performed according to Brown and Clarke (2006) by the second author. Codes were generated from the students' description of each photograph, then categories were formulated from EE literature and themes were generated, then reviewed and refined together with the first author. In this paper we present findings from the "What I like" question. We adopted the definition of the urban environment of Russ & Krasny (2015) as a combination of biophysical, socio-cultural and built components. The biophysical sub-theme included living organisms (animals and plants), physical resources like water, and ecosystems like the sea. People, relationships, social activities and environmental art formed the social environment subtheme. Buildings, historical monuments, public squares, etc. were grouped under the built environment sub-theme. Green spaces like parks or municipal gardens, squares and playgrounds were considered as biophysical elements if the description focused on plants. If they were described as meeting places or places of play, they were categorized as social environment. We would like to emphasize that the analysis was based on students' own descriptions and interpretations of their photographs. This is a difference between photovoice and other qualitative methods that use photography where photographs are interpreted by the researchers (Briggs et al., 2014). Participants' photographs can have personal meanings that can be revealed only from the verbal description of the participant (Woodgate et al., 2017). Another difference between photovoice and other research methods that use photography is that the photographs are presented in groups, giving opportunity for group discussion, exchange of knowledge and sharing of experiences (Briggs et al., 2014).

## 23.3 Results

In total, 111 photographs were presented by the participants for the question "*What I like/What I don't in Rethymno's environment*". Students' interpretations of their photographs were analysed in two respects: what was photographed and why?

### 23.3.1 What Students Liked in the Local Urban Environment

Thematic analysis produced two main themes in students' preferences: elements of the urban environment and sustainable practices in the city. As seen in Table 23.2, biophysical elements of the city were the most photographed subtheme followed by sociocultural elements and the built environment.

Regarding biophysical elements, students photographed plants, animals, the sea, the sky, water, the beach and the municipal garden.

All students photographed plants. Individual plants in narrow streets, neighbourhoods, flowerbeds, pots, and in green spaces like the city's municipal garden (Fig. 23.1). Most students did not know the plant names, and used descriptions instead. For example "*Rethymno is full of these yellow flowers*" (student 1); "*In my first visit here I saw a plant, not a magnolia, a climbing plant with purple flowers*" (student 2), who in another instance referred to a tree in a photograph as "*not an orange tree*". Students also photographed palm trees along the beach road and in public squares (Fig. 23.2).

There is a native palm tree in Crete that forms two natural forests included in all tourist promotions and campaigns. However, these palm trees in the city are of the Canarian type having been planted in the 1980s by the municipality. The practice of planting palm trees for decorative purposes in public spaces is being criticized in the ecology classes as an unsustainable practice, with economic and environmental costs in view of the invasive red palm weevil that infests and destroys them. Nevertheless, these students liked the palm trees in the city contrary to what they had heard in ecology class (Fig. 23.2). In their own words" "*They are attractive, make the landscape beautiful"(student 11); "I like the landscape with palm trees, it reminds me of the summer, I Know that they are not for Greece, but I like them*" (student 4); "*I like this square with palm trees. I know that our professor in ecology class told us that is not a native plant, it is from abroad, maybe other plants would be better suited in a Mediterranean landscape, but I like this square, I think it is a beautiful effort*" (student 6); "*Palm trees, I know they are from elsewhere, but in my opinion it is innovative that the authorities have given a different tone to Rethymno*" (student 7).

**Table 23.2** Numbers of photographs per theme and student group (some photos are listed in more than one categories)

| | Theme 1: Elements of the urban environment | | | Theme 2: sustainable practices |
|---|---|---|---|---|
| | Subtheme:biophysical | Subtheme:built | Subtheme:social | |
| GROUP A | 18/30 (60%) | 7/30 (23%) | 9/30 (30%) | 2/30(6.6%) |
| GROUP B | 8/10 (80%) | 4/10 (40%) | 3/10 (30%) | 1/10 (10%) |
| GROUP C | 45/71 (63%) | 16/71 (22.5%) | 12/71 (16.9%) | 8/71 (11.2%) |

**Fig. 23.1** All students photographed plants, individual ones (left, centre) or in green spaces like the municipal garden (right)

**Fig. 23.2** Students liked the planted exotic palm trees on the beach road

Few animals were photographed, just random encounters such as a bee on a flower, a butterfly, domesticated ducks and stray cats. However, no positive comments were made on their presence in the city or other characteristics (Fig. 23.3). In fact some negative comments were made: "*I do not like cats but this one was calm.*" (student 1), "*I am not fond of them, this one outside my work place was a little cute, despite being black and ugly*" (student 7); "*There was a bee among the flowers, I was scared and ran away*" (student 9).

In general, animals are more difficult to observe in the city and to photograph than plants. Small animals like insects and spiders are demanding objects for photography. However, all these students except one missed entire groups of animals like birds and fish. It should be noted that dolphins were included in the animal category from the photograph of a statue of dolphins, a student's favorite animals since childhood (see below Fig. 23.5). Another point of attraction for the students was the sea. The city has an extended sandy beach, a marina, a historical Venetian port and a rocky shore. Students referred to its calmness, colour and vastness. These

**Fig. 23.3** Few animals were photographed

are stereotypic images projecting relaxation and carefree mood characteristic of student life. However, no reference was made to the sea as an ecosystem providing food and other services.

The built environment was photographed less often: students liked characteristic sights of the city like the picturesque Venetian harbour with the lighthouse, the narrow streets in the historical centre, and a Venetian fountain, although without knowing their names or historical context. For example, student 1 described a Turkish minaret as "a tower". The historic Venetian harbour was referred to by all as the "old harbour", while the elaborate Venetian fountain with lion-head springs, one of the most characteristic tourist icon in the city centre, was photographed only by students 1 and 4 and referred to simply as "the faucets". The impressive Venetian fortress that dominates the city was photographed by only two students, from the distance and from below. Among the three groups, student 4, near graduation, photographed more plants than the rest, knew their names and was the only student to photograph two public squares with plants he used as a meeting place with his friends. He was the only one to photograph plants in the backyard of his block of flats, a neglected urban space.

Regarding the socio-cultural environment, two students photographed people at work (a grocer, and two fishermen) without any reference to food. One student engaged in conversation with workers pruning trees. Some photovoice studies place limitations on photographying people's faces (Woodgate, et al. 2017), while others require signed consent from human subjects (Bellino & Adams, 2014). Although no such limitations were placed here however, some students seemed to be aware of such considerations (*"I wanted to photograph people on their bicycles but it was difficult to photograph so as not to show their faces so I photographed one bicycle instead"* (student 11); "*This was next to a café where two gentlemen were eating but I did not want to photograph them, they looked fearsome"* (student 6). Street art was also included in this sub-theme. Student 3 thought that "*A nice drawing makes the streets beautiful*", differentiating however street art from graffiti. This prompted a debate about whether graffiti is art or vandalism. Some symbolic photographs of emotions like friendship, togetherness and loneliness, were also included in the social environment. It should be noted that there were no photos from the University campus as this is located 4.5 km away from the city.

The second theme in students' preferences was sustainable practices in the city, like sustainable mobility (bicycles), examples of renewable energy and recycling

(Fig. 23.4). Student 8 photographed solar panels and solar heaters on rooftops, examples of energy conservation; student 2 photographed "a wall of kindness", a designated place where food and other items can be left for the have-nots. Several photographed recycling cans stating that they liked this practice in Rethymno. The issue of solid waste management initiated a discussion in group C since there is a particular management system in the historical centre that students did not know, and it was explained by the facilitator.

### 23.3.2 Students' Reasons for Taking the Photographs

Thematic analysis of students' reasons for taking particular photographs produced the three themes shown in Table 23.3.

As stated previously, ten participants came to study in Rethymno from other parts of the country. Their places of origin were still in their minds, making comparisons, assessing differences or similarities with Rethymno. Students took photos of the familiar, what they had at home (like "the wall of happiness", a familiar practice from home for student 3), or the different and the unusual that they did not have at home (like the palm trees and public squares for student 5, the sea for student 1, and the extended sandy beach and the picturesque harbour for student 3). Students also referred to positive memories from childhood, family and vacations. One example of a childhood memory in relation to a photograph is the statue of dolphins, taken by student 2 (Fig. 23.5). *"I adored dolphins since I was a child because my father, a seaman, used to tell me stories with dolphins. So I loved dolphins. When I decided I will study here, I searched the internet and saw this statue; with this project, I thought I will definitely photograph it".*

**Fig. 23.4** Photographs showing sustainable practices in the city. Bicycle path, solar panels, recycling

**Table 23.3** Reasons for photography and number of students that gave them

| | |
|---|---|
| Comparisons with place of origin | 8 / 11students |
| Positive memories from childhood, family and vacations | 4 /11 students |
| Acknowledgement of ecosystem services | 7 / 11 students |

**Fig. 23.5** Photography due to a positive memory from childhood

Finally, students referred to cultural ecosystem services to justify their photographs of the biophysical elements of the city. Ecosystem services are various benefits provided to humans by ecosystems: provisioning (food, water), regulating (climate control), supporting (primary production, nutrient cycling), and cultural (spiritual, recreational, educational) (Krasny et al., 2014). Examples were the feelings of relaxation from the sight of the sea, the beautiful colours of the sky and the sea, and the aesthetic calmness and tranquility that the sea evokes, as well as leisure activities such as walking and swimming: "*I like the sea very much, I adore the sea. When I do not feel all right I come to the sea to calm down, to feel tranquil* (student 3); "*the sky here has this lovely blue colour, summery like, unlike ours up north.*" (student 6). They also stated that they appreciated plants for their provision of oxygen, for the shade they cast (supporting services), and for their pleasant smell as well as the beauty and colours of flowering plants: "*I was attracted to this plant by the shadow it casts and its smell*" (student 1); "*There is colour in this plant, it's not just green*" (student 2); "*These flowers add colour to the beach road, they are a beautiful sight for us and the tourists*" (student 4); "*I like these plants, they give me hope in the urban environment. In contrast, in my hometown, there is a lot of traffic and there are not enough plants to absorb carbon dioxide effectively and produce oxygen*" (student 7).

## 23.4 Discussion

### *23.4.1 Students' Perceptions of the Urban Environment*

Giving individuals a camera gives them the opportunity to help others see the world through their eyes (Wang, 1999). In this study we were able to see the urban environment through the eyes of our students. Our students appreciated mainly the biophysical elements, particularly plants. This was an unexpected finding in view of all the literature on plant blindness (Wandersee & Schussler, 1999), that is the inability to see or notice plants in one's environment, recognize their importance in the biosphere and to humans, appreciate their aesthetic and unique features, a condition affecting according to the authors and subsequent reports, young people of different ages in different countries. However, our students neglected other forms of urban biodiversity like birds, fish and insects. Urban biodiversity has been neglected even by scientists until fairly recently, but now is considered an important part of global biodiversity (Muller & Werner, 2010). Our students' appreciation of elements of the urban environment, appreciation of some ecosystem services (cultural) and sustainable practices in the city, reveal they ascribe some ecological meaning to the city, which is an important factor for environmental participation and action (Russ & Krasny, 2017). As shown in other studies (Stedman, et al., 2004), students choice of photographs was guided by their memories and life histories. Student 8, who comes from the capital city Athens photographed people, rooftops and the sea. Student 3, who comes from a neighbourhood with large unemployment and the country's larger harbour, photographed "the wall of kindness", a place to donate for the have-nots, a familiar practice from her own town. Meanwhile, student 2 who comes from an island, photographed the statue of dolphins, an animal associated with her childhood and father. As photographer Ernst Haas has said "*I see what I think, I see what I feel, because I am what I see, with the camera I can make others see it".(Ernst Haas estate,* 2021*)*

### *23.4.2 Implications for Environmental Education and Teacher Education*

With increased urbanization, urban EE is becoming critically important to foster ecological literacy on urban biodiversity and urban ecosystems, and to make cities more sustainable through citizen participation and action (Russ & Krasny, 2017). Capitalising on students' appreciation of urban plants (as shown here), place-based instruction on local flora could increase our students' and prospective elementary school teachers' knowledge of plants as well as other forms of urban biodiversity like birds (Lindemann-Matthies, 2005). Place-based instruction and EE should also address the coast and the sea, as ecosystems, which in the Greek secondary educational system are largely under-represented, as well as other types of ecosystem services besides the cultural ones cited here (provisional, supporting and regulating types). The students' perception of the urban environment should be completed to incorporate the built and social environment as well as relationships between them.

### *23.4.3 Some Comments on Using the Photovoice Method*

To our knowledge this was the first application of photovoice in environmental education in Greece, and we believe it is therefore useful to make some comments on using this method. Our study showed that photovoice was a good tool for our students to document and reflect on the urban environment. Photovoice gave students the opportunity to examine and document the local environment, share stories of lived experiences and memories, exchange information on local issues, like the solid waste management system in the city's historical centre and debate whether urban graffiti is art or vandalism. Some photographs of participants had personal meanings that were revealed by the participants' presentations and interpretations. That was the case with the photograph of the statue of dolphins in Fig. 23.5. A depiction of an ancient coin, an emblem of the city, was photographed by student 2 not for her appreciation of the arts, but for the dolphins, her favourite animals, associated with childhood and her father. Therefore what photos appear to be and what they really represent may be very different things, and it is essential to uncover the intended meanings of the participants (Stedman et al., 2004).

As described in the methodology, we encountered difficulties in recruiting participants. We consider this not a random and isolated occurrence, however unusual it may seem in this age of social media. Three years ago we noticed the same phenomenon as members of the review committee in a student photographic contest organized by the University entitled "My life at the University". Participation was extremely low (0.1%) and the content of photographs was "the self", a paradoxical finding in a place of collective existence like the University. We interpreted this as a generalized lack of interest in photography other than the glorification of the self, a common trend on social media. On the other hand, photovoice requires some time commitment (attending the photography seminar, photography and attend discussion meetings), which our college students may not have, since due to the country's economic crisis many have to work to support themselves. A solution would be to incorporate photovoice in the context of a university course, particularly seminars that require compulsory presence. Bellino & Adams (2014) argued that by making environmental education more local and by giving students the tools to investigate the environment they live in, young people can develop the skill to take a critical lens though which to observe their surroundings.

However, in our experience, students' engagement with dialogue and critical reflections do not happen automatically, effort is required on behalf of the facilitator. Our application lacked a collective action at the end to reach policymakers, an essential element of the method (Wang, 1999). In a recent review of four photovoice applications in environmental and sustainability contexts, Derr & Simmons (2020) concluded that photovoice is a promising methodology, however they emphasized the importance of maintaining its original emancipatory intents. Similarly, in this study, students did not consider taking any action on their own accord. A good practice for our future applications will be to follow Adams et al. (2014) and show participants photovoice projects from the literature, with examples of collective action such as organisation of a photography exhibition for the general public or production of a photobook, and present it to policy makers.

**Acknowledgments** We thank all the students who participated in our study, Stelios Mantadakis for delivering the digital photography seminar to groups B and C, and the Department of Primary Education University of Crete for providing the digital cameras. We also thank two anonymous reviewers for their constructive comments.

## References

Adams, S., Brooks, K., & Green, M. (2014). Using photovoice to empower k-12 teachers and students through authentic literacy engagements. *Writing & Pedagogy, 6*(3), 649–664.

Agarwal, N., Moya, E., Yura Yasui, N., & Seymour, C. (2015). Participatory action research with college students with disabilities. *Photovoice for an Inclusive Campus/Journal of Postsecondary Education and Disabilitu, 28*(2), 243–250.

Bellino, M., & Adams, J. (2014). Reimaging environmental education: Urban youth's perce-ptions and investigations of their communities. *RevistaBrasileira de PesquisaemEducacaoemCiencias, 14*(2), 27–38.

Briggs, L., Stedman, R., & Krasny, M. (2014). Photo-elicitation methods in studies of children's sense of place. *Children, Youth and Environments, 24*(3), 153–172.

Brown, V., & Clarke, V. (2006). Using thematic analysis in psychology. *Qualitative Research in Psychology, 3*(2), 77–101.

Derr, V., & Simmons, J. (2020). A review of photovoice applications in environments, sustainability and conservation contexts: is the method maintaining its emancipatory intents? *Environmental Education Research, 46*(3), 359–380.

Ernst Hass Estate. (2021). *Writings by Ernst Haas*. Retrieved from http://ernst-haas.com/writings-by-haas/

Farley, L., Brooks, K., & Pope, K. (2017). Engaging students in Praxis using photovoice research. *Multicultural Education, Winter, 2017*, 49–52.

Krasny, M., Russ, A., Tidball, K., & Elmqvist, T. (2014). Civic ecology practices: Participatory approaches to generating and measuring ecosystem services in cities. *Ecosystem Services, 7*, 177–186.

Leou, M., & Kalaitsidaki, M. (2017). The city as classroom. In A. Russ & M. Krasny (Eds.), *Urban environmental education review* (pp. 215–223). Cornell University Press.

Lindemann-Matthies, P. (2005). "Loveable" mammals and "lifeless" plants: How children's interest in common local organisms can be enhanced through observation of nature. *International Journal Science Education, 27*, 655–677.

Muller, N., & Werner, P. (2010). Urban biodiversity and the case for implementing the convention on biological diversity in towns and cities. In N. Muller, P. Werner, & G. Kelsey (Eds.), *Urban biodiversity and design* (pp. 1–32). Blackwell Publishing.

Russ, A. (2015). Introduction. In A. Russ (Ed.), *Urban environmental education review* (p. 6). Cornell University Civic Ecology Lab, NAAEE and EE Capacity project.

Russ, A., & Krasny, M. (2015). Urban environmental education trends. In A. Russ (Ed.), *Urban environmental education review* (pp. 12–25). Cornell University Civic Ecology Lab, NAAEE and EE Capacity project.

Russ, A., & Krasny, M. (2017). Introduction. In A. Russ & M. Krasny (Eds.), *Urban environmental education review* (pp. 1–9). Cornell University Press.

Sobel, D. (2004). *Place-based education: Connecting classrooms and community*. Orion Society Press.

Stedman, R., Beckley, T., Wallace, S., & Ambard, M. (2004). A picture and 1000 words: Using resident-employed photography to understand attachment to high amenity places. *Journal of Leisure Research, 36*(4), 580–606.

Wandersee, J., & Schussler, E. (1999). Preventing plant blindness. *The American Biology Teacher, 61*(2), 82–86.

Wang, C. (1999). Photovoice: A participatory action research strategy applied to women's health. *Journal of Women's Health, 8*(2), 185–192.

Wang, C., & Burris, M. (1997). Photovoice: Concept, methodology, and use forparticipatory needs assessment. *Health Education & Behavior, 24*, 369–387.

Withrow-Clark, R., Konrad, J., & Siddal, K. (2015). Sense of place and interpretation. In A. Russ (Ed.), *Urban environmental education* (pp. 51–54). Cornell University Civic Ecology Lab, NAAEE. EE Capacity.

Woodgate, R., Zurba, M., & Tennent, P. (2017). Worth a thousand words? Advantages, challenges and opportunities in working with photovoice as a qualitative research method with youth and their families. *Forum Qualitative Sozialforschung, 18*(1), Retrieved from http://www.qualitative-research.net/index.php/fqs/article/view/2659/4045

The manufacturer's authorised representative in the EU is Springer Nature Customer Service Centre GmbH, Europaplatz 3, 69115 Heidelberg, Germany. If you have any concerns regarding our products, please contact ProductSafety@springernature.com

Printed and bound by CPI Group (UK) Ltd, Croydon, CR0 4YY
28/09/2026
02224094-0001